Introduction to Drug Disposition and Pharmacokinetics

Introduction to Drug Disposition and Pharmacokinetics

STEPHEN H. CURRY
ADispell Inc., Rochester, New York, USA

ROBIN WHELPTON
Queen Mary University of London, UK

WILEY

This edition first published 2017

Registered Office
John Wiley & Sons, Ltd, The Atrium, Southern Gate, Chichester, West Sussex, PO19 8SQ, United Kingdom

For details of our global editorial offices, for customer services and for information about how to apply for permission to reuse the copyright material in this book please see our website at www.wiley.com.

Library of Congress Cataloging-in-Publication Data applied for

ISBN: 9781119261049

A catalogue record for this book is available from the British Library.

Set in 10/12pt Times by SPi Global, Pondicherry, India

Printed and bound in Malaysia by Vivar Printing Sdn Bhd

10 9 8 7 6 5 4 3 2 1

Contents

Preface

The application of knowledge of drug disposition, and skills in pharmacokinetics, is crucial to the development of new drugs and to a better understanding of how to achieve maximum benefit from existing ones. No matter how efficacious a potential new drug may be in the laboratory, the development of the new agent will come to nought if it cannot be delivered to its site of action and maintained there for sufficient duration for it to have the desired effect. Similarly, if it transpires that one (or more) of its metabolites is toxic then further development will cease. Also, knowledge of the concepts and principles of pharmacokinetics allows better clinical use of drugs and for this reason the topic is considered to some degree in most pharmacological texts.

We have written this book for those wishing, or needing, to study the topic in more depth than covered by most books of pharmacology. As an introductory text to drug disposition and pharmacokinetics, the book takes the reader from basic concepts to a point where those who wish to will be able to perform pharmacokinetic calculations and be ready to read more advanced texts and research papers. Exercises and further references have been included on the companion website for such readers.

The book will be of benefit to students of medicine, pharmacy, biomedical sciences and veterinary science, particularly those who have elected to study the topic in more detail, such as via electives and special study modules. It will be of benefit to those involved in drug discovery and product development, pharmaceutical and medicinal chemists, as well as budding toxicologists and forensic scientists who require the appropriate knowledge to interpret their findings. It will also provide a starting point for clinical pharmacologists, and may find use amongst professionals on the fringes of biomedical science, such as analytical chemists and patent lawyers. The book has been arranged to take the reader from a brief introduction to largely qualitative descriptions of drug disposition, and then to kinetic modelling to explain and predict drug effects, and to assist in the development of analytical skills that are needed in evaluating and mathematically describing pharmacokinetics. From there, we describe the special factors encountered in patient populations. The journey finishes with consideration of how aspects of drug interactions and toxicity are explainable, and indeed predictable, from consideration of the pharmacokinetic principles described in the previous chapters. The book concludes with a short history of the topic from its origins to where it is today and reflects upon the future, including the contribution that the reader might make. This chapter may be read as an introduction, but we feel the reader will gain most benefit by reading it with the knowledge of the previous chapters.

We envisage that some students will find the first chapters helpful in their early years and find the later ones more relevant if they gain more clinical experience, or find themselves working alongside clinicians. For some readers the book will provide all that they require but for others the book will be a valuable introduction to the topic, allowing them to progress to more advanced aspects of the subject.

We are grateful to the publishers for agreeing to the use of full-colour illustrations throughout as this enhances the learning experience and enables readers to assimilate what they need to know much more quickly. For example, many of the principal pharmacokinetic relationships can be demonstrated empirically by the movement of dye into and out of volumes of water, an approach we have used to illustrate the validity of several models, and one that has proved popular with our own students. We encourage the reader to visualize such movement, and we have reinforced this by including colour plates. Specific examples, some from our own experience, and others from the literature, often of seminal importance to the development of the subject, are included. Some of these examples refer to drugs that might be considered as being 'old', and indeed some may no longer be available in some countries, but many, propranolol, warfarin, digoxin, aspirin, and theophylline for instance, are still important members of their respective classes. Some of our examples, such as phenacetin and phenylbutazone, have been withdrawn or are no longer used in humans, but were the subject of investigations that informed our understanding, particularly of interactions and toxicity. We have not felt it necessary to search for examples involving more recently introduced drugs when drugs with more familiar names have already provided the examples that were needed. To do so would only increase the burden of drug names for the reader to learn.

In regard to drug names, some years ago the World Health Organization (WHO) started the introduction of a system of recommended international non-proprietary names (rINNs), which we have followed. Thus, different names and spellings may be encountered in different texts until the system is adopted universally. Generally it is clear as to what the new spelling refers, amfetamine and amphetamine, for instance, but when it is unclear we have added the alternative name in parentheses. This will be particularly helpful for our North American readers. Also, there is at present no internationally agreed system for pharmacokinetic symbols, but we have elected to use as simple a system as is possible, only adding additional sub- and superscripts when the meaning is not clear from the context.

Chapter 1 introduces the reader to the fundamental concepts related to the growth and decay of drug concentrations in plasma occurring alongside the growth and decay of effects, developed further in later chapters, and then continues with a brief presentation of the general chemical principles underlying the key mechanisms and processes described in the later chapters. To some extent, this reviews the relevant scientific language needed. Chapters 2 and 3 are largely qualitative descriptions of how drugs are administered and the physiochemical properties that influence their absorption, distribution, metabolism and excretion. Pharmacokinetic modelling of drugs and their metabolites can be found in Chapters 4–6. Chapter 7 examines extraction and clearance in more detail as these are the cornerstones of modern approaches to the subject and are vital to understanding physiologically-based pharmacokinetic modelling. Chapter 8 is devoted to the integration of pharmacokinetics and pharmacodynamics (PK-PD), while Chapter 9 describes the pharmacokinetics of what we have referred to as 'large molecules', a topic that is becoming increasingly important with the recent introduction of therapeutic enzymes and monoclonal antibodies. The next three chapters (10–12) deal with what can be referred to as 'special populations' or 'special considerations', sex, disease, age and genetics in particular. The plan in these chapters is that the reader will develop the ability to relate the many variables that affect drug response in the whole animal to the standard pharmacokinetic patterns shown in Figure 1.1 and described in Chapter 4, and later. Chapter 13 exemplifies the importance of pharmacokinetics in considering aspects of drug interactions and toxicity. Thus our sequence is from scientific preparation, through relevant science, to an introduction to clinical applications.

Additional material can be found on the companion website. This includes multiple choice questions, calculations and simulation exercises. We have also included suggestions for individual research or, preferably, for group discussions. Some of the material for these may be found in our previous book, *Drug Disposition and Pharmacokinetics: From Principles to Applications*. This also contains chapters on extrapolation from animals to humans and therapeutic drug monitoring, which we elected not to include in this book.

Whatever your reason for choosing this book we hope that reading it proves to be both a beneficial and an enjoyable experience.

Stephen H. Curry
Rochester, NY, USA

Robin Whelpton
Paignton, Devon, UK

Companion Website Directions

The companion website associated with this book (http://www.wiley.com/go/curryand whelpton/IDDP) contains a number of questions, exercises, simulations and more detailed explanations of selected topics which we feel readers will find helpful. We have taken this approach to encourage learning beyond that which is possible using the book on its own. Thus the website materials will assist you in improving your competence in the simple calculations that are essential to the subject, and also lead you to utilize other literature to broaden the educational experience. Hopefully, it will stimulate thinking, both by the reader on his or her own, and in the group discussions that we have found to be very helpful in the discovery of 'grey' areas of the subject, and which can facilitate understanding of controversial issues.

The resources are displayed in two ways:

1. By chapter: Use the drop-down menu at the top to view resources for that chapter.
2. By resource: Click the name of the resource in the top menu to see all content for that resource.

The types of resource include:

- *Multiple choice questions*: so that readers can quickly check their knowledge and comprehension; these can also be used as discussion topics.
- *Simulations*: so that the effects of changing pharmacokinetic parameters may be observed.
- *Short questions:* requiring brief explanations and sketches.
- *Topics for group discussion or private research:* these require students to think and research using more advanced texts and original scientific papers, and to reflect generally on what they have learnt.

Answers to the questions and, where appropriate, worked examples are provided, so that individuals using the book as a self-teaching aid can monitor their progress. Course leaders should be able to devise suitable questions for enhancing learning, and for testing their students.

We have suggested initial values for the simulations, which have been designed to reinforce the material in the book and should prove useful to all readers (see Appendix 4). Additionally, curve fitting using the method of residuals is demonstrated using a spreadsheet that can be used as a template for other sets of data. There is a detailed presentation explaining the relationship between apparent volumes of distribution in multiple-compartment models.

Should students require more examples, these can be accessed via the companion website that accompanies *Drug Distribution and Pharmacokinetics: From Principles to Applications*:

www.wiley.com/go/curryandwhelpton/IDDP

1

Introduction: Basic Concepts

> **Learning objectives**
>
> This chapter was written for those unfamiliar with certain aspects of pharmacology and chemistry, including physical chemistry, and for those who feel a little revision would be helpful. By the end of the chapter the reader should be able to:
>
> - use the Henderson–Hasselbalch equation to calculate the ionization of weak acids and bases
> - plot concentration–time data to determine first-order and zero-order rate constants
> - explain the effect of ionization on the partitioning of weak electrolytes between buffers and octanol.

1.1 Introduction

Pharmacology can be divided into two major areas, pharmacodynamics (PD) – the study of what a drug does to the body – and pharmacokinetics (PK) – the study of what the body does to the drug; hardly rigorous definitions but they suffice. Drug disposition is a collective term used to describe drug absorption, distribution, metabolism and excretion whilst pharmacokinetics is the study of the rates of these processes. By subjecting the observed changes, for example in plasma concentrations as a function of time, to mathematical

Introduction to Drug Disposition and Pharmacokinetics, First Edition. Stephen H. Curry and Robin Whelpton.

Companion website: www.wiley.com/go/curryandwhelpton/IDDP

equations (models) pharmacokinetic parameters such as elimination half-life ($t_{½}$), volume of distribution (V) and plasma clearance (CL) can be derived. Pharmacokinetic modelling is important for the:

- selection of the right drug for pharmaceutical development
- evaluation of drug delivery systems
- design of drug dosage regimens
- appropriate choice and use of drugs in the clinic.

A detailed knowledge of mathematics is not required to understand pharmacokinetics and it is certainly not necessary to be able to differentiate or integrate complex equations. The few examples in this book are standard differentials or integrals that can be quickly learnt if they are not known already. To understand the equations in this book requires little more than a basic knowledge of algebra, laws of indices and logarithms, a brief explanation of which can be found in Appendix 1. Furthermore, the astute reader will quickly realize that, although seemingly different, many equations take the same form, making learning easier. For example, drug binding to macromolecules, whether they be receptors, plasma proteins, transporters or enzymes, can be described using the same basic equation. Similarly, the equation describing the time course of formation and excretion of a drug metabolite is very much like that describing the plasma concentrations during the absorption and elimination of a drug.

The role of pharmacokinetics is illustrated in Figure 1.1. There is an optimum range of concentrations over which a drug has beneficial effects, but little or no toxicity – this range is the therapeutic range, sometimes referred to as the therapeutic window. There is a threshold concentration below which the drug in ineffective and a higher threshold above which adverse effects become problematic. If a single dose of a drug, for example aspirin taken to relieve a headache, is consumed, the concentration in the plasma will rise until the aspirin becomes effective. After a period of time the processes which remove aspirin from the body will reduce the concentration until the drug is no longer effective (Figure 1.1, curve (a)).

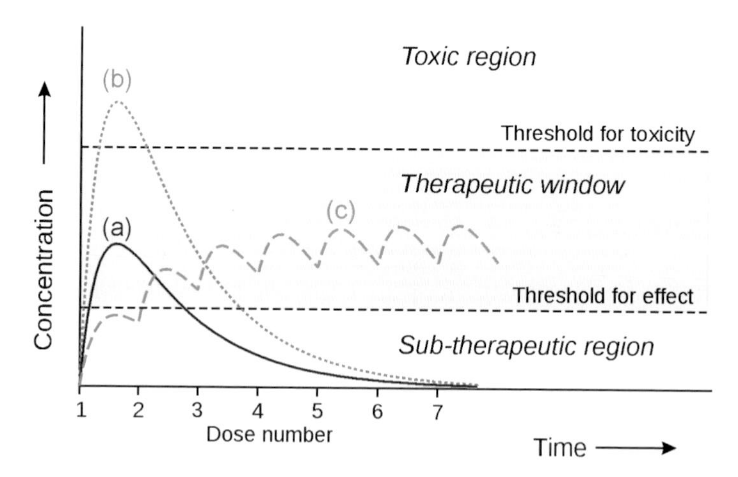

Figure 1.1 Typical concentration–time curves after oral administration of a drug: (a) single dose of drug; (b) a single dose twice the size of the previous one; (c) the same drug given as divided doses. The dose and frequency of dosing for (c) were calculated to ensure the concentrations remained in the therapeutic window.

The short duration of action may be fine for treating a headache but if the aspirin is to treat rheumatoid arthritis a much longer duration of action is required. Simply increasing the size of the dose is not the answer because eventually the plasma concentrations will enter the toxic region (Figure 1.1, curve (b)). However, by giving the aspirin as smaller divided doses at regular intervals the plasma concentrations can be maintained within the therapeutic window (Figure 1.1, curve (c)). The three curves depicted in Figure 1.1 were produced using relatively simple pharmacokinetic equations which will be explained later.

1.2 Drugs and drug nomenclature

A drug is a substance that is taken, or administered, to produce an effect, usually a desirable one. These effects are assessed as physiological, biochemical or behavioural changes. There are two major groups of chemicals studied and used as drugs. First, there is a group of pharmacologically interesting endogenous substances, for example epinephrine, insulin and oxytocin. Second, there are the non-endogenous or 'foreign' chemicals (xenobiotics), which are mostly products of the laboratories of the pharmaceutical industry. Early medicines, some of which have been used for at least 5000 years, relied heavily on a variety of mixtures prepared from botanical and inorganic materials. Amongst the plant materials, the alkaloids, morphine from opium, cocaine from coca leaves and atropine from the deadly nightshade (belladonna) are still used today. Insulin, once obtained from pigs (porcine insulin), is more usually genetically engineered using a laboratory strain of *Escherichia coli* bacteria to produce human insulin. A few inorganic chemicals are used as drugs, including lithium carbonate (Li_2CO_3) and sodium hydrogen carbonate (sodium bicarbonate, $NaHCO_3$).

1.2.1 Drug nomenclature

Wherever possible specific drug examples are given throughout this book, but unfortunately drug names can lead to confusion. Generally a drug will have at least three names: a full chemical name, a proprietary name, i.e. a trade name registered to a pharmaceutical company, and a non-proprietary name (INN) and/or an approved or adopted name. Names that may be encountered include the British Approved Name (BAN), the European Pharmacopoeia (EuP) name, the United States Adopted Name (USAN), the United States Pharmacopoeia (USP) name and the Japanese Approved Name (JAN). The World Health Organization (WHO) is introducing a system of recommended INNs and it is hoped that this will became the norm for naming drugs, replacing alternative systems (http://www.who.int/medicines/services/inn/innguidance/en/, accessed 17 February 2016). For example, lidocaine is classed as a rINN, USAN and JAN, replacing the name lignocaine that was once a BAN. Generally, the alternatives obviously refer to the same drug, e.g. ciclosporin, cyclosporin and cyclosporine. There are some notable exceptions, for example pethidine is known as meperidine in the US and paracetamol as acetaminophen. Even a simple molecule like paracetamol may have several chemical names but the number of proprietary names or products containing paracetamol is even greater, including Panadol, Calpol, Tylenol and Anadin Extra. It is therefore necessary to use an unequivocal approved name whenever possible, but alternative names and spellings are likely to be encountered, some examples of which are given in Table 1.1. Useful sites for checking, names, synonyms,

Table 1.1 Differences in rINN and USAN nomenclature

rINN:BAN	USAN:USP	Alternative spellings*
Aciclovir	Acyclovir	Acyclovir
Amfetamine	Amphetamine	Amphetamine
Bendroflumethiazide	Bendroflumethiazide	Bendrofluazide
Benzylpenicillin	Penicillin G	Benzyl penicillin
Cefalexin	Cephalexin	Cephalexin
Ciclosporin	Cyclosporine	Cyclosporin
Epinephrine	Epinephrine	Adrenaline
Furosemide	Furosemide	Frusemide
Glycerol	Glycerin	
Glyceryl trinitrate	Nitroglycerin	
Indometacin	Indomethacin	Indomethacin
Isoprenaline	Isoproterenol	
Lidocaine	Lidocaine	Lignocaine
Metamfetamine	Methamphetamine	Methamphetamine
Norepinephrine	Norepinephrine	Noradrenaline
Paracetamol	Acetaminophen	
Pethidine	Meperidine	
Phenoxymethylpenicillin	Penicillin V	Phenoxymethyl penicillin
Rifampicin	Rifampin	
Salbutamol	Albuterol	
Sulfadimidine	Sulfamethazine	Sulphadimidine

*Chiefly previous BAN entries.

chemical properties and the like include http://chem.sis.nlm.nih.gov/chemidplus/and www.chemicalize.org/ (accessed 17 February 2016).

1.3 Law of mass action

The reversible binding of drugs to macromolecules such as receptors and plasma proteins is described by the law of mass action: 'The rate at which a chemical reaction proceeds is proportional to the active masses (usually molar concentrations) of the reacting substances'. This concept is easily understood if the assumption is made that for the reaction to occur, collision between the reacting molecules must take place. It follows that the rate of reaction will be proportional to the number of collisions and the number of collisions will be proportional to the molar concentrations of the reacting molecules. If a substance X is transformed into substance Y,

$$X \rightarrow Y$$

the rate of reaction $= k[X]$, where k is the rate constant and [X] represents the molar concentration of X at that time. If two substances A and B are reacting to form two other substances C and D, and if the concentrations of the reactants at any particular moment are [A] and [B] then:

$$A + B \rightarrow C + D$$

and the rate of reaction $= k[A][B]$.

1.3.1 Reversible reactions and equilibrium constants

Consider the reaction:

$$A + B \rightleftharpoons C + D$$

The rate of the forward reaction is:

$$\text{forward rate} = k_1[A][B] \tag{1.1}$$

whilst the backward rate is:

$$\text{backward rate} = k_{-1}[C][D] \tag{1.2}$$

where k_1 and k_{-1} are the rate constants of the forward and backward reactions, respectively. When equilibrium is reached the forward and backward rates are equal, so:

$$k_1[A][B] = k_{-1}[C][D] \tag{1.3}$$

The equilibrium constant K is the ratio of the forward and backward rate constants, and rearranging Equation 1.3 gives:

$$K = \frac{k_1}{k_{-1}} = \frac{[C][D]}{[A][B]} \tag{1.4}$$

The term *dissociation constant* is used when describing the equilibrium of a substance which dissociates into smaller units, as in the case, for example, of an acid (Section 1.4). The term is also applied to the binding of a drug, D, to a macromolecule such as a receptor, R, or plasma protein (Sections 2.5 and 8.2). The complex DR dissociates:

$$DR \rightleftharpoons D + R$$

so:

$$K = \frac{[D][R]}{[DR]} \tag{1.5}$$

An *association* constant is the inverse of a dissociation constant.

1.3.1.1 Sequential reactions

When a product D arises as a result of several sequential reactions (Figure 1.2), it cannot be formed any faster than the rate of at which its precursor C is formed, which in turn cannot be formed any faster than its precursor B. The rates of each of these steps are determined by the rate constants k_1, k_2 and k_3, therefore, the rate at which D is formed will be the rate of the slowest step, i.e. the reaction with the lowest value of rate constant. Say, for example, k_2 is the lowest rate constant, then the rate of formation of D is determined by k_2 and the reaction $B \rightarrow C$ is said to be the *rate-limiting* or *rate-determining* step. This concept is fundamental to understanding sustained-release preparations (Chapter 4) and also drug metabolism when it occurs in more than one step (Chapter 6).

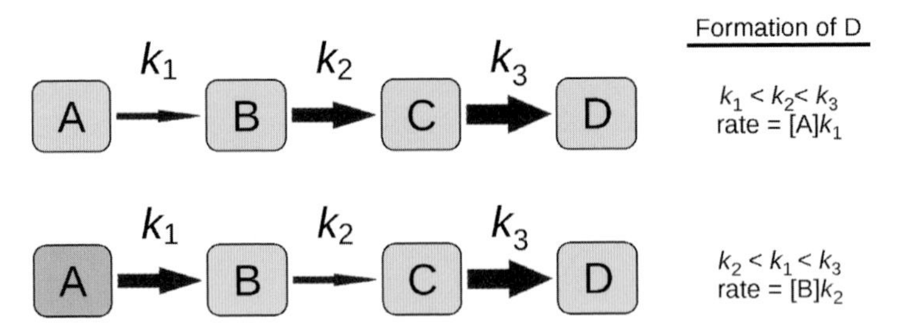

Figure 1.2 *Illustration of the rate-determining step in sequential reactions. The thinnest arrow represents the smallest rate constant and is therefore the rate-determining step.*

1.3.2 Reaction order and molecularity

The order of a reaction is the number, n, of concentration terms affecting the rate of the reaction, whereas molecularity is the number of molecules taking part in the reaction. The order of a reaction is measured experimentally and because it is often close to an integer, 0, 1, or 2, reactions may be referred to as zero-, first- or second-order, respectively. The reaction

$$X \rightarrow Y$$

is clearly monomolecular, and may be either zero- or first-order depending on whether the rate is proportional to$[X]^0$ or $[X]^1$. The reactions

$$2X \rightarrow Y$$

and

$$A + B \rightarrow C + D$$

are both bimolecular and second-order providing the rate is proportional to $[X]^2$ in the first case and to [A][B] in the second. Note how the total reaction order is the sum of the indices of each reactant: rate $\propto [A]^1[B]^1$, so $n=2$. However, if one of the reactants, say A, is present in such a large excess that there is no detectable change in its concentration, then the rate will be dependent only on the concentration of the other reactant, B. Thus, the rate is proportional to $[A]^0[B]^1$. The reaction is first-order (rate $\propto$ [B]) but it is still bimolecular. Hydrolysis of an ester in dilute aqueous solution is a commonly encountered example of a bimolecular reaction which is first-order with respect to the concentration of ester and zero-order with respect to the concentration of water, giving an overall reaction order of unity.

Enzyme-catalysed reactions have reaction orders between 1 and 0 with respect to the drug concentration. This is because the Michaelis–Menten equation (Section 4.7) limits to zero-order when the substrate is in excess and the enzyme is saturated so that increasing the drug concentration will have no further effect on the reaction rate. When the concentration of enzyme is in vast excess compared to the substrate concentration, the enzyme concentration is not rate determining and the reaction is first order. Thus, the reaction order of an enzyme-catalysed reaction changes as the reaction proceeds and substrate is consumed.

1.3.3 Decay curves and half-lives

As discussed above, the rate of a chemical reaction is determined by the concentrations of the reactants and from the foregoing it is clear that a general equation relating the rate of decline in concentration ($-dC/dt$), rate constant (λ) and concentration (C) can be written:

$$-\frac{dC}{dt} = \lambda C^n \tag{1.6}$$

Note the use of λ to denote the rate constant when it refers to decay; the symbol is used for radioactive decay, when it is known as the decay constant. Use of λ to denote *elimination* rate constants is becoming more prevalent in pharmacokinetic publications.

1.3.3.1 First-order decay

Because first-order kinetics are of prime importance in pharmacokinetics, we shall deal with these first. For a first-order reaction, $n=1$ and

$$-\frac{dC}{dt} = \lambda C \tag{1.7}$$

Thus, the rate of the reaction is directly proportional to the concentration of substance present. As the reaction proceeds and the concentration of the substance falls, the rate of the reaction decreases. This is exponential decay, analogous to radioactive decay, where the probability of a disintegration is proportional to the number of unstable nuclei present. The first-order rate constant has units of reciprocal time (e.g. h^{-1}). Integrating Equation 1.7 gives:

$$C = C_0 \exp(-\lambda t) \tag{1.8}$$

which is the equation of a curve that asymptotes to 0 from the initial concentration, C_0 (Figure 1.3(a)). Taking natural logarithms of Equation 1.8 gives:

$$\ln C = \ln C_0 - \lambda t \tag{1.9}$$

which is the equation of a straight line of slope, $-\lambda$ (Figure 1.3(b)). Before the advent of inexpensive calculators and the availability of spreadsheets, common logarithms were often used to plot log C against t when the slope was $-\lambda/2.303$. Another way of presenting the data is to plot C on a logarithmic scale. This approach was often used when computers where not readily available. The half-life can be read easily from such graphs and λ can be calculated via Equation 1.10.

The half-life ($t_{½}$) is the time for the initial concentration (C_0) to fall to $C_0/2$, and substitution in Equation 1.9 gives:

$$t_{½} = \frac{\ln 2}{\lambda} = \frac{0.693}{\lambda} \tag{1.10}$$

because ln 2 = 0.693. This important relationship, where $t_{½}$ is constant (independent of the initial concentration) and inversely proportional to λ, is *unique* to first-order reactions.

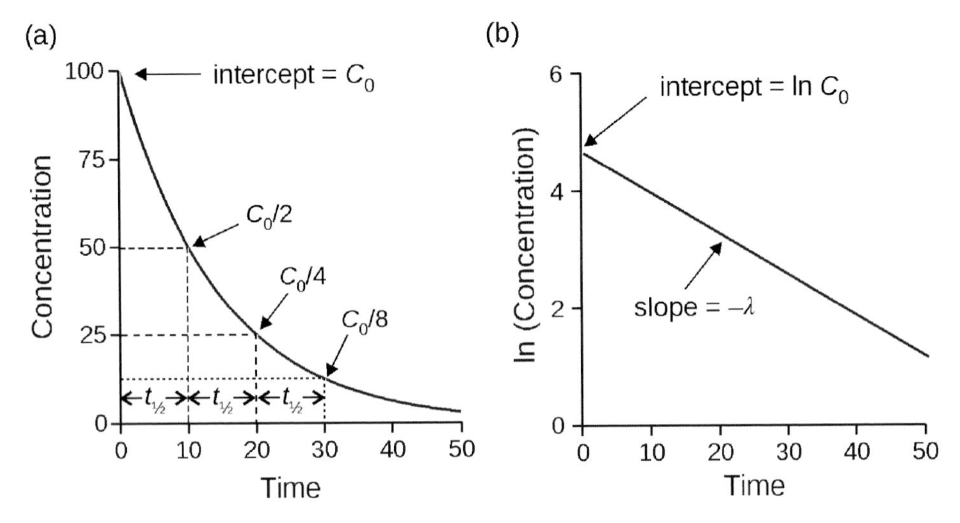

Figure 1.3 Curves for first-order decay plotted as (a) C versus t and (b) ln C versus t.

Because $t_{1/2}$ is constant, 50% is eliminated in $1 \times t_{1/2}$, 75% in $2 \times t_{1/2}$ and so on. Thus, when five half-lives have elapsed less than 5% of the substance remains, and after seven half-lives less than 1% remains.

1.3.3.2 *Zero-order decay*

For a zero-order reaction, $n = 0$ and:

$$-\frac{dC}{dt} = \lambda C^0 = \lambda \tag{1.11}$$

Because $C^0 = 1$, it is clear that a zero-order reaction proceeds at a *constant rate*, and the zero-order rate constant must have units of *rate* (e.g. g L^{-1} h^{-1}). Integrating Equation 1.11:

$$C = C_0 - \lambda t \tag{1.12}$$

gives the equation of a straight line of slope, $-\lambda$, when concentration is plotted against time (Figure 1.4(a)). The ln C plot is a convex curve because initially the *proportion* of drug eliminated is less when the concentration is higher (Figure 1.4(b)).

The half-life can be obtained as before, and substituting $t = t_{1/2}$ and $C = C_0$ gives:

$$t_{1/2} = \frac{C_0}{2\lambda} \tag{1.13}$$

The zero-order half-life is inversely proportional to λ, as would be expected, but $t_{1/2}$ is also directly proportional to the initial concentration. In other words, the greater the amount of drug present initially, the longer the time taken to reduce the amount present by 50%, as would be expected. The term 'concentration-dependent half-life' has been applied to this situation.

Equations such as Equations 1.8 and 1.12 are referred to as *linear* equations. Note that in this context it is important not to confuse 'linear' with 'straight-line'. While it is true

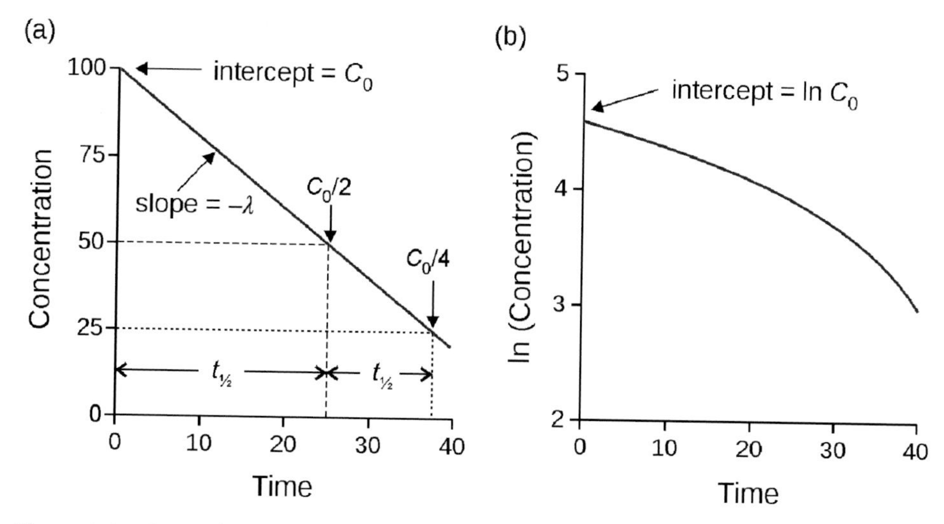

Figure 1.4 Curves for zero-order decay plotted as (a) C versus t and (b) ln C versus t. Note how for an initial concentration of 100, $t_{½} = 25$, but for an initial concentration of 50, $t_{½} = 12.5$.

that the equation of a straight-line is a linear equation, exponential equations are also linear. On the other hand, non-linear equations are those where the variable to be solved cannot be written as a linear combination of independent variables. The Michaelis–Menten equation is such an example.

1.3.3.3 Importance of half-life in pharmacokinetics

Half-life is a very useful parameter in pharmacokinetics. It is much easier to compare the duration of action of drugs in terms of their relative half-lives rather than rate constants, and the rate of attainment of steady-state concentrations during multiple dosing and the fluctuations in peak and trough levels is a function of $t_{½}$ (Figure 4.16). However, it is important to recognize at the outset that the half-life of a drug is, in fact, dependent on two other pharmacokinetic parameters, apparent volume of distribution (V) and clearance (CL). The apparent volume of distribution, as its name implies, is a quantitative measure of the extent to which a drug is distributed in the body (Section 2.4.1.1) whist clearance can be thought of as an indicator of how efficiently the body's eliminating organs remove the drug, therefore the larger the value of CL, the shorter will be $t_{½}$, and any change in half-life will be as a result of changes in either V or CL, or both (Section 4.2.1).

1.4 Ionization

The degree of ionization of a molecule can have a major influence on its disposition and pharmacokinetics. The term *strength* when applied to an acid or base refers to its tendency to ionize. The term should not be confused with *concentrated*. Strong acids and bases can be considered to be 100% ionized at any practical pH value. Weak electrolytes, such as amines and carboxylic acids, are only partially ionized in aqueous solutions:

$$R{-}NH_2 + H^{\oplus} \rightleftharpoons R{-}NH_3^{\oplus}$$

$$R{-}COOH \rightleftharpoons R{-}COO^{\ominus} + H^{\oplus}$$

The degree of ionization is determined by the pK_a of the ionizing group and the pH of the aqueous environment. The pK_a of the compound, a measure of its inherent acidity or basicity, is numerically equal to the pH at which the compound is 50% ionized.

For an acid, AH, dissolved in water:

$$AH \rightleftharpoons H^+ + A^-$$

The acid dissociation constant is:

$$K_a = \frac{[H^+][A^-]}{[AH]} \tag{1.14}$$

Clearly the more the equilibrium is to the right, the greater is the hydrogen ion concentration, with a subsequent reduction in the concentration of unionized acid, so the larger will be the value of K_a. Note that because pK_a is the negative logarithm of K_a (analogous to pH), strong acids have *low* values of pK_a. Taking logarithms (see Appendix 1 for details) of Equation 1.14 gives:

$$\log K_a = \log[H^+] + \log[A^-] - \log[AH] \tag{1.15}$$

and on rearrangement:

$$-\log[H^+] = -\log K_a + \log\frac{[A^-]}{[AH]} \tag{1.16}$$

Because $-\log[H^+]$ is the pH of the solution:

$$pH = pK_a + \log\frac{[A^-]}{[AH]} = pK_a + \log\frac{[base]}{[acid]} \tag{1.17}$$

where pK_a=$-\log K_a$, by analogy with pH. Note that when $[A^-]=[AH]$ the ratio is 1 and because log(1)=0, the pK_a=pH, as stated earlier.

The range of pK_a values extends below 1 and above 14, but for the majority of drugs values are between 2 and 13. Benzylpenicillin (pK_a=2.3) is an example of a relatively strong acid and metformin (pK_a=12.4) is an example of a relatively strong base (Figure 1.5). It should be noted that it is not possible from a knowledge of the pK_a alone to say whether a substance is an acid or a base. It is necessary to know how the molecule ionizes. Pentobarbital, pK_a=8.0, forms sodium salts and so must be an acid, albeit a rather weak one. The electron-withdrawing oxygen atoms result in the hydrogen atom being acidic. Diazepam, pK_a=3.3, must be a base because it can be extracted from organic solvents into hydrochloric acid. Imines are weak bases because of delocalization of the nitrogen lone pair of electrons around the C=N double bond. In oxazepam, the electron-withdrawing effect of the oxygen in the hydroxyl group reduces the pK_a (=1.7) of the imine compared to that in diazepam. Molecules can have more than one ionizable group, for example salicylic acid has a carboxylic acid (pK_a = 3.0) and a weaker acidic phenol group (pK_a = 13.4). The amide in oxazepam is very weakly acidic (pK_a = 11.6), making this compound amphoteric, that is, both acidic and basic. Sulfonamides are usually more acidic than amides and if the primary aromatic amine is not acetylated they are amphoteric, as illustrated by sulfadimidine. Similarly, morphine is amphoteric, having a tertiary amine group (pK_a = 8.0) and an acidic phenol (pK_a = 9.9) (Figure 1.5).

Not all drugs ionize. The volatile and gaseous anaesthetics are usually neutral compounds, for example enflurane ($CHClF_2$-CF_2C-O-CF_2H), which is an ether. Alcohols such as ethanol and chloral hydrate ($CCl_3CH(OH)_2$) are usually referred to as being neutral as they do not ionize at physiological pH values.

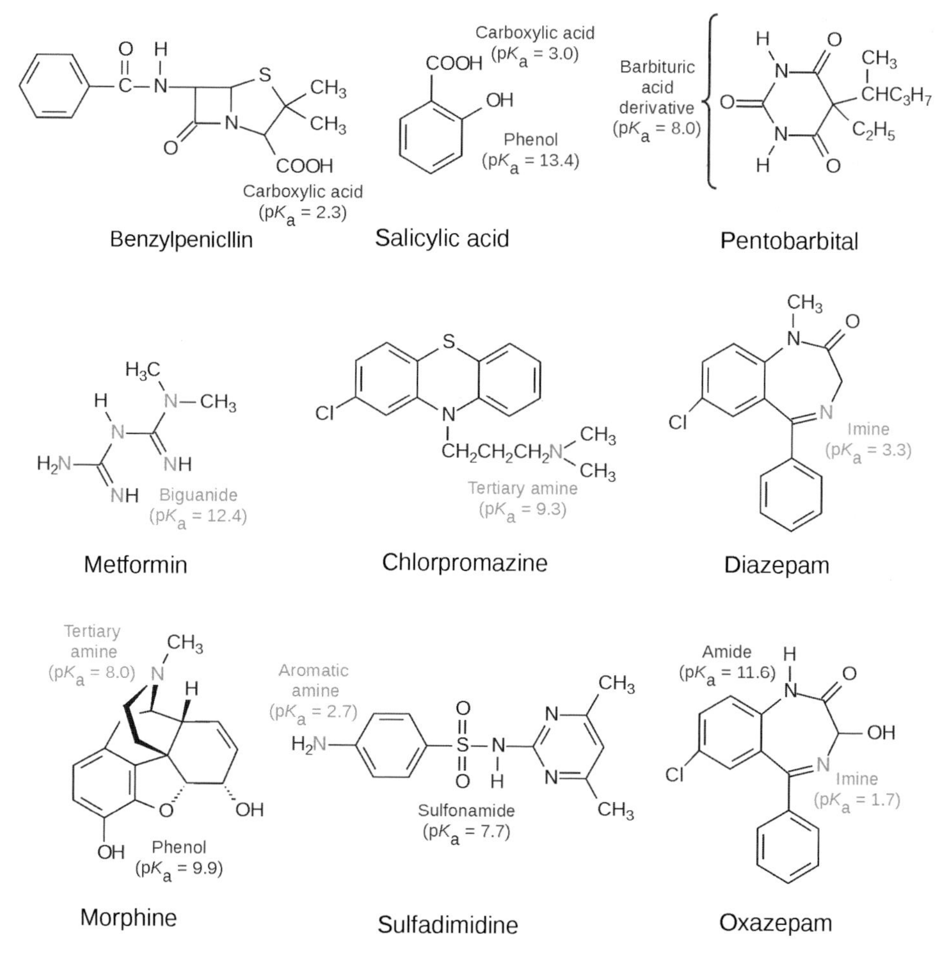

Figure 1.5 Examples of ionizable groups in selected drug examples. Note how the pK_a values range (strong to weak) from 2.3 to 11.6 for the acids and 12.4 to 1.7 for the basic groups. Acidic hydrogen atoms are shown in red and basic nitrogen atoms in blue.

1.4.1 Henderson–Hasselbalch equation

Equation 1.17 is a form of the Henderson–Hasselbalch equation, which is important in determining the degree of ionization of weak electrolytes and calculating the pH of buffer solutions. If the degree of ionization is α, then the degree non-ionized is $(1 - \alpha)$ and, for an acid:

$$\mathrm{pH} = \mathrm{p}K_\mathrm{a} + \log\frac{\alpha}{1-\alpha} \tag{1.18}$$

Taking antilogarithms and rearranging allows the degree of ionization to be calculated:

$$\alpha = \frac{10^{(\mathrm{pH}-\mathrm{p}K_\mathrm{a})}}{1+10^{(\mathrm{pH}-\mathrm{p}K_\mathrm{a})}} \tag{1.19}$$

The equivalent equation for a base is:

$$\alpha = \frac{10^{(\mathrm{p}K_\mathrm{a}-\mathrm{pH})}}{1+10^{(\mathrm{p}K_\mathrm{a}-\mathrm{pH})}} \tag{1.20}$$

Although Equations 1.19 and 1.20 may look complex, they are easy to use. Using the ionization of aspirin as an example, the $\mathrm{p}K_\mathrm{a}$ of aspirin is ~ 3.4, so at the pH of plasma (pH 7.4)

$$\mathrm{pH} - \mathrm{p}K_\mathrm{a} = 7.4 - 3.4 = 4$$

$$\alpha = \frac{10^4}{1+10^4} = \frac{10000}{10001} = 0.9999$$

In other words aspirin is 99.99% ionized at the pH of plasma, or the ratio of ionized to non-ionized is 10,000:1. In gastric contents, pH 1.4, aspirin will be largely non-ionized; $1.4 - 3.4 = -2$, so the ratio of ionized to non-ionized is $1:10^{-2}$, that is, there are 100 non-ionized molecules for every ionized one.

1.5 Partition coefficients

The ability of a drug to dissolve in, and so cross, lipid cell membranes can be a major factor in its disposition. This ability can be assessed from its partition coefficient. When an aqueous solution of a substance is shaken with an immiscible solvent (e.g. diethyl ether) the substance is extracted into the solvent until equilibrium between the concentration in the organic phase and the aqueous phase is established. For dilute solutions the ratio of concentrations is known as the distribution, or partition coefficient, P:

$$P = \frac{\text{concentration in organic phase}}{\text{concentration in aqueous phase}} \tag{1.21}$$

Organic molecules with large numbers of paraffin chains, aromatic rings and halogens tend to have large values of P, whilst the introduction of polar groups such as hydroxyl or carbonyl groups generally reduces the partition coefficient. Drugs with high partition coefficients are lipophilic or hydrophobic, whereas those that are very water soluble and are poorly extracted by organic solvents are hydrophilic. Lipophilicity can have a major influence on how a drug is distributed in the body, its tendency to bind to macromolecules such as proteins and, as a consequence, drug activity. A relationship between partition coefficient and pharmacological activity was demonstrated as early as 1901, but it was in the 1960s that Corwin Hansch used regression analysis to correlate biological activity with partition coefficient. He chose *n*-octanol as the organic phase and this has become the standard for such studies (Figure 1.6). Because *P* can vary between <1 (poorly extracted by the organic phase) to several hundred thousand, values are usually converted to log *P*, to encompass the large range (Appendix 1).

1.5.1 Effect of ionization on partitioning

Generally, ionized molecules cannot be extracted into organic solvents, or at least not appreciably. Thus, for weak electrolytes the amount extracted will be dependent on the degree of ionization, which of course is a function of the pH of the aqueous solution and the pK_a of the ionizing group, as discussed above (Section 1.4), and the partition coefficient. If the total concentration (ionized + non-ionized) of solute in the aqueous phase is measured and used to calculate an apparent partition coefficient *D*, then the partition coefficient *P*, can be calculated. For an acid:

$$P = D\left[1 + 10^{(\mathrm{pH} - \mathrm{p}K_a)}\right] \tag{1.22}$$

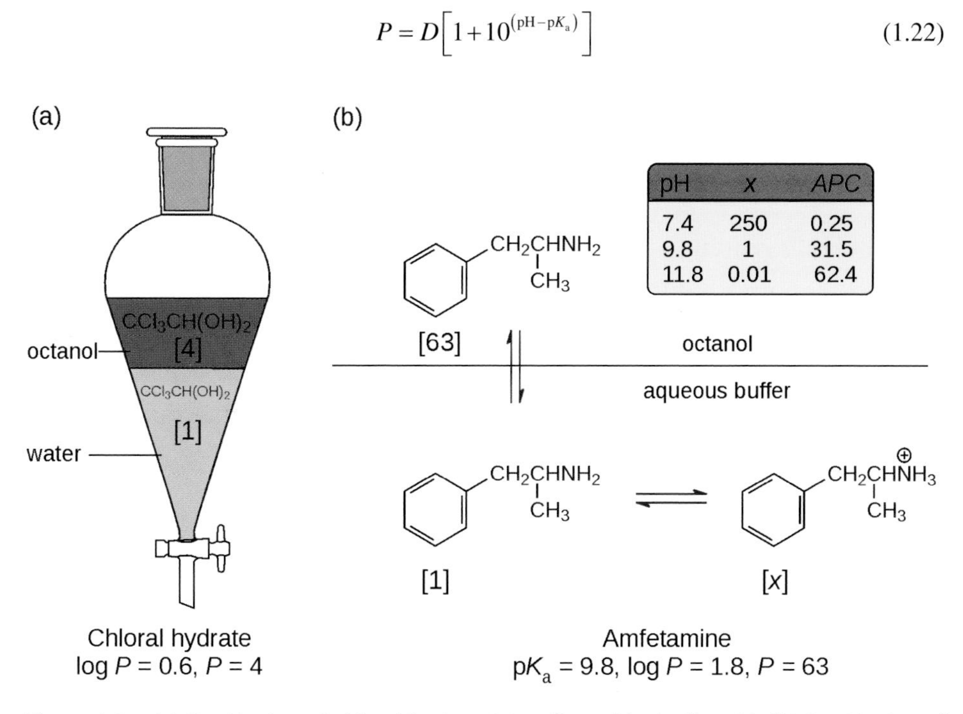

Figure 1.6 (a) Partitioning of chloral hydrate is unaffected by buffer pH. (b) Partitioning of non-ionized amfetamine remains constant, 63:1. However, the ratio of ionized to non-ionized is affected by buffer pH and as a consequence affects the apparent partition coefficient (APC) and the proportion extracted (inset). Note how when pH = pK_a, x = 1 and there are equal concentrations of ionized and non-ionized amfetamine in the aqueous phase.

and for a base:

$$P = D\left[1 + 10^{(\mathrm{p}K_a - \mathrm{pH})}\right] \tag{1.23}$$

When pH = pK_a then, because $10^0 = 1$, $P = 2D$. When the pH is very much less than the pK_a, in the case of acids, or very much larger than the pK_a, in the case of bases, there will be no appreciable ionization and then D will be a good estimate of P (Figure 1.6(b)).

Unless stated otherwise, log P is taken to represent the logarithm of the true partition coefficient, that is, when there is no ionization of the drug. However, for some weak electrolytes, biological activity may correlate better with the partition coefficient between octanol and pH 7.4 buffer solution. These values are referred to as log D or log $D_{7.4}$.

Differences in the pH of different physiological environments, for example plasma and gastric contents, can have a major influence on the way drugs are absorbed and distributed (Chapter 2).

Summary

This chapter has introduced rates, rate constants and half-lives, all crucially important to understanding the pharmacokinetics described in Chapters 4 and 5, and elsewhere in this book. Rate-determining reactions will be encountered when considering sustained-release formulations (Chapter 4) and the kinetics of metabolism (Chapter 6). The influence of the degree of ionization on partitioning of weak electrolytes will be helpful in understanding Chapters 2 and 3.

1.6 Further reading

Allen LV. (ed.) *Remington: The Science and Practice of Pharmacy*. 22nd edn. London: Pharmaceutical Press, 2012.

Atkins P, de Paula J. *Atkins' Physical Chemistry*. 10th edn. Oxford: Oxford University Press, 2014.

Brunton L, Chabner BA, Knollmann BC. *Goodman and Gilman's the Pharmacological Basis of Therapeutics*. 12th edn. New York: McGraw-Hill, 2011. A classic text, many other books of pharmacology are available.

Clayden J, Greeves N, Warren S. *Organic Chemistry*. 2nd edn. Oxford: Oxford University Press, 2012.

Curry SH, Whelpton R. *Disposition and Pharmacokinetics: from Principles to Applications*. Chichester: Wiley-Blackwell, 2011. Chapter 1. Additional information on sources and classification of drugs and stereochemistry.

Leo A, Hansch C, Edkins D. Partition coefficients and their uses. *Chem Rev.* 1971; 36: 1539–44.

Marieb EN, Hoehn K. *Human Anatomy and Physiology*. 10th edn. Pearson Education, 2015.

2

Drug Administration and Distribution

Learning objectives

By the end of this chapter the reader should be able to:

- describe the mechanisms by which substances cross biological membranes
- explain the pH-partition hypothesis and its role in the absorption and sequestration of weak acids and bases
- compare the advantages and disadvantages of administering drugs by the routes described in this chapter
- discuss apparent volume of distribution and explain its relationship with known anatomical volumes
- explain the mechanisms by which drugs are sequestered in tissues
- discuss plasma protein binding and explain its importance in pharmacology.

2.1 Introduction

In order to achieve its effect, a drug must first be presented in a suitable form at an appropriate site of *administration*. It must then be *absorbed* from the site of administration and *distributed* through the body to its site of action. For the effect to wear off the drug must nearly always be *metabolized* and/or *excreted*. These processes are often given

Introduction to Drug Disposition and Pharmacokinetics, First Edition. Stephen H. Curry and Robin Whelpton.

Companion website: www.wiley.com/go/curryandwhelpton/IDDP

the acronym ADME, or occasionally LADME, where L stands for liberation of drug from its dosage form, which can be an important factor in the usefulness of the drug. Finally, drug residues are *removed* from the body. Removal refers to loss of material, unchanged drug and/or metabolic products, in urine and/or faeces, once this material has been excreted into the bladder or bowel by the kidneys and liver. Absorption and distribution comprise the *disposition* (placement around the body) of a compound. Metabolism and excretion comprise the *fate* of a compound.

The most common pathway for an orally administered drug is as indicated by the red arrows in Figure 2.1. This pathway involves metabolism and excretion of both unchanged drug and metabolites. A drug that is excreted in its unmetabolized form will bypass metabolism (orange). An intravenously administered drug undergoes no absorption (purple) but will be distributed, metabolized and excreted in the same way as it would if it had been orally administered. An oral dose may not be absorbed into the systemic circulation because it cannot cross the gastric mucosa (blue) or because it is rapidly converted to its metabolites in the intestinal mucosa or the portal circulation so that only the drug metabolites reach the systemic circulation; this is referred to as 'pre-systemic' or first-pass metabolism (Section 2.4.1.1). Excretion products in the intestine may be reabsorbed (green), for example as with enterohepatic cycling (Section 3.3.8.1), rather than being removed via the faeces.

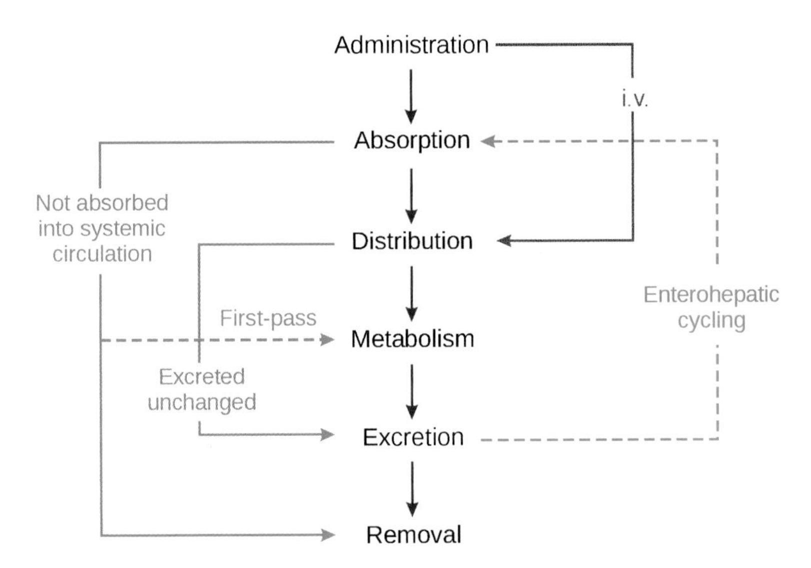

Figure 2.1 General scheme showing the relationship between the various events involved in drug disposition and fate (see text for details).

2.2 Drug transfer across biological membranes

Absorption, distribution and excretion of drugs involve transfer of drug molecules across various membranes, such as the gastrointestinal (GI) epithelium, the renal tubular epithelium, the blood–brain barrier and the placental membrane. Transfer of substances across biological membranes can occur by one or more of five possible mechanisms:

- *Passive diffusion:* through the membrane, down a concentration gradient.
- *Filtration:* through channel proteins and pores in the membrane.
- *Facilitated diffusion:* a carrier-mediated process that does not require energy and where the net flow is down a concentration gradient.
- *Active transport:* involving carrier processes, requiring energy and occurring against the concentration gradient.
- *Pinocytosis:* microscopic invaginations of the cell wall engulf drops of extracellular fluid and solutes are carried through in the resulting vacuoles of water.

Historically, passive diffusion has been viewed as by far the most important mechanism for foreign molecules (Section 2.2.1). Filtration can be transcellular, via pores in the plasma membrane (Figure 2.2) or paracellular (i.e. between cells) and is important for the transfer of small molecules into interstitial fluid via the fenestrations in peripheral capillaries (Section 2.2.3). Filtration plays a major part in the urinary excretion of drug molecules.

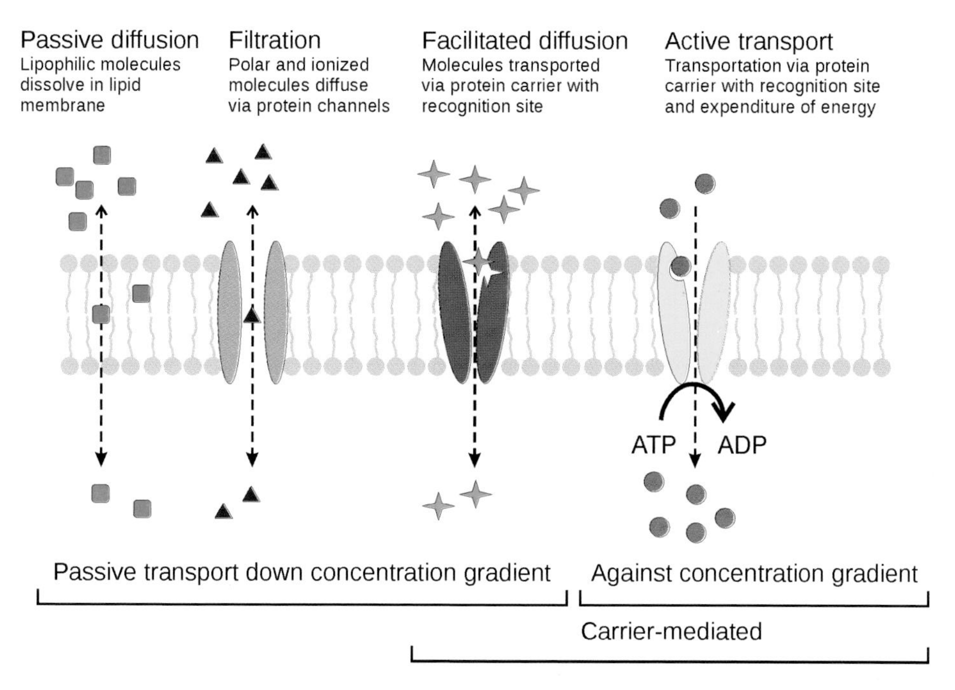

Figure 2.2 *Schematic representation of some of the mechanisms by which substances are transferred across lipid membranes.*

There has been increased interest in the roles of active transport and facilitated diffusion in the disposition of drugs (Section 2.2.2) and the carrier proteins are being investigated as targets for new drugs. The transfer of small foreign compounds across membranes by pinocytosis is largely unknown, but it is believed to be the mechanism by which macromolecules such as botulinum toxin are absorbed. Pinocytosis and other forms of endocytosis are receiving a great deal of interest as a way of absorbing drugs as nanoparticles, micelles and the like. The ileum differs from the duodenum and the jejunum in that it has regions of aggregated lymphoid nodules known as Peyer's patches, and macromolecules, including antigens, can be absorbed via associated specialized cells, known as microfold, or M, cells, which one day may prove to be a route by which insulin can be administered orally (Lopes *et al.*, 2014). Anything that is absorbed in this way drains into the lymphatic system.

2.2.1 Passive diffusion

The rate of diffusion of a drug, dQ/dt, is a function of the concentration gradient across the membrane, ΔC, the surface area over which transfer occurs, A, the thickness of the membrane, Δx, and the diffusion coefficient, D, which is characteristic of the particular drug. The diffusion constant of a compound is a function of its solubility in the membrane, its molecular weight and its steric configuration, therefore:

$$\frac{dQ}{dt} = RDA\frac{\Delta C}{\Delta x} \tag{2.1}$$

where R is the partition coefficient of the drug between the membrane and the aqueous phase. Because for a given drug diffusing through a particular membrane A, D, R and the distance over which diffusion is occurring are constants, these can be combined into one constant, known as permeability, P, so that:

$$\frac{dQ}{dt} = P\Delta C = P(C_1 - C_2) \tag{2.2}$$

It is important to note that not only is the rate of diffusion dependent of the nature of the drug and the membrane and, of course, the concentration gradient, the area over which transfer is taking place is of major importance. The micro villi of the small intestine and the kidney proximal convoluted tubular cells provide a large surface area over which transfer can take place.

2.2.1.1 pH-partition hypothesis

With passive diffusion drug molecules have to dissolve in the membrane and so low molecular weight, lipophilic species diffuse freely, but polar, particularly ionized, molecules do not. Whether a weak electrolyte is ionized or not will be a major determinant of whether it will diffuse across a biological membrane. The ratio of ionized to non-ionized forms is a function of the pH of the aqueous environment and the acid dissociation constant, $K_{a,}$ as explained in Section 1.6. This means that in any aqueous solution both ionized and non-ionized forms are present to varying degrees.

Calculation of the equilibrium distribution of aspirin between stomach contents and plasma water illustrates the importance of pH partitioning (Figure 2.3). It will be readily appreciated that, at equilibrium, for the *non-ionized* species (which alone passes the lipid membrane) the concentration ratios will be the same:

$$\frac{\text{concentration in membrane}}{\text{concentration in gastric acid}} = \frac{\text{concentration in membrane}}{\text{concentration in plasma water}}$$

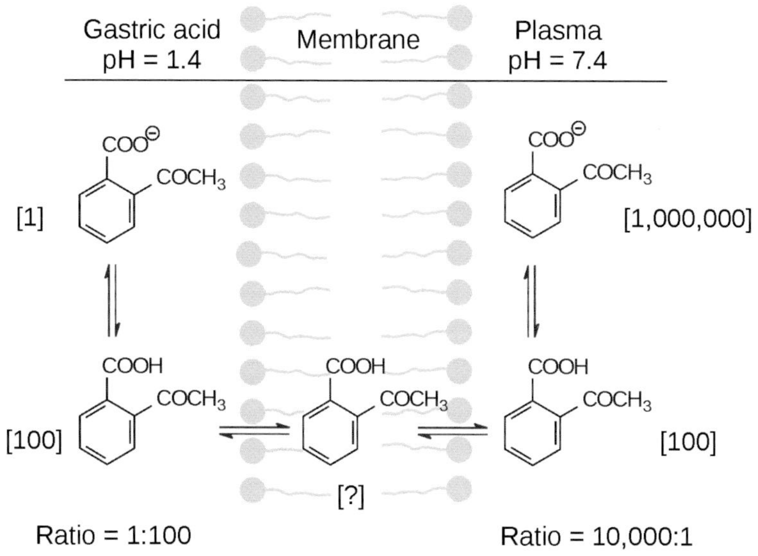

Figure 2.3 *Equilibrium distribution of aspirin* (pK_a = 3.4) *between two solutions of pH 1.4 and pH 7.4 separated by a simple membrane. Relative concentrations are shown in* [ ].

Thus for the non-ionized species the concentrations on each side of the membrane will be identical. For convenience, in Figure 2.3 we have designated this concentration as being 100 units, [100]. However, from the Henderson–Hasselbalch equation (Section 1.6.1) the ratio of ionized to non-ionized aspirin at pH 1.4 is 1:100. In other words, for every 100 units non-ionized there is 1 unit of ionized aspirin. In the alkaline environment of the plasma, aspirin will be highly ionized, the ratio being 10,000:1, so at pH 7.4, for every 100 molecules of non-ionized aspirin there will be 10,000 × 100 = 1,000,000 molecules ionized, therefore at equilibrium the ratio of total material on the two sides will be:

$$\frac{\text{total concentration in acid}}{\text{total concentration in plasma}} = \frac{100+1}{1{,}000{,}000+100} = \frac{101}{1{,}000{,}100} \approx \frac{1}{9900}$$

that is, almost 10,000 to 1 in favour of plasma. If drug molecules are introduced into any part of the system, they will transfer between the various media, including the membrane, until this concentration ratio is achieved. However, when aspirin is ingested, after absorption into the plasma it will be carried away in the bloodstream. Any binding to plasma proteins will also increase the concentration gradient in favour of absorption because there will be fewer non-ionized, non-bound molecules in the plasma.

2.2.2 Carrier-mediated transport

There are two important superfamilies of drug transporters, solute carriers (SLC) and ABC transporters, the latter functioning chiefly as efflux pumps (Table 2.1).

The SLC superfamily of solute carriers is the second largest family of membrane proteins after G protein-coupled receptors. The family includes the organic cation transporters

Table 2.1 Selected examples of SLC and ABC transporters

Transporter	Gene	Selected localizations	Selected substrates	Selected inhibitors
OATP1A2	*SLCO1A2*	Brain Enterocytes (luminal) Kidney PCT (luminal)	Atenolol Ciprofloxacin Fexofenadine	Grapefruit juice Orange juice Verapamil
OATP2B1	*SLCO2B1*	Enterocytes (luminal) Hepatocytes (sinusoidal)	Aliskiren Atorvastatin Pravastatin	Ciclosporin Grapefruit juice Orange juice
OATP1B1	*SLCO1B1*	Hepatocytes (sinusoidal)	Pravastatin Rifampicin Simvastatin	Ciclosporin Lopinavir Rifampicin
OATP1B3 (OATP8)	*SLCO1B3*	Hepatocytes (sinusoidal) Kidney PCT (basolateral)	Paclitaxel Pravastatin Rifampicin	Clarithromycin Ciclosporin Rifampicin
OAT1	*SLC22A6*	Kidney PCT (basolateral)	Captopril Cidofovir Furosemide	Indometacin Probenecid
OAT3	*SLC22A8*	Kidney PCT (basolateral)	Benzylpenicillin Furosemide Methotrexate Pravastatin	Indometacin Probenecid Salicylate
OCT2	*SLC22A2*	Brain Lung Kidney PCT (basolateral)	Cimetidine Cisplatin Metformin Ranitidine	Cimetidine Procainamide Ranitidine Trimethoprim
MATE1	*SLC47A1*	Hepatocytes (canalicular) Kidney PCT (basolateral) Skeletal muscle	Cefalexin Metformin Procainamide	Cimetidine Pyrimethamine Quinidine
PepT1	*SLC15A1*	Enterocytes (luminal) Kidney PCT (luminal)	Penicillins Cephalosporins Enalapril Valaciclovir	Glibenclamide Tolbutamide
P-gp	*ABCB1*	Adrenal gland Blood–brain barrier Enterocytes (luminal) Hepatocytes (canalicular) Placenta Kidney PCT (luminal)	Digoxin Docetaxel Fexofenadine Ritonavir Vinblastine	Clarithromycin Ciclosporin Itraconazole Quinidine Ritonavir Verapamil
MRP2	*ABCC2*	Enterocytes (luminal) Hepatocytes (canalicular) Kidney PCT (luminal)	Glucuronide and GSH conjugates Bromosulfophthalein	Ciclosporin Probenecid
BCRP	*ABCG2*	Enterocytes (luminal) Hepatocytes (canalicular) Mammary glands	Doxorubicin Methotrexate 5-Flurouracil	Ritonavir Omeprazole

BCRP: breast cancer resistance protein; MATE: multidrug and toxin extrusion protein; MRP: multidrug resistance-associated protein; OAT: organic anion transporter; OATP: organic anion-transporting polypeptide; OCT: organic cation transporter; PepT: peptide transporter; P-gp: P-glycoprotein.

(OCTs), which transport small cations such as tetramethylammonium, choline, histamine and norepinephrine, and the organic anion-transporting polypeptides (OATPs), of which there are several human isoforms (Shugarts & Benet, 2009). These are widely distributed and are found in brain, liver, kidney, small intestine and many other tissues. Many SLC proteins are responsible for facilitated diffusion, the net flow of solute being down the concentration gradient, whereas others are coupled to ion transport and function by secondary active transport. For example, multi-drug and toxin extrusion protein (MATE1) is dependent on an oppositely directed proton gradient as the driving force, and may act as either an uptake or efflux transporter, depending on the relative pH values of intracellular and extracellular fluids. SLC proteins are important in the transfer of conjugates of glucuronic acid, sulfate and glutathione across cell membranes. OATPs in the small intestine may be important for the absorption of drugs, for example OATPs transport fexofenadine from the intestinal lumen. SLC transporters located on the basolateral side of proximal tubular cells (PCT) of the kidney transport pencillins and cephalosporins (OAT1 and OAT3) whilst OCT1 and OCT2 transport organic bases. The natural substrates for PepT1, which is located on the luminal membrane of enterocytes, are di- and tripeptides, but β-lactams and angiotensin-converting enzyme (ACE) inhibitors are also substrates. PepT1 has been the focus for development of new drugs, particularly drugs that have been modified to contain peptide-like structures so that they can exploit this active uptake. The antiviral drug aciclovir has an oral availability of ~10–30%, whereas the bioavailability of its valyl ester, valaciclovir (valacyclovir), is ~63% (Cao *et al.*, 2012).

The adenosine triphosphate (ATP)-binding cassette family of transport proteins (ABC transporters) are a superfamily of efflux pumps (Yang, 2013) that transport substrates from the intracellular to the extracellular side of the cell membrane. They derive their energy from binding ATP, which allows transport of substrates against their concentration gradients. The most studied member of the family is P-glycoprotein (permeability protein, P-gp), which was first recognized in tumour cells that were resistant to a number of cytotoxic agents as a result of over-expression of what was then referred to as the multidrug resistance (*MDR1*) gene, now known as the *ABCB1* gene. Other members of this family include multidrug resistance-associated proteins (MRP), which are important in the biliary secretion of glucuronides and efflux of anions from PCTs, and breast cancer resistance protein (BCRP), which is widely located and appears to have a protective role as an efflux pump in the intestine and blood–brain barrier.

The transporters of both superfamilies are subject to induction, inhibition and genetic differences, and so are sites for drug–drug interactions and variability in drug response (Müller & Fromm, 2011).

2.2.3 Nature of the membrane

Drug transfer will also be affected by the nature of the membrane over which transfer is occurring. Membranes vary depending on their location and function. There are three basic types of capillary (Figure 2.4). Small molecular weight drugs will readily diffuse through those peripheral capillaries that have gaps or *fenestrations*. Human serum albumin (M_r ~69,000) is too large to diffuse and as a consequence is excluded from interstitial fluid. At the glomerulus substances up to about the size of albumin pass freely, while the gaps in the liver sinusoids are large enough for macromolecules such as lipoproteins to be filtered. Other membranes, notably those of the GI tract and the placenta, do not have

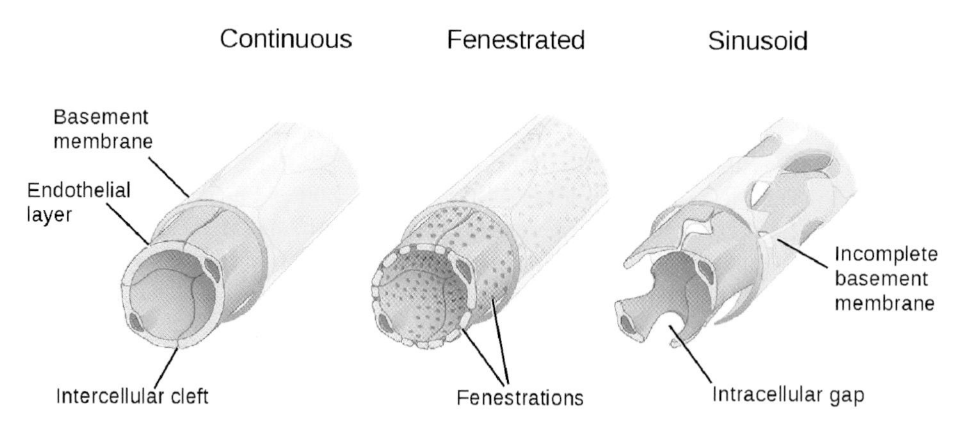

Figure 2.4 Representation of three types of capillary. Reproduced from '2104 Three Major Capillary Types' by OpenStax College – Anatomy & Physiology, Connexions website. http://cnx.org/content/col11496/1.6/, 19 June 2013. Licensed under CC BY 3.0 via Commons.

fenestrations. The observation from early studies that certain dyes did not enter the brain led to the concept of the blood–brain barrier (BBB). It is now known that the brain is separated from the peripheral circulation by specialized capillaries with tightly-packed endothelial cells. For these membranes, drugs must traverse them either by passive diffusion or specialized carrier-mediated transport.

2.3 Drug administration

The route of drug administration will be determined by the nature of the drug and the indication for its use.

2.3.1 Oral administration

This route is popular as it is generally convenient and does not require medical skills or sterile conditions. Thus it is appropriate for outpatient use and medicines bought over the counter (OTC). For people who have difficulty swallowing tablets or capsules, the elderly or infants for example, the drug may be may be given as a solution or suspension in liquid.

However, the GI tract is a harsh environment. Gastric pH is low and drugs that are unstable at low pH values (acid labile) such as benzylpenicillin (penicillin G) and methicillin are inactivated. The acid environment catalyses opening of the β-lactam ring in these pencillins, but these drugs may sometimes be used in the very young or the elderly in whom gastric pH is higher, but generally they are given by injection. Similarly, the presence of proteases makes the oral route unsuitable for many proteins and peptides such as insulin and oxytocin.

Absorption occurs chiefly by passive diffusion of lipophilic molecules and carrier-mediated transport of drugs that are endogenous, levodopa for example, or those that are structurally similar to endogenous compounds, such as the cytotoxic agent 5-fluorouracil. With passive diffusion the rate of absorption is proportional to the concentration or the amount of drug to be absorbed and the *faction* or *percentage* absorbed in a given interval remains constant. With carrier-mediated mechanisms, be it active transport or facilitated diffusion, there is a limited capacity and the transporter can be saturated as shown in Figure 2.5. Note how increasing concentrations of uracil saturate the carrier so the percentage absorbed decreases, whereas for salicylic acid, which is absorbed by passive diffusion, the percentage absorbed is independent of the concentration. Indeed the amount of salicylic acid absorbed is directly proportional to the initial drug concentration (Figure 2.5(b)), indicating first-order absorption. The relationship for uracil, however, is hyperbolic and asymptotes to a maximum value, in much the same way as is seen with protein binding (Section 2.5) or ligand-receptor interactions (Section 8.2).

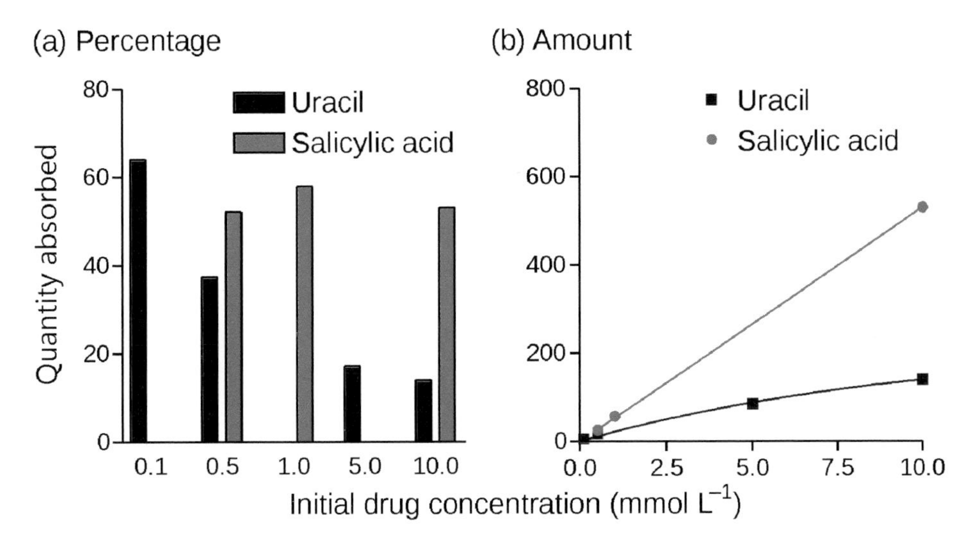

Figure 2.5 Absorption of uracil and salicylic acid from the rat small intestine as function of initial drug concentration as (a) percentage and (b) amount absorbed (mmol). Data for uracil were fitted to a hyperbola and those for salicylic acid to a straight line.

Being largely unionized in acid, aspirin is absorbed from the stomach (Section 2.2.1.1), but most of the absorption occurs in the small intestine where the large surface area compensates for the less favourable degree of ionization. Weak bases, which are highly ionized in gastric acid, cannot be absorbed until they have left the stomach, so delayed gastric emptying can delay the effect of such drugs. Quaternary ammonium compounds (Figure 2.6), which are permanently ionized, would not be expected to be absorbed to an appreciable extent by passive diffusion and this appears to be the case. Tubocurarine, the purified alkaloid from the arrowhead poison curare, is not absorbed and the oral absorption of pyridostigmine is very low and erratic, such that when an oral dose is used to treat myasthenia gravis the dose is typically 30 to 60 times larger than an equivalent intravenous or intramuscular one, for example to reverse the effects of tubocurarine. Of toxicological importance is the absorption of the diquaternary ammonium herbicide, paraquat, which is interesting as extensive absorption would not be expected for such an ionized molecule. However, its absorption is saturable, which supports the hypothesis that one or more carriers are involved, possibly one for choline and one for putrescene.

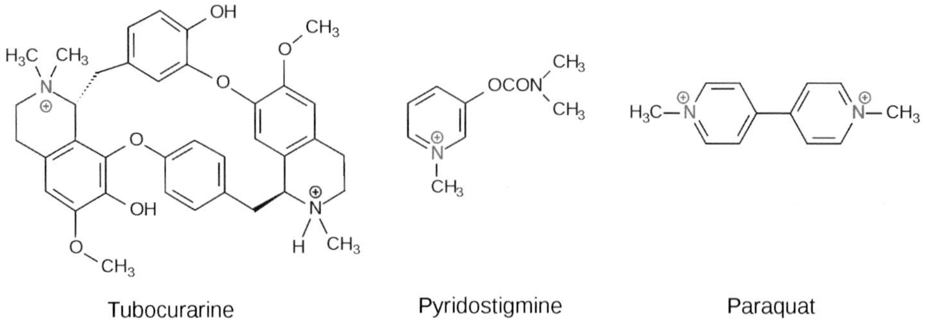

Tubocurarine Pyridostigmine Paraquat

Figure 2.6 Examples of quaternary ammonium compounds. Quaternary nitrogen atoms are shown in blue, the tertiary amine in tubocurarine is shown protonated because that is how it is at physiological pH values.

2.3.1.1 Pre-systemic metabolism

Drugs may be metabolized before they reach the systemic circulation, and this pre-systemic metabolism may markedly reduce their bioavailability (Section 4.4.3). Materials absorbed from the stomach and intestine are carried, via the mesenteric capillary network and the hepatic portal vein, to the liver (see Figure 2.8). If the drug is largely metabolized as it passes through the liver, then little of it will reach the systemic circulation. The consequences of this *first-pass metabolism*, so-called because it represents the first time that the drug has passed thought the liver, will depend on whether or not the metabolites are pharmacologically active. In the case of glyceryl trinitrate (GTN, nitroglycerin), which is almost totally metabolized, the di- and mononitrate metabolites have very reduced activity and generally this drug is considered to be inactive when taken orally. GTN is given by more suitable, alternative routes. The extent to which a drug undergoes pre-systemic metabolism can be obtained by comparing the plasma concentration–time curves and areas under the curves (AUC), after an oral (p.o.) and an intravenous (i.v.) dose of the drug (Equation 4.32). This is illustrated for chlorpromazine in Figure 2.7, where the systemic availability of chlorpromazine is markedly reduced after oral administration, as evidenced by the reduced AUC_{po} compared to the area under the intravenous curve, AUC_{iv}. Note that the values have been adjusted to take account of the fact that the intravenous dose was only ¼ of the oral dose. The low oral bioavailability is known to be largely due to pre-systemic metabolism because studies with radioactive drug show that large quantities of radioactive metabolites are measurable in the systemic circulation.

First-pass metabolism may influence the relative proportions of metabolites produced when compared with other routes of administration. This is the case for propranolol; when it is given orally, pharmacologically active 4′-hydroxypropranolol is produced, but little of this metabolite is measurable when propranolol is given intravenously. It has been suggested that this is because the high concentrations of propranolol reaching the liver after oral administration saturate the pathway that produces naphthoxylacetic acid, a major metabolite after i.v. injection. Comparison of the AUC values after oral and intravenous doses support this supposition. The amount of drug reaching the systemic circulation is very much reduced when given orally and, furthermore, there appears to be a threshold dose of approximately 20 mg below which little or no propranolol is measurable in the

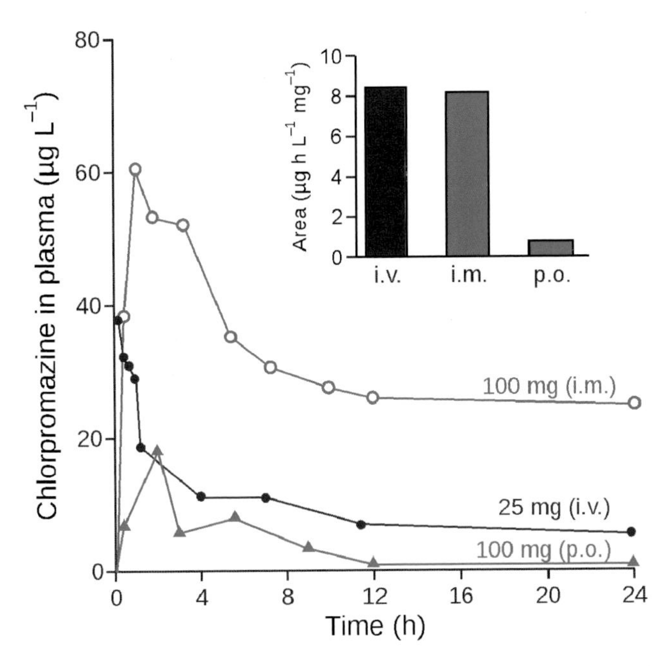

Figure 2.7 *Plasma concentrations of chlorpromazine after three routes of administration and the areas under the curves after normalization for dosage (inset).*

plasma, suggesting that at low doses all of the dose may be metabolized and that higher doses saturate some of the drug-metabolizing pathways (Shand & Rangno, 1972). Saturation of first-pass metabolism may apply to other drugs that are normally extensively metabolized as they flow through the liver.

Drugs may be metabolized before they even reach the liver. GI tract mucosal cells contain several drug-metabolizing enzymes, probably the most important of which is cytochrome P450 3A4 (CYP3A4; Section 3.2.1.1). As described above, pre-systemic metabolism may render a drug inactive, or larger doses of a drug may have to be administered when it is given orally. On the other hand, prodrugs, that is drugs that are inactive until they have been metabolized (Section 3.2.1.1), will be activated by pre-systemic metabolism. Dose for dose prodrugs should be more pharmacologically active when given orally than by other routes as they will be activated by first-pass metabolism.

2.3.1.2 Efflux proteins

Not only are drugs metabolized by intestinal cells, drug that is not metabolized may be returned to the gut lumen by efflux. P-Glycoprotein and other proteins, such as MRP2 and BCRP (Table 2.1), transport substrates from the intracellular to the extracellular side of cell membranes. The P-gp gene is highly expressed in the apical membrane of enterocytes lining the GI tract, renal proximal tubular cells, the canalicular membrane of hepatocytes and other important blood–tissue barriers such as those of the brain, testes and placenta. Its location suggests that it has evolved to transport potentially toxic substances out of cells. P-gp is inducible and many of the observations that were once ascribed to enzyme induction or inhibition (Chapter 13) may in fact be due to changes in P-gp activity.

2.3.1.3 Gastrointestinal motility and splanchnic blood flow

Absorption is facilitated by thorough mixing of the drug within the GI tract. Mixing increases the efficiency with which the drug is brought into contact with surfaces available for absorption. In some instances a reduction in GI motility, for example by opiates or antimuscarinic drugs, may reduce absorption, including their own. On the other hand, excessive motility and peristalsis will reduce the transit time and this may be critical for drugs that are slowly absorbed or are only absorbed from particular regions of the GI tract. The most favourable site for absorption for the majority of drugs is the small intestine with its large surface area due the presence of microvilli, sometimes referred to as the 'brush-border' because of its appearance. The transit through this part of the GI tract is usually about 3–4 h. The surface area for absorption decreases from the duodenum to the rectum, although the transit time in the large bowel is in general about 12–24 h, possibly longer. It is thought that some drugs are absorbed from particular regions, for example part of an oral dose of drug may be absorbed for 3–4 h with little more appearing in the blood after that time; the remaining portion of the dose being expelled in the faeces. Increased intestinal motility may reduce the absorption of such a drug because the time it is in the optimal region for absorption is reduced. However, a drug that is rapidly and extensively absorbed from the duodenum will be less affected. As a consequence it is sometimes difficult to predict how changes in GI motility will affect the oral availability of a drug.

Splanchnic blood flow will affect the rate of removal of the drug from the site of absorption, as bulk flow and transport of drugs by plasma proteins is the major mechanism by which drugs are carried away from their sites of absorption and around the body. The blood flow to the GI tract, which represents approximately 30% of the cardiac output, is lower during fasting than after feeding. Weight for weight, the mucosa of the small intestine receive the largest proportion of the flow, followed by those of the colon and then the stomach.

Gastric emptying, or rather the lack of it, such as the pyloric stenosis that often follows surgery, can have a major influence on oral availability. Tablets and capsules, particularly enteric-coated tablets, which are designed not to release their contents until they reach the intestine, will be trapped in the stomach.

2.3.1.4 Food and drugs

It is often recommended that drugs should be taken at meal times. In part this is as a reminder to aid compliance (adherence) and under some circumstances to reduce gastric irritation, as, for example, with aspirin, which can cause gastric bleeding. However, food may have a major, but not always predictable, effect on oral availability. Food generally delays gastric emptying but, as discussed above, it increases splanchnic blood flow and increases secretion of gastric acid. The reduced pH and greater time spent in the stomach will reduce the availability of acid labile drugs, but drugs that increase gastric pH, for example H_2-receptor antagonists, may reduce the absorption of drugs such as ketoconazole that are more soluble in acid (Table 2.2). It is often assumed that food delays absorption but does not necessarily reduce it. The absorption of griseofulvin, a very poorly soluble antifungal drug, is increased when taken with a 'fatty' meal. Similarly, the absorption of the antiretroviral drug saquinavir can be doubled when taken with a fatty meal as the release of bile salts aids dissolution and facilitates absorption.

Table 2.2 Examples of the effects of food and drugs on the oral availability of drugs

Mechanism	Affected drug	Effect on absorption	Notes
Gastric acid - low pH (induced by food)	Benzylpenicillin Methicillin Azithromycin Isoniazid	Reduced	Acid labile compounds destroyed with loss of activity Occurs to a lesser extent with ampicillin
Gastric acid – high pH (fasting, H_2-antagonists)	Amprenavir Ketoconazole	Reduced	Reduced solubility at higher pH values
Dairy products Ferrous sulfate Antacids	Bisphosphonates Tetracyclines Ciprofloxacin Penicillamine	Reduced	Chelation with divalent metal ions, Ca^{2+}, Fe^{2+}, Mg^{2+}
Fatty meals	Griseofulvin Mefloquine Saquinavir	Enhanced	May be due to increased solubility in bile salts released in response to fatty meal
Dietary fibre	Digoxin	Reduced	Adsorption to the fibre
Reduced gastric emptying	Many drugs Enteric-coated tablets	Delayed	May be drug-induced, e.g. opiates used during surgery
Grapefruit and other juices/ketoconazole	Terfenadine	Increased	Inhibition of pre-systemic metabolism by CYP3A4
Apple juice	Fexofenadine	Decreased	Inhibition of uptake by OATP12A
Indometacin	Sulfasalazine	Increased	Inhibition of efflux by MRP2

The constituents of a meal may interfere with the processes of absorption, such as the components of grapefruit, or they may interact directly with the drug. Tetracyclines chelate divalent metal ions to form unabsorbable complexes, and so their absorption is reduced by milk (high in calcium), magnesium-containing antacids and ferrous sulfate. The effects of grapefruit juice and some drugs can be complex. The interaction between terfenadine and grapefruit juice and drugs such as ketoconazole led to this antihistamine being withdrawn. The drug is normally metabolized almost entirely to fexofenadine by CYP3A4 in the enterocytes but inhibition of this enzyme resulted in absorption of the parent drug, which can exhibit cardiotoxicity in some individuals. Paradoxically, grapefruit and other fruit juices reduce the oral bioavailability of fexofenadine; this has been ascribed to inhibition of uptake by OATP1A2.

From the foregoing, it should be obvious that concomitant use of other drugs can affect oral absorption in a number of ways, including changing gastric pH, gut motility and by forming unabsorbable complexes. Yet despite the main problems (Figure 2.8), the oral route is the preferred one for outpatient and OTC medicines.

2.3.2 Sublingual administration

Sublingual (literarily 'under the tongue') administration is when tablets are chewed or crushed and held in the mouth for absorption from the buccal cavity. This occurs by diffusion and although the surface area for absorption is small, there is a rich blood supply and

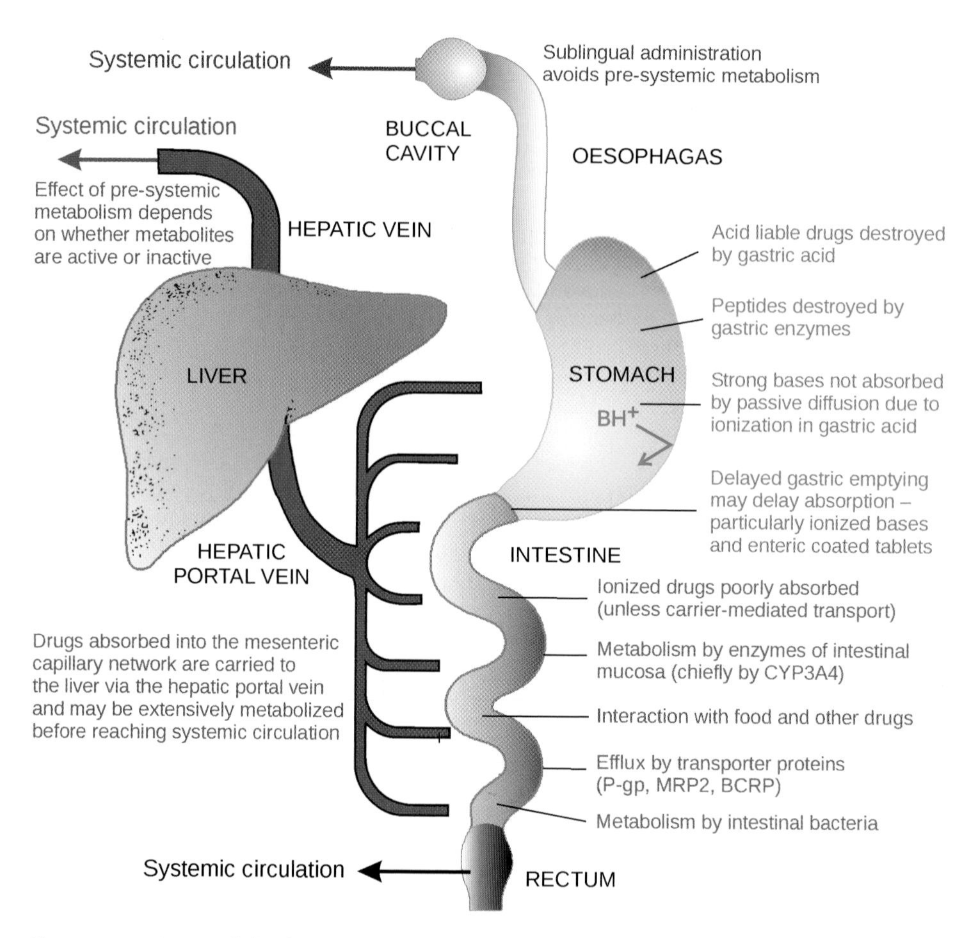

Figure 2.8 *Some of the factors that may reduce oral availability. Other factors include GI motility, absorptive surface area and splanchnic blood flow.*

absorption occurs rapidly. Furthermore, materials absorbed across the buccal membrane avoid the hepatic portal system so this route is useful for drugs such as glyceryl trinitrate (GTN) with which a rapid effect is desired and which can be destroyed in the GI tract before absorption or by pre-systemic metabolism (Figure 2.8). Because GTN is volatile and may be lost from tablets, limiting its shelf-life, it is available as an aerosol for spaying into the mouth as a metered dose. Buprenorphine is available as a lozenge that can be held in the mouth to give sustained-release of this potent opioid.

2.3.3 Rectal administration

Substances absorbed from the lower part of the rectum are carried to the vena cava and so avoid first-pass metabolism. However, this is not the chief reason for choosing this route of administration. The surface area for absorption is not especially large, but the blood supply is extremely efficient and absorption can be quite rapid. This is considered sufficiently advantageous to warrant extensive use of this route of administration for a wide range of drugs in certain countries. Apart from when a local effect is required, in the UK suppositories tend to be used for more specialized purposes, such as the antiemetic, prochlorperazine – there is little point in taking a drug orally when one is vomiting. Diazepam suppositories are available for use in epileptic infants, when insertion of an intravenous cannula might be considered dangerous. Slow-release aminophylline is given rectally, often at night, to ease breathing in asthmatic children. The disadvantages of this route, apart from patient acceptability, include local irritation and inflammation, and the possibility that the patient may need to defaecate shortly after insertion of the suppository.

2.3.4 Intravenous and intra-arterial injections

Injecting a drug directly into the circulation avoids any problems of absorption and usually produces the most rapid onset of effects of any route. Peak blood concentrations occur immediately after a rapid 'bolus' injection, but if this presents a problem or a sustained effect is required, then the drug can be given as a slow intravenous infusion, over several minutes, hours or even days. For the longer periods, portable (ambulatory) pumps are available. These have been used to deliver insulin and opiates (usually in terminally ill patients), and in the treatment of iron poisoning and certain cancers, amongst other applications.

Intravenous injections are not used simply to overcome problems that may be encountered using other routes, but because the i.v. route is the most appropriate. Examples of such drugs include the i.v. general anaesthetics (thiopental, propofol), muscle relaxants (tubocurarine, suxamethonium) and neostigmine (to reverse the effects of tubocurarine-like drugs); all these drug examples are used during surgery and so in the hospital environment it is not particularly inconvenient to use this method of administration.

Intra-arterial injections are more specialized and are typically used in the treatment of certain tumours. By injecting a cytotoxic drug into an artery the drug is carried directly to the tumour.

The disadvantages of i.v. injections include the fact that sterile preparations and equipment are required. A high degree of skill is required, particularly to prevent extravasation, that is, injection near to the vein or leakage from it, which can lead to serious tissue damage.

2.3.5 Intramuscular and subcutaneous injections

Some of the problems of low oral bioavailability can be avoided by intramuscular (i.m.) or subcutaneous (s.c.) injections. Unlike bolus i.v. injections there is no immediate peak plasma concentration as the drug has to be absorbed from the injection site. Drugs, including ionized ones, enter the systemic circulation via fenestrations in the capillary walls. Size does not appear to be a limiting factor up to approximately $M_r = 5000$, although the absorption is flow dependant and increasing local blood flow, for example by warming and massaging the injection site, can increase the rate of absorption. The rate of absorption can be delayed by co-injection of a vasoconstrictor to reduce blood flow to the area, for example the use of epinephrine to prolong the effect of a local anaesthetic.

The rate of absorption may be faster or slower than that following oral administration. Absorption after i.m. injection of chlordiazepoxide or diazepam may be delayed because the drugs precipitate at the injection site. Some i.m. preparations may be formulated to provide sustained release from the injection site, for example microcrystalline salts of penicillin G (i.m.) and various insulin preparations (s.c.) (Section 4.4.4).

Sterile preparations and equipment are required but as they require less skill, patients or their carers can be trained to perform i.m or s.c. injections.

2.3.6 Transdermal application

The epidermis behaves as a lipoprotein barrier while the dermis is porous and permeable to almost anything. Consequently, lipophilic molecules penetrate the skin easily and rapidly, whilst polar ionized molecules penetrate poorly and slowly. For many years it has been known that chlorinated solvents and some organic nitro compounds are potentially toxic because they are rapidly absorbed through the skin. More recently, several pharmacological preparations designed for absorption across the skin have been introduced. These may be in the form of ointments and creams to be rubbed on to the skin or patches to be stuck on. Some of the patches incorporate a rate-limiting membrane to ensure a steady, sustained release of the drug. Generally the drugs have to be potent as well as lipophilic because large doses would be problematic. Examples of drugs applied to the skin for systemic effects include glyceryl trinitrate, hyoscine, buprenorphine, steroids (contraceptives and hormone replacement) and nicotine.

2.3.7 Insufflation

The nose, with its rich blood supply, highly fenestrated capillaries and an epithelium with gaps around the goblet cells, allows absorption of drugs that cannot be given orally. Peptide hormones, insulin, calcitonin and desmopressin can be given as nasal sprays. Furthermore, it has been suggested that some drugs can enter the CNS directly via the nose and this route is being investigated for molecules which do not normally enter the brain. However, the volume that can be sprayed into the nose is limited.

2.3.8 Inhalation

Volatile and gaseous general anaesthetics are generally given via the lungs, their large surface area giving rapid absorption and onset of effect. Because most of these anaesthetics are also excreted via the lungs, the level of anaesthesia can be controlled by adjusting the partial pressure of the drug in the apparatus used to administer it.

Other drugs frequently given by inhalation are those to relieve bronchiolar constriction in asthmatic patients. The β-adrenoceptor agonists salbutamol (albuterol) and terbutaline, and the antimuscarinic drug ipratropium, are examples. Applying these agents directly to the lungs gives rapid relief, reduces the dose of drug required and so lessens the severity of any systemic adverse effects. Disodium cromoglicate (cromolyn sodium) is very poorly absorbed when given orally but is absorbed from the lungs after inhalation as a fine dry powder.

2.3.9 Other routes of administration

Drugs may be applied to various other sites, usually for a local effect. Intravaginal applications of antifungal creams are usually for local effects against *Candida albicans* ('thrush'). Prostaglandin pessaries may be used to induce labour. Lipophilic drugs such as physostigmine, pilocarpine and timolol are absorbed across the cornea and used to treat glaucoma. Tropicamide eye-drops may be used to dilate the pupil to aid ophthalmic examinations. Sometimes sufficient drug may be absorbed to cause systemic effects. Timolol eye drops, for example, have been known to cause bradycardia.

Drugs are generally well absorbed from the peritoneal cavity and although this route is rarely used in human beings, intraperitoneal injections (i.p.) are a convenient method for dosing laboratory animals.

2.4 Drug distribution

The majority of drugs have to be distributed to their site(s) of action. Only rare examples such as anticoagulants, heparin and the like, which have their effects in the bloodstream, do not. Drugs are carried by the circulation, often bound to plasma proteins from where they equilibrate with their sites of action or other storage sites (Figure 2.9).

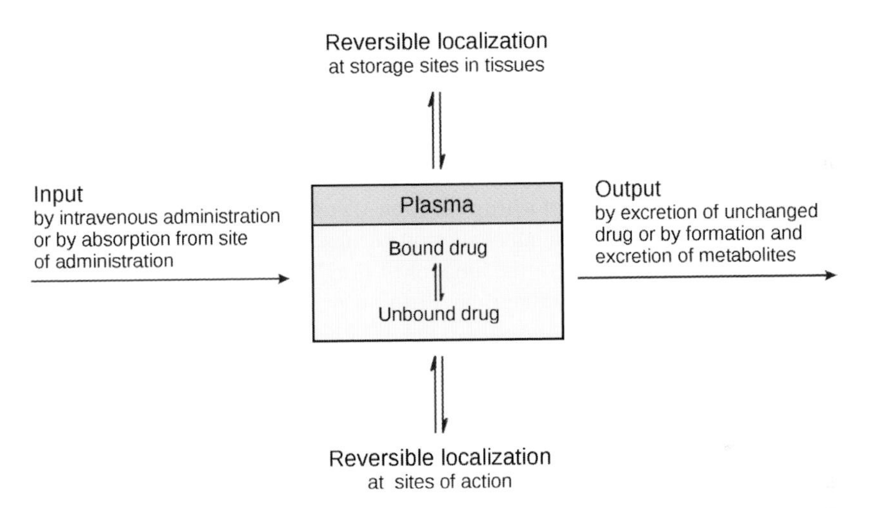

Figure 2.9 Scheme illustrating the key role of plasma in transporting drugs to sites of action, other storage sites and sites of elimination.

2.4.1 Extent of distribution

The extent and rate at which a drug is distributed is dependent on its physiochemical characteristics. Small lipophilic drugs that readily penetrate membranes are generally widely distributed, whereas polar ionized ones and macromolecules are often contained within particular anatomical volumes. The body is made up of approximately 60% water, 18% protein, 15% fat and 7% minerals. Body water can be subdivided into that in the cells (intracellular water (ICF), 40% of total body weight) and the remaining extracellular water (ECF, 20%), which can be subdivided further into interstitial fluid (15%) and plasma (5%). The blood volume is 9% of total body water (TBW), the 4% of body water associated with the red cells being part of the intracellular volume (Figure 2.10).

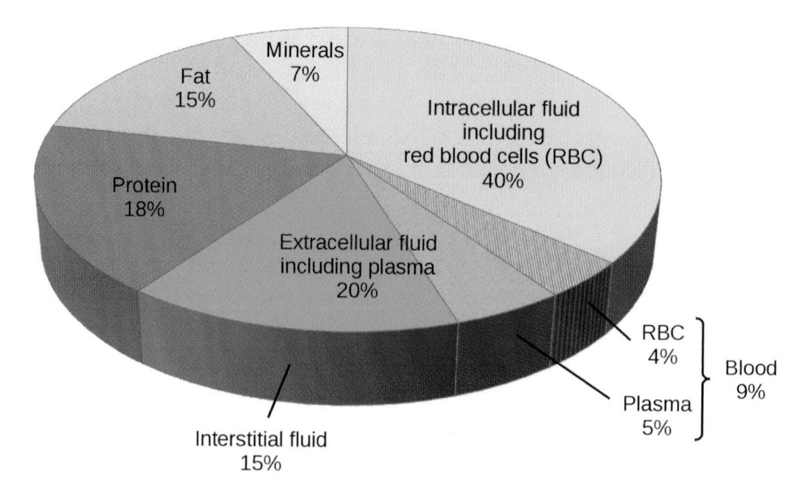

Figure 2.10 Approximate composition of the body.

2.4.1.1 Apparent volume of distribution

The extent of distribution of a drug may be assessed by a parameter known as the apparent volume of distribution (V). It is an important pharmacokinetic parameter because not only does it give an indication of the extent to which a drug is distributed, it has a major influence on the elimination half-life of a drug (Section 4.2). Values of V are normally obtained from determination of the plasma concentration because it can be defined as 'the volume of fluid that is required to dissolve the amount of drug in the body (A_t) to give the same concentration as that in plasma at that time, t'. In other words:

$$V = \frac{A_t}{C_t} \tag{2.3}$$

where C_t is the plasma concentration at time t. Immediately following a rapid intravenous injection of drug, the amount in the body is, for all intents and purposes, the dose, D, that has been injected, therefore if the plasma concentration was known at this time then the volume could be calculated. The practical situation is a little more complicated because time has to be allowed for the drug to mix within the circulation and be distributed to whichever tissues the drug can penetrate. It is therefore usual to measure the plasma concentrations over a period of time and to plot them, or ln C, against time (Figure 1.3) and to extrapolate to $t = 0$ to obtain a value of C_0, from which V can be calculated:

$$V = \frac{D}{C_0} \tag{2.4}$$

Although C_0 may be referred to as the concentration at time zero, it is a *theoretical* plasma concentration that would occur if it *were possible* for the drug to be distributed instantaneously, which of course it cannot. This is discussed further in Chapter 4. To allow for differences in body weight amongst different subjects, V may be expressed as litres per kilogram of body weight (L kg^{-1}).

Table 2.3 *Examples of apparent volumes of distribution*

Compound	V (L kg^{-1})*	Notes
Evans blue[†]	0.05	Dye to measure plasma volume
Heparin	0.06	Macromolecule – cannot enter interstitial fluid
Inulin	0.21	Used to measure ECF
Penicillin G	0.2	Does not penetrate cells
Tubocurarine	0.2	Quaternary ammonium compound
Deuterium oxide (D_2O)	0.55–0.65	Isotopically labelled water – to measure TBW
Ethanol	0.65	Distributes in TBW
Phenazone (antipyrine)	0.6	Distributes in TBW, used to assess enzyme induction
Digoxin	5	Binds to Na^+/K^+ATPase
Chlorpromazine	20	Lipophilic, binds to proteins and other cell constituents
Amiodarone	62	Little found in the CNS
Mepacrine (quinacrine)	500	Intercalates in DNA

* Normalized to body weight.
[†] Compounds in italics may be used to measure anatomical volumes.

Compounds such as Evans blue, inulin and isotopically labelled water can be used to measure the volumes of plasma, ECF and TBW, respectively. Evans blue, named after the American chemist Herbert McLean Evans, binds so avidly to albumin that it does not leave the plasma, and the intensity of the colour can be assessed spectroscopically to estimate the concentration. Inulin, a water-soluble polysaccharide that does not enter cells, can be used to the estimate ECF. Isotopically labelled water, whether ^2H_2O or ^3H_2O, distributes in TBW. Apparent volumes of distribution can give some indication of where a drug may be distributed (Table 2.3) but often this is not possible from knowledge of V alone.

Heparin (a polymer, M_r = 12,000–15,000) is too large to pass through the fenestrations in the peripheral capillaries and so is confined to plasma. Several drugs, including the penicillins and tubocurarine, do not readily enter cells but are small enough to filter into interstitial fluid and are distributed in a volume equal to ECF. Drugs which can cross lipid cell membranes but are not concentrated (sequestered) in cells or bound to plasma proteins have volumes approximately equal to those of TBW. Ethanol and the now obsolete antipyretic drug phenazone (antipyrine) are examples of such compounds. This property is why phenazone can be used to investigate enzyme induction. Centrally-acting drugs generally have to be lipophilic enough to cross the blood–brain barrier and often have apparent volumes of distribution >1 L kg^{-1}. However, the converse is not true, for example a drug such as digoxin (Table 2.3) has a large value of V because it binds to cardiac and skeletal muscle Na^+/K^+ ATPase.

2.4.2 Mechanisms of sequestration

Sequestration of drugs in various parts of the body arises because of differences in pH, binding to macromolecules, dissolution in lipids, transportation (usually against the concentration gradient) and what may be termed 'irreversible' binding.

2.4.2.1 *pH differences*

Local differences in pH may lead to high concentrations of weak electrolytes in one area relative to another because of differences in the degree of ionization. This is predictable from pH partition considerations (Section 2.2.1.1). For example the ionized to non-ionized ratio of salicylic acid ($pK_a = 3.0$) at intracellular pH (6.8) is 6300:1 whereas at the pH of plasma (7.4) the ratio is 25,000:1. This represents a ratio of ~4:1 in favour of plasma water and is true for any weak acid. The converse is the case for bases, when the difference in intracellular and plasma pH means that bases will be distributed ~4:1 in favour of the more acidic fluid.

2.4.2.2 *Binding to macromolecules*

Drugs may be concentrated by binding to several types of macromolecules: plasma proteins, tissue proteins, including enzymes, and nucleic acids. The antimalarial drugs mepacrine (quinacrine) and chloroquine are concentrated in tissues because of binding to DNA. Chlortalidone (chlorthalidone), a thiazide-like diuretic, binds to red cell carbonic anhydrase, whilst anticholinesterases such as neostigmine and pyridostigmine bind to red cell acetylcholinesterase. Binding to plasma proteins, which can lead to important differences in distribution and hence pharmacological activity, is discussed in Section 2.5. The possible effects of protein binding on drug kinetics are considered in Chapter 7.

2.4.2.3 *Dissolution in lipids*

Lipophilic drugs are often concentrated in lipid cell membranes and fat deposits. This is thought to be simple partitioning, analogous to solvent:water partitioning. The distribution of thiopental in adipose tissue is an important determinant of its duration of action (Section 2.4.3.1).

2.4.2.4 *Active transport*

As mentioned earlier (Section 2.2.2), compounds that are structurally similar to endogenous molecules may be substrates for transport proteins. Guanethidine (and probably other adrenergic neuron blocking drugs) is concentrated in cardiac tissue by active uptake. Amfetamine and other indirectly-acting sympathomimetic drugs are actively transported into aminergic nerves. Paraquat is concentrated in the lungs because the distance between the quaternary nitrogen atoms (~8 nm) makes it a very good substrate for a putrescine transport protein.

2.4.2.5 *Special processes*

This refers to binding processes which, for all intents and purposes, can be considered as irreversible. Examples include deposition of tetracyclines in bone and teeth and of drugs in hair. These areas have very poor blood supply so that penetration is slow and loss of drug by diffusion back to the blood is in effect irreversible. It is probable that drug residues are laid down as hair and teeth are formed in areas of rich blood supply, and the deposits are carried away from these areas as the tissues grow.

Other examples of irreversible binding include covalent binding in tissue with a good blood supply. Such localization usually involves only a small proportion of the total amount of drug in the body but it is particularly important as a mechanism of drug toxicity (Chapter 13).

2.4.3 Kinetics of distribution

Drugs carried in the bloodstream will penetrate those tissues which they can, net transfer being down the concentration gradient, a process sometimes referred to as 'random walk'. A drug placed at any point within the system will diffuse backwards and forwards until characteristic equilibrium concentration ratios are reached (Figure 2.11). Within equilibrium of course diffusion continues, but the *relative* concentrations at the various points do not change. Loss of drug from the system occurs by metabolism, and by excretion, from the extracellular fluid, so the initial concentration gradient is reversed and now the net movement is from tissues to plasma.

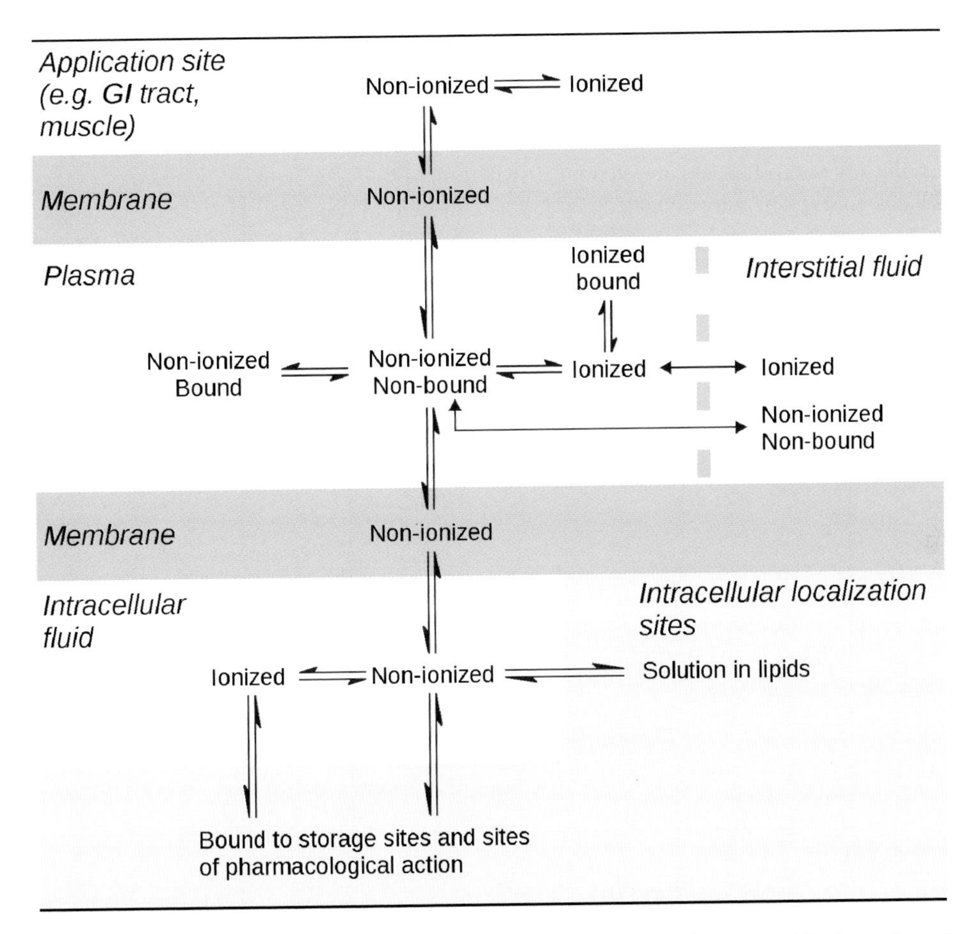

Figure 2.11 Simplified diagram for the equilibrium distribution of an ionizable drug of small relative molecular mass. The drug readily filters into interstitial fluid so the unbound concentrations are equal to those in plasma.

The result of tissue localization is a pattern throughout the body of concentrations in each tissue higher than concentrations in plasma by varying amounts. The highest concentrations are commonly found in the liver, lung and spleen, but the significance of this, except perhaps in regard to drug metabolism and excretion by the liver, remains obscure. The speed with which a particular ratio is achieved is governed by the nature of the drug and the tissues in which it is distributed. A highly ionized drug like pyridostigmine rapidly enters interstitial fluid via capillary fenestrations and, because it does not enter cells, the equilibration time is very short. Because lipophilic drugs rapidly diffuse across cell walls, the rate-limiting step for equilibration is delivery of the drug to the tissue, that is, it is flow-limited. Thus the rate of equilibration will be a function of the vascularity of the particular tissue and is rapidly established with well-perfused tissues such as kidney, liver, lung and brain. Muscle is intermediate, in that a rising ratio can often be detected. Poorly perfused adipose tissue can require many hours for equilibrium to be achieved. Presumably drugs of intermediate lipophilicity will show a mixture of flow-limited and diffusion-limited equilibration. A number of drugs have been studied in detail with regard to tissue distribution.

2.4.3.1 *Tissue distribution of thiopental*

Thiopental is a lipophilic barbiturate that is used as a short-acting general anaesthetic. Brodie and his colleagues invested the distribution of this drug in the 1950s as part of their studies to explain why repeated, or higher, doses of thiopental gave disproportionate increases in duration of action (Brodie *et al.*, 1952, 1956; Brodie & Hogben, 1957; Brodie, 1967; Mark *et al.*, 1957). Their work made a major contribution to the understanding of the kinetics of drug distribution. After an intravenous injection in a dog, plasma and liver thiopental concentrations fell rapidly (Figure 2.12(a)). The log(concentration)–time plots

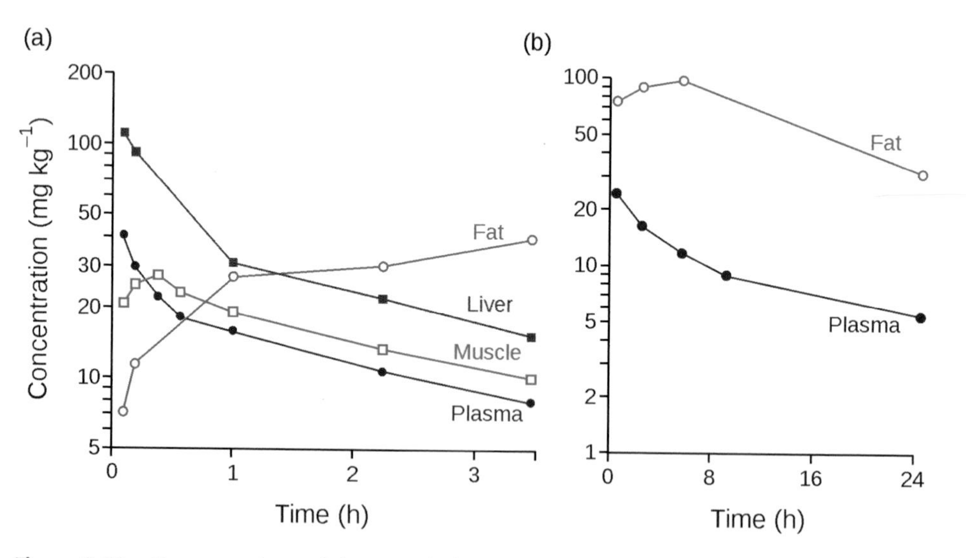

Figure 2.12 *Concentrations of thiopental (logarithmic scale) in the plasma and various tissues of a dog after intravenous administration of 25 mg kg^{-1}. (a) 0–3.5 h; (b) 0–24 h. Redrawn with permission from Brodie* et al., *1952.*

were parallel throughout the sampling period, showing that the liver and plasma concentrations had equilibrated by the time of the first sampling point.

The concentration in skeletal muscle rose over the first 30 min and had equilibrated by about 1 h, after which the muscle to plasma ratios remained constant. Thiopental concentrations in fat continued to rise for several hours, equilibrated and then declined in parallel to the plasma concentration when plotted on a semi-logarithmic plot (Figure 2.12(b)). A slow rate of return from fat to plasma is responsible for the slow decline in plasma concentrations at later times because the drug cannot be eliminated from the plasma any faster than it reaches plasma from the adipose tissue, that is, the rate of return is rate determining. The peak concentrations in cerebrospinal fluid were recorded at 10 min, after which the concentrations were similar to those in plasma water (Figure 2.13(a)). Clearly, the short duration of action of thiopental (about 15–20 min) cannot be explained by it being rapidly removed from the body. However, in a separate study it was shown that brain concentrations rapidly equilibrated with those in the plasma (Figure 2.13(b)) and thus the steep decline in plasma concentrations is accompanied by a similar sharp fall in brain concentrations. Hence, the rapid onset and short duration of action of thiopental can be explained in terms of the kinetics of its distribution. After an intravenous injection, this lipophilic drug rapidly crosses the blood–brain barrier to enter the brain, giving an almost immediate loss of consciousness. Over the next few minutes, the plasma concentration declines, not due to elimination of the drug, but because of loss of the drug from the plasma to the less well-perfused tissues. Because of the rapid transfer between brain and plasma, brain concentrations quickly fall and the patient regains consciousness. In other words, the short duration of action of thiopental is due to redistribution of the drug from the brain, via the plasma, to less well-perfused tissues, probably muscle and then fat. This phenomenon probably occurs with other lipophilic centrally acting drugs such as the related barbiturate methohexital and the opioid fentanyl.

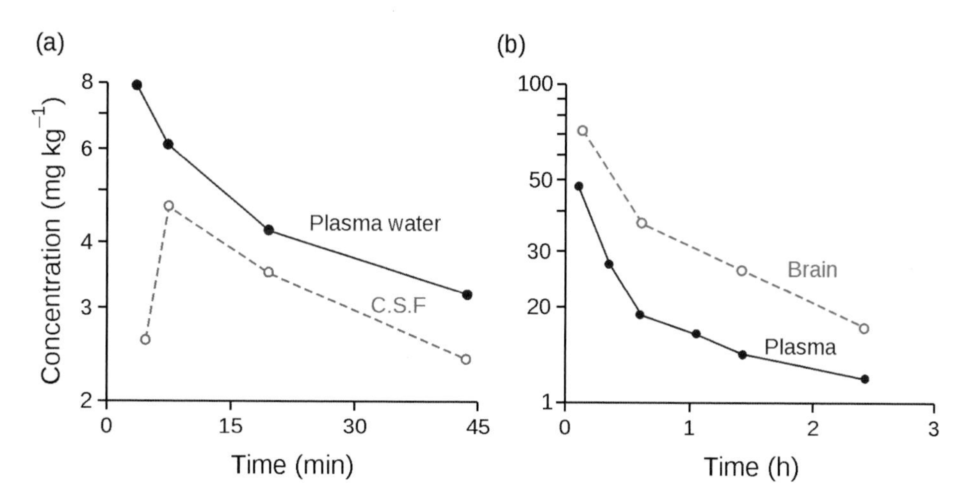

Figure 2.13 *(a) Concentrations of thiopental in plasma water and cerebrospinal fluid (C.S.F.) of a dog given 25 mg kg^{-1} intravenously. (b) Thiopental concentrations in the brain and plasma of a dog after intravenous administration of 40 mg kg^{-1}. Redrawn with permission from Brodie* et al., *1952.*

There are drugs which do not fit the pattern of distribution described above. Guanethidine, for example, illustrates how the tissue distribution of one drug may be more complex than another, particularly when active transport is involved and the simple model of passive diffusion does not apply.

2.4.4 Studying tissue distribution: more modern approaches

At one time tissue distribution studies of the type described above for thiopental were commonplace. The objective was, first, to determine which tissues showed selective uptake, in the hope of discovering sites of action, and, second, to aid the understanding of the time course of both the drug in the body and its pharmacological effect. The thiopental studies described above provided a fundamental basis for the later concepts of pharmacokinetic compartment modelling, and also influenced dosing practices with intravenous anaesthetics. Studies of this type were time-consuming, and they have largely been superseded by such techniques as microdialysis, and by imaging methods such as whole body autoradiography (Curry & Whelpton, 2011).

2.4.4.1 Positron emission tomography

Positron emission tomography (PET) scanning permits the production of images in tissues of live organisms, including human subjects. Synthetic radioactive isotopes (e.g. ^{11}C, ^{13}N, ^{15}O and ^{18}F) with atomic masses less than the naturally occurring stable isotopes have half-life values of 2–110 min and emit positrons that interact with electrons to emit gamma radiation that can be detected outside the body. The isotopes are generated in a cyclotron and incorporated into the drug molecules immediately before the administration of the drug. The disposition of [^{11}C]-triamcinolone after intranasal administration is shown in Figure 2.14(a). This technique is quantitative, and plots of percentage of administered dose against time up to 2 h in various regions were generated, illustrating how large amounts of data can be collected (Figure 2.14(b)). Half-lives for the labelled drug in these regions could be calculated.

PET scanning allows discrete regions of tissues to be investigated as exemplified above and by Figure 2.15. Dopamine receptors within the brain were labelled using the precursor [^{18}F]-fluoroDOPA to study the effect of drugs that modify dopamine function in exerting their beneficial effects on psychiatric illness. The figure shows eight images, four generated within a healthy volunteer and four generated in a patient. Two areas of the brain were examined, and the disposition of [^{18}F]-fluorodopamine was studied at two different time points. The images clearly show that clozapine reduces dopamine binding in the cortex of the brain.

2.5 Plasma protein binding

Binding of drugs to plasma proteins can have a major role in the overall distribution of a drug and may influence both its pharmacological activity and its kinetics. Most binding interactions are reversible, probably due to ionic and hydrophobic bonding, where lipophilic molecules associate with a hydrophobic part of the protein. Acids tend to bind to albumin and bases bind to α_1-acid glycoprotein and albumin. Covalent bonding, when it

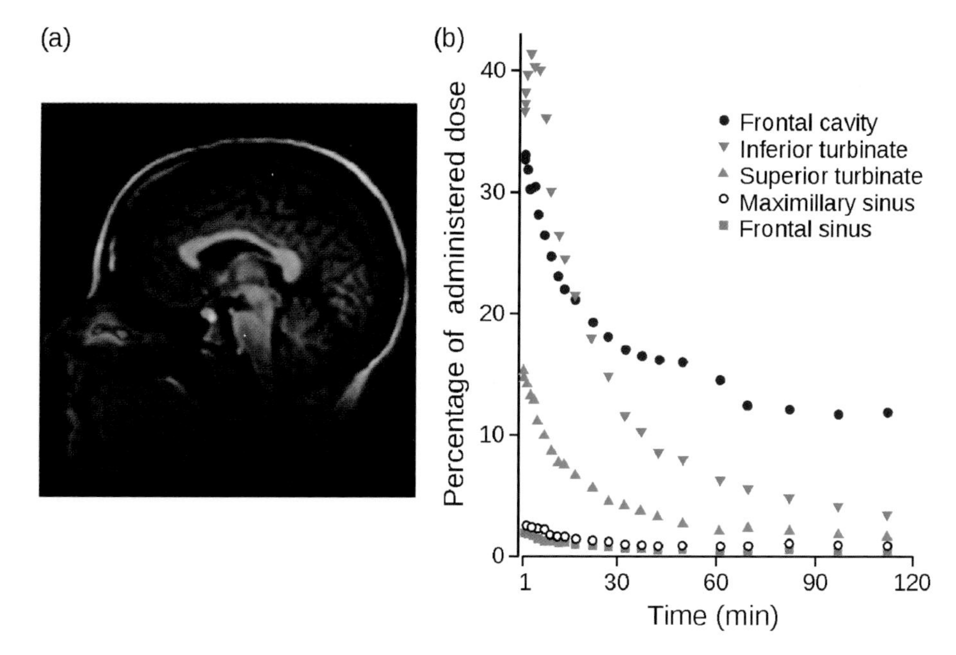

Figure 2.14 *(a) PET scan of [^{11}C]-triamcinolone (see text for details). (b) Time course of [^{11}C]-triamcinolone in selected regions. Each point is the mean from three volunteers plotted at the midpoint of each PET scan. Adapted with permission from Berridge* et al.*, 1998.*

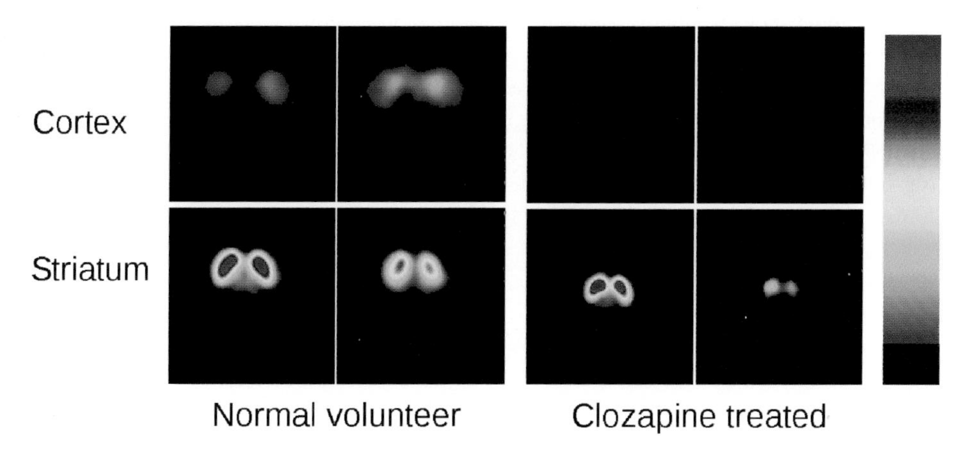

Figure 2.15 *PET scans of the cortex and striatum after administration of [^{18}F]-fluoroDOPA, the precursor of [^{18}F]-fluorodopamine.*

occurs, may result in antibody production and hypersensitivity reactions (Chapter 13). The extent of binding can be described by the fraction or percentage of drug bound, β, or less commonly by the unbound fraction, α. The term 'free' is best avoided as the term is used by some to describe the non-ionized form or the non-conjugated form of a drug, for example morphine rather than morphine glucuronide, potentially resulting in untold confusion.

The protein binding can be treated in the same way as any other reversible binding isotherm:

$$\text{drug} + \text{protein} \rightleftharpoons \text{drug} - \text{protein complex}$$

If the molar concentration of bound drug is D_b and total concentration of protein is P_t, then, assuming one binding site per protein molecule, the concentration of protein without drug bound to it is $(P_t - D_b)$, so:

$$D_f + (P_t - D_b) \rightleftharpoons D_b$$

where D_f is the concentration of unbound drug. The association constant, K, is:

$$K = \frac{D_b}{D_f(P_t - D_b)} \tag{2.5}$$

Rearrangement gives:

$$\frac{D_b}{P_t} = \frac{KD_f}{(1 + KD_f)} = r \tag{2.6}$$

where r is the number of moles bound per total number of moles of protein. If there are n equivalent binding sites, that is, sites with the same binding constants, this can be included (Equation 2.7). Equation 2.6 has been rearranged in a number of ways so that K and, usually, n can be estimated. The double reciprocal plot of Klotz (Figure 2.16(a)) and the Scatchard plot (Figure 2.16(b)) require the molar concentration of the protein to be known, which of course means one has to know the characteristics of the protein to which the drug is binding. This usually means that the binding experiment uses purified protein. Plotting D_b/D_f against D_b overcomes this because K can be estimated without knowing the

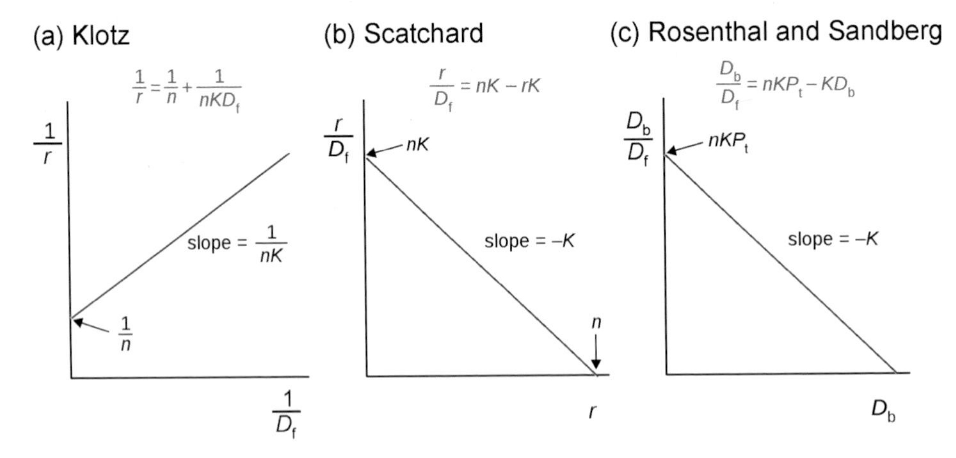

Figure 2.16 *Three approaches for solving protein binding data.*

molecular weight of the protein, and is more useful when measuring binding in plasma. If P_t is known then n can be derived from the y intercept of this plot (Figure 2.16(c)). This plot is also frequently used for receptor–ligand binding studies, and it is often referred to simply as a Scatchard plot although it was first proposed by Sandberg and Rosenthal.

Curvature of binding plots is indicative of the existence of more than one class of site and the number of moles bound is the result of binding to all the sites. So for two classes of binding sites:

$$r = \frac{n_1 K_1 D_f}{1 + K_1 D_f} + \frac{n_2 K_2 D_f}{1 + K_2 D_f} \tag{2.7}$$

Often a change in the fraction of drug bound in plasma is barely observable within the therapeutic range and when the concentration is increased over several orders of magnitude the change may be small (Figure 2.17).

Experimentally, proteins appear to be able to 'mop-up' some drugs from aqueous solutions or suspensions, even when in some cases the solubility in water has been exceeded. Scatchard plots are often curved when binding is studied over a large range of concentrations, suggesting that there is high-affinity binding (specific binding), with a small value of n_1 and relatively large K_1, and non-specific binding, which prevails at high concentrations where K_2 is small but n_2 is large (Equation 2.7). Thus, for many drugs there appears to be excess binding capacity, but this may not be the case for less potent drugs that are used at higher doses. For example, therapeutic concentrations of phenylbutazone (approximately 50–100 mg L^{-1} = 0.16–0.32 mmol L^{-1}) are not very far removed from the molar concentration of albumin in plasma (0.58 mol L^{-1}). A phenylbutazone concentration of ~180 mg L^{-1} represents a 1:1 ratio of drug:albumin and the fraction bound shows a marked increase in unbound concentration at concentrations higher than this (Figure 2.18).

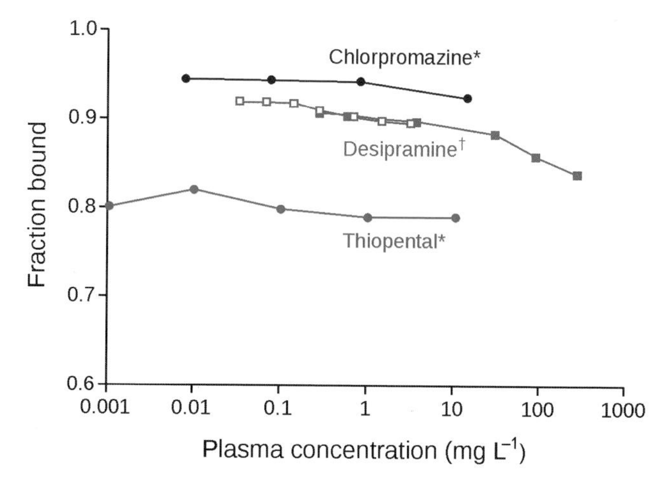

Figure 2.17 Fraction bound to human plasma protein for desipramine, chlorpromazine and thiopental (†two individuals, Borga et al., *1969; *blood bank plasma, Curry, 1970). Redrawn with permission of the copyright holders.*

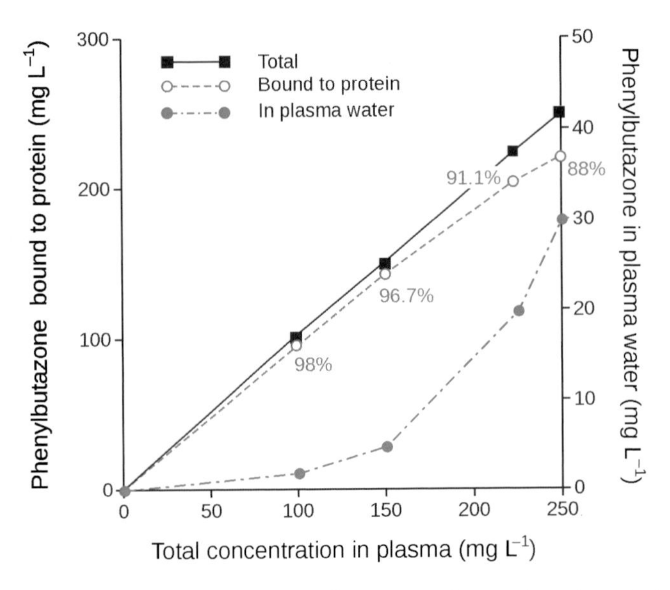

Figure 2.18 Binding of phenylbutazone to plasma protein as a function of phenylbutazone concentration. The percentages indicate the binding at 100, 150, 225 and 250 mg L^{-1}. Redrawn with permission from Brodie & Hogben, 1957.

2.5.1 Pharmacological importance of binding to plasma proteins

Clearly plasma protein binding has a major influence on the distribution of drugs. Extensive binding to plasma proteins reduces the apparent volume of distribution because a larger proportion of the amount of drug in the body will be in the plasma. It is usual to measure the 'total' concentration of drug (i.e. bound + unbound) in plasma.

Binding to plasma proteins provides an efficient way of transporting drugs in the circulation, sometimes at concentrations that exceed their solubility in plasma water. Binding has an important role in absorption as it maintains a favourable concentration gradient for the unbound drug. It is generally assumed that plasma protein binding reduces the proportion of a dose of drug available to its receptors and so it can have a major influence on drug activity. Changes in the faction bound may occur because of:

- displacement by a second drug – this may be clinically important with salicylate and valproic acid, displacing drugs that attain molar concentrations similar to that of the binding protein
- changes in protein concentration, often as result of disease (Chapter 12)
- concentration-dependent binding.

Many *in vitro* studies have demonstrated displacement of one drug by another, but *in vivo* the situation is more complex. The 'total' concentration of a displaced drug in plasma will be reduced as some of the liberated drug diffuses into tissues as new equilibria are established. The increased concentration of unbound drug may lead to greater, possibly toxic, effects. Hence, measurement of the 'total' (bound + unbound) concentration of a drug in plasma may be misleading under certain circumstances. When phenytoin was displaced by salicylate, for example, the percentage unbound increased from 7.14 to 10.66%, and this was accompanied by a significant decrease in total serum phenytoin concentration from 13.5 to 10.3 mg L^{-1}. The salivary phenytoin concentration rose from 0.97 to 1.13 mg L^{-1} (Leonard *et al.*, 1981).

The effect of protein binding on drug elimination is more complicated than it might at first appear. It is reasonable to assume that binding will reduce glomerular filtration, as the composition of the filtrate is in essence that of plasma water. However, other factors such as urine flow rate and reabsorption of drug from renal tubular fluid need to be considered (Section 3.3.1). The effects on drug metabolism are more complex. Briefly, binding reduces the rate of elimination of those drugs that are poorly extracted by the liver but not those that are extensively metabolized. In fact, for these drugs plasma protein binding can be considered as an efficient mechanism for delivering the drug to its site of metabolism. This is discussed in more detail in Chapter 7.

Summary

Knowledge of the processes by which substances cross biological membranes and the physicochemical properties and structures that are required for drugs to exploit those mechanisms is crucial to understanding:

- the most appropriate method of administration of a particular drug
- how differences in pH values affect the absorption and distribution of weak electrolytes
- why drugs are distributed to different parts of the body and to different degrees
- how plasma protein binding can affect the distribution and pharmacological activity of some drugs.

2.6 Further reading

Curry SH, Whelpton R. *Drug Disposition and Pharmacokinetics: From Principles to Applications*. Chichester: Wiley-Blackwell, 2011. Chapter 2 – Further examples of modern approaches to tissue distribution and experimental methods for measuring protein binding.

2.7 References

Berridge MS, Heald DL, Muswick GJ, Leisure GP, Voelker KW, Miraldi F. Biodistribution and kinetics of nasal carbon-11-triamcinolone acetonide. *J Nucl Med* 1998; 39: 1972–7.

Borga O, Azarnoff DL, Forshell GP, Sjoqvist F. Plasma protein binding of tricyclic antidepressants in man. *Biochem Pharmacol* 1969; 18: 2135–43.

Brodie BB. Physicochemical and biochemical aspects of pharmacology. *Jama* 1967; 202: 600–9.

Brodie BB, Bernstein E, Mark LC. The role of body fat in limiting the duration of action of thiopental. *J Pharmacol Exp Ther* 1952; 105: 421–6.

Brodie BB, Burns JJ, Mark LC, Papper EM. Clinical application of studies of the physiologic disposition of thiopental. *N Y State J Med* 1956; 56: 2819–22.

Brodie BB, Hogben CA. Some physico-chemical factors in drug action. *J Pharm Pharmacol* 1957; 9: 345–80.

Cao F, Gao Y, Ping O. Advances in research of PepT1-targeted prodrug. *Asian J Pharm Sci* 2012; 7: 110–22.

Curry SH. Plasma protein binding of chlorpromazine. *J Pharm Pharmacol* 1970; 22: 193–7.

Curry SH, Whelpton R. *Drug Disposition and Pharmacokinetics: From Principles to Applications*. Chichester: Wiley-Blackwell, 2011.

Leonard RF, Knott PJ, Rankin GO, Robinson DS, Melnick DE. Phenytoin-salicylate interaction. *Clin Pharmacol Ther* 1981; 29: 56–60.

Lopes MA, Abrahim BA, Cabral LM, Rodrigues CR, Seiça RM, de Baptista Veiga FJ, Ribeiro AJ. Intestinal absorption of insulin nanoparticles: contribution of M cells. *Nanomedicine* 2014; 10: 1139–51.

Mark LC, Burns JJ, Campomanes CI, Ngai SH, Trousof N, Papper EM, Brodie BB. The passage of thiopental into brain. *J Pharmacol Exp Ther* 1957; 119: 35–8.

Müller F, Fromm MF. Transporter-mediated drug-drug interactions. *Pharmacogenomics.* 2011; 12: 1017–37.

Shand DG, Rangno RE. The disposition of propranolol. I. Elimination during oral absorption in man. *Pharmacology* 1972; 7: 159–68.

Shugarts S, Benet LZ. The role of transporters in the pharmacokinetics of orally administered drugs. *Pharm Res* 2009; 26: 2039–54.

Yang Z. The roles of membrane transporters on the oral drug absorption. *J Mol Pharm Org Process Res* 2013; 1: e102.

3

Drug Metabolism and Excretion

Learning objectives

By the end of the chapter the reader should be able to:

- explain the role of metabolism in the elimination of drugs
- describe the relationship between phase 1 and phase 2 metabolism
- list examples of phase 1 oxidations, reductions and hydrolyses
- briefly explain the isoforms of cytochrome P450
- discuss why 'metabolism' is not synonymous with 'detoxification'
- compare renal and hepatic elimination of foreign materials
- explain how manipulating urine pH may be used to increase or decrease the renal clearance of some drugs
- discuss recycling processes and their effects on pharmacological activity.

3.1 Introduction

The action of a drug may be terminated by redistribution away from its site of action, as discussed in Section 2.4.3. However, most drugs have a limited duration of action because they are eliminated by metabolism and/or excretion. Some drugs, particularly polar, water-soluble ones, may be excreted in the urine unchanged but lipophilic drugs tend to

Introduction to Drug Disposition and Pharmacokinetics, First Edition. Stephen H. Curry and Robin Whelpton.

Companion website: www.wiley.com/go/curryandwhelpton/IDDP

be highly protein bound, which reduces glomerular filtration and any of the drug in kidney tubular fluid tends to be reabsorbed back into the body. Consequently, lipophilic drugs would have very long elimination half-lives if they were not metabolized into polar molecules that are more readily excreted. Furthermore, drug metabolites can also contribute to the pharmacological effect(s) of a drug.

An appreciation of how drugs are metabolized and excreted is important to understanding the duration of action and clinical effect. Predicting which metabolites might be formed is important not only for disposition studies but for analytical method development, as it is generally necessary to be able to quantify the parent drug rather than metabolite(s). Knowledge of which enzymes are involved can aid selection of an appropriate species for animal studies. Similarly, understanding the routes and mechanisms by which drugs and their metabolites are excreted may inform the choice of fluids for analysis and pharmacokinetic studies.

3.2 Metabolism

The liver is undoubtedly the major site of metabolism for the majority of drugs but most tissues have some metabolizing capacity. Other notable sites of metabolism include:

- plasma and other body fluids – hydrolysis, mainly of esters
- nerve terminals – metabolism of endogenous transmitters and drugs with similar structures
- mucosal cells – responsible for pre-systemic metabolism of many drugs
- kidney, lung and muscle – metabolize some drugs
- brain – capable of glucuronidation of morphine
- intestinal bacteria – metabolism of several drugs and hydrolysis of glucuronide metabolites.

Drug metabolism has been divided into two, sometimes three, distinct phases. Phase 1 reactions either insert, usually via oxidation, or reveal, by hydrolysis or desalkylation, reactive or functional groups that can undergo phase 2 reactions. For this reason, phase 1 reactions may be referred to as 'functionalization' reactions. In phase 2 reactions, molecules with suitable functional groups are conjugated with endogenous compounds such as glucuronic acid, sulfate, acetate and amino acids. Conjugation may be with products of phase 1 metabolism or if the drug has suitably reactive groups conjugation may occur directly. Phenacetin is desalkylated to paracetamol (acetaminophen), which is then conjugated with either glucuronate or sulfate:

$NHCOCH_3$ / OC_2H_5 — Phenacetin → (Phase 1, Desalkylation) → $NHCOCH_3$ / OH — Paracetamol → (Phase 2, Glucuronidation) → $NHCOCH_3$ / $O.C_6H_9O_6$ — Paracetamol glucuronide

Clearly, paracetamol can be conjugated directly without prior functionalization. The term 'phase 3 metabolism' is sometimes used to describe the further metabolism of phase 2 conjugates, such as occurs with the glutathione conjugate of paracetamol.

3.2.1 Phase 1 metabolism

These reactions include reductions and hydrolyses, but the most common reactions are the oxidations catalysed by the mixed-function oxidase (MFO) system, so-called because of the apparent lack of specificity for individual substrates. These enzymes are isolated by subjecting tissue homogenates to high centrifugal forces (100,000 *g*) after the cell debris has been removed by centrifugation at 9,000 *g* (Figure 3.1). The pellet comprises microsomes, small particles of smooth endoplasmic reticulum (SER), whilst the supernatant layer (soluble fraction) contains enzymes from the cytosol, such as alcohol dehydrogenase (ADH) and *N*-acetyltransferase type 2 (NAT2).

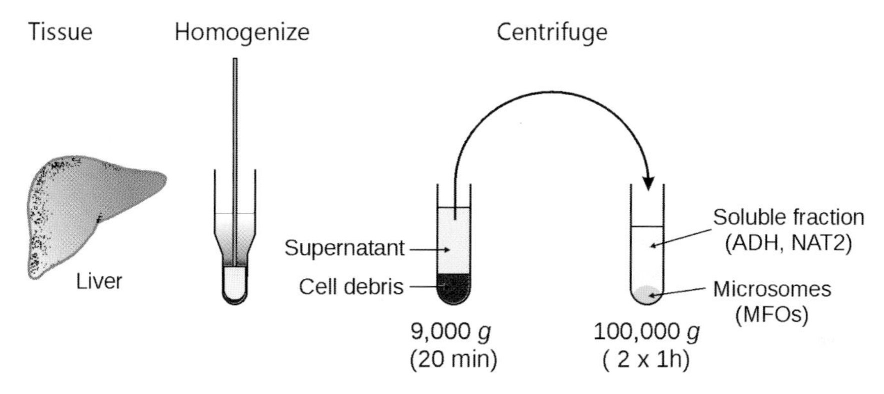

Figure 3.1 Preparation of a microsomal pellet containing mixed function oxidases.

Microsomal oxidations involve a relatively complex chain of redox reactions and require (i) NADPH, (ii) a flavoprotein (NADPH-cytochrome P450 reductase) or cytochrome b_5, (iii) molecular oxygen and (iv) a haem-containing protein, cytochrome P450, so-called because in its reduced form it binds with carbon monoxide to produce a characteristic spectrum with a peak at 450 nm. Cytochromes, which comprise a superfamily with nearly 60 genes having been identified in humans, catalyse the final step (Figure 3.2). Drug (DH) combines with the cytochrome, and the iron(III) in the complex is reduced to iron(II) by acquiring an electron from NADPH-P450 reductase. The reduced complex combines with oxygen $[(DH)Fe^{2+}O_2)]$ and combination with a proton and a further electron (from NADPH-cytochrome P450 reductase or cytochrome b_5) produces a $[DH\text{-}Fe^{2+}OOH]$ complex. The addition of a proton liberates water and a Fe(III) oxene complex $[DH(FeO)^{3+}]$, which extracts hydrogen from the drug with the formation of a pair of free radicals. Finally, the oxidized drug is released from the complex with the regeneration of the P450 enzyme. The last stage of the reaction involves free radicals and the same products can sometimes be produced by treating the drug with Fenton's reagent, a source of hydroxyl radicals.

3.2.1.1 Cytochrome P450 superfamily

Although the Human Genome Project has identified 57 human genes that encode various forms of cytochrome P450, only three families, CYP1, CYP2 and CYP3, appear to be important in drug metabolism. The enzymes are classified on the basis of cDNA cloning according to similarities in amino acid sequence. A family contains genes that have at least a 40% sequence homology. Members of a subfamily (denoted by a letter) must have at least 55% identity. The individual gene products are identified by the final number.

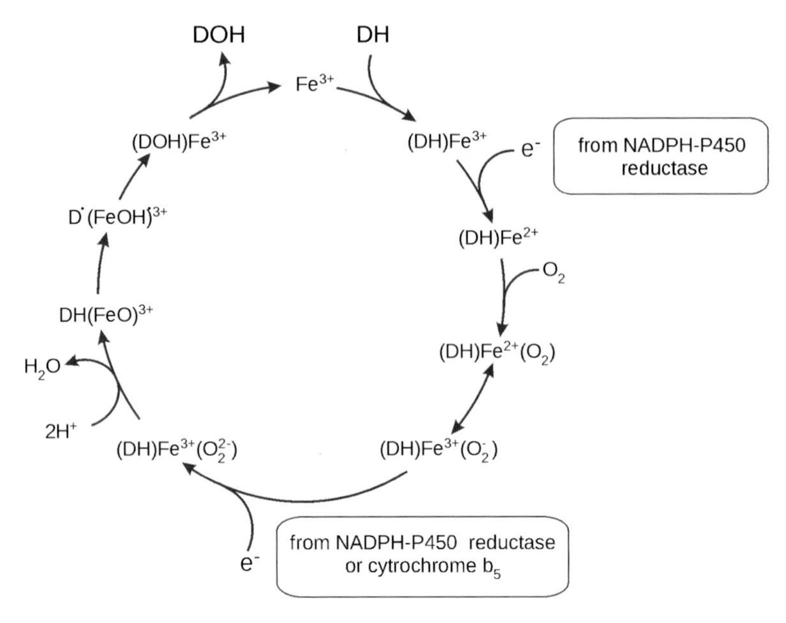

Figure 3.2 The final stage of incorporation of molecular oxygen into drug (DH) showing the various oxidation states of iron in the haem-containing cytochrome.

The substrate specificity of the CYP isoforms varies, as do their inhibitors and the agents that can induce their synthesis (Table 3.1). Note how several inhibitors and inducers appear several times. Ketoconazole and related drugs work by inhibiting fungal cytochromes whilst tranylcypromine binds covalently to monoamine oxidase and probably binds to other enzymes as well. It therefore might be predicted that these drugs would have the effects that they do. The H_2-receptor antagonist cimetidine inhibits several CYP isoforms. Some drugs are both substrates and (competitive) inhibitors. The well-known enzyme-inducing agents phenobarbital, rifampicin (rifampin) and phenytoin feature large in Table 3.1. Note also that components of foods, particularly grapefruit, and herbal remedies, for example Saint John's wort, can inhibit or induce enzymes. Enzyme induction and inhibition are discussed in Chapter 13.

The distribution of the various CYPs also varies. CYP3A4, which metabolizes a large number of drugs, constitutes approximately 30% of the total CYP content of the liver and 70% of the total CYP in GI epithelial cells. Many of its substrates are also transported by P-gp, with which it appears to work in concert, so that if a molecule escapes metabolism it is likely to be expelled into the GI lumen by the efflux protein. CYP3A5 is the most abundant CYP3A form in the kidney. CYP3A7 is a foetal form of the enzyme that is believed to be important in the metabolism of endogenous steroids and *trans*-retinoic acid. It rapidly declines during the first week of life and is rarely expressed in adults. Although CYP2D6 represents only 2% of hepatic CYP, it metabolizes a large number of substrates, including many antipsychotic and antidepressant drugs, β-blockers and the anti-hypertensive debrisoquine. It is also responsible for activating the prodrug tramadol, and for the *O*-demethylation of codeine to morphine. It has been suggested that those with a deficit of CYP2D6 fail to obtain pain relief when taking these drugs (Section 10.5.1).

Table 3.1 *Some examples of drug-metabolizing cytochromes*

Isoform	Substrates	Inhibitors	Inducers
CYP1A2	Caffeine, clozapine, fluvoxamine, olanzapine theophylline	Ciprofloxacin, cimetidine	Phenobarbital, phenytoin, rifampicin, tobacco smoke
CYP2A6	Cotinine, coumarin, nicotine	Ketoconazole, star fruit juice, grapefruit juice	Dexamethasone, rifampicin, phenobarbital
CYP2B6	Bupropion, cyclo-phosphamide*, pethidine (meperidine)	Fluoxetine, orphenadrine	Carbamazepine, cyclophosphamide, phenobarbital
CYP2C9	Losartan*, NSAIDs, phenytoin, tolbutamide, *S*-warfarin	Amiodarone, ketoconazole, tranylcypromine, valproic acid	Phenobarbital, phenytoin, rifampicin, Saint John's wort
CYP2C19	Diazepam, omeprazole, phenytoin	Omeprazole, moclobemide, sulfaphenazole	Phenobarbital, phenytoin, rifampicin, Saint John's wort
CYP2D6	Codeine*, debrisoquine, haloperidol, phenothiazines, SSRIs†	Cimetidine, fluoxetine, quinidine, terbenafine	Glutethimide
CYP2E1	Enflurane, ethanol, halothane, paracetamol	Disulfiram, ethanol (acute), miconazole	Ethanol (chronic), isoniazid, rifampicin
CYP3A4	Clozapine, nifedipine, simvastatin*	Cimetidine, erythromycin, ketoconazole, SSRIs, grapefruit juice	Carbamazepine, phenobarbital, rifampicin, Saint John's wort
CYP3A5	Medazepam, caffeine		Dexamethasone

* Prodrug.
† Selective serotonin reuptake inhibitors.

3.2.1.2 *Other oxidases*

Not all oxidases are cytochromes. Monoamine oxidase (MAO) is a mitochondrial enzyme located in aminergic nerve terminals, the liver and intestinal mucosa. Not only does it deaminate endogenous neurotransmitters, norepinephrine, dopamine and serotonin, it is normally responsible for the first-pass metabolism of indirectly-acting sympathomimetic amines (IASA) such as tyramine and ephedrine. When MAO is inhibited, for example by antidepressant drugs like tranylcypromine, the absorption of tyramine and subsequent displacement of elevated concentrations of norepinephrine can result in a dangerous hypertensive crisis. Amfetamine is an IASA but it is a poor substrate for MAO because it has an α-methyl substituent, but it may inhibit this enzyme.

Flavin-containing monooxygenases (FMOs) are microsomal enzymes that catalyse the NADPH-dependent oxidation of a large number of sulfur-, selenium- and nitrogen-containing compounds. Examples include the *N*-oxidation of tertiary amines and stereo-specific oxidation of sulfides. A genetic failure to express one isoform, FMO3, results in an inability to metabolize trimethylamine, which leads to large quantities being excreted in the urine and sweat, a very distressing condition for the sufferers, who smell of putre-fying fish.

Other phase 1 enzymes include the soluble enzyme alcohol dehydrogenase (ADH), the mammalian form of which oxidizes several alcohols as well as ethanol, including methanol, ethylene glycol and 2,2,2-trichloroethanol, the active metabolite of chloral hydrate. Xanthine oxidase metabolizes 6-mercaptopurine, a metabolite of azathioprine. The conversion of xanthine to uric acid may be inhibited by allopurinol, a drug sometimes used to treat gout.

3.2.1.3 Hydrolyses

Blood contains several esterases, including butyrylcholinesterase (BChE), a soluble enzyme (also known as pseudo- or plasma cholinesterase), and acetylcholinesterase (AChE), which is bound to the membranes of red cells. Amides may be hydrolysed by plasma but the liver contains higher amidase activity as well as esterases. Esters are usually rapidly hydrolysed; examples include acyl and carbamoyl esters (Figure 3.3).

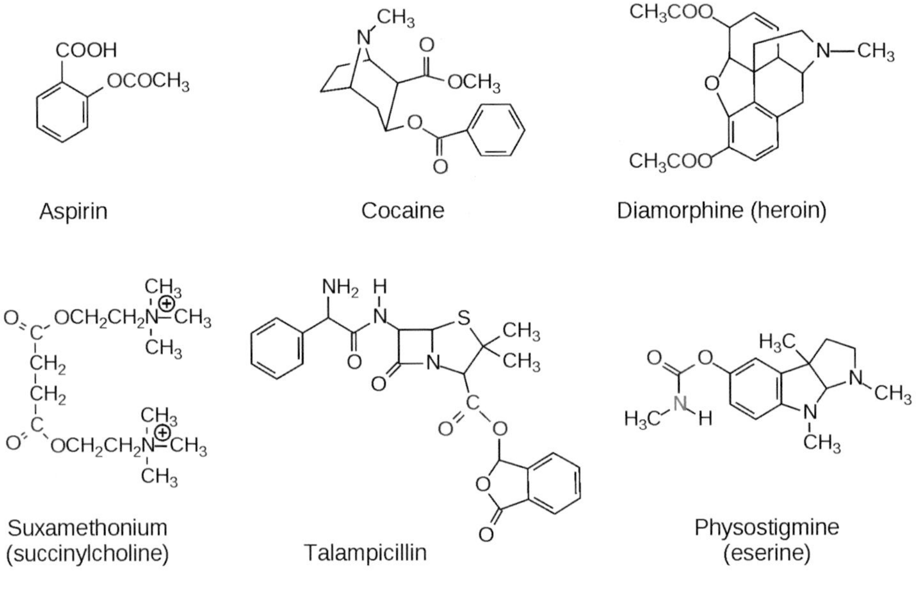

Aspirin   Cocaine   Diamorphine (heroin)

Suxamethonium (succinylcholine)   Talampicillin   Physostigmine (eserine)

Figure 3.3 Examples of esters that are hydrolysed by plasma esterases. Esters are coded in red. The nitrogen of the carbamate ester of physostigmine is shown in blue.

Amides and hydrazides may be hydrolysed (Figure 3.4). Procainamide is hydrolysed to *p*-aminobenzoic acid. The amide local anaesthetics lidocaine (lignocaine) and prilocaine are hydrolysed more slowly than the local anaesthetics that are esters, for example procaine and benzocaine. The *o*-toluidine (2-methylaniline) produced from prilocaine is thought to be responsible for the methaemoglobinaemia observed as an unwanted effect of this anaesthetic. Hydrolysis of hydrazides may lead to reactive hydrazines.

3.2.1.4 Reductions

Azo-, nitro- and dehalogenation reductions are catalysed by microsomal reductases. Several pathways are probably involved that are dependent on NADPH, cytochrome *c* reductase and flavoprotein enzymes. It is thought that azo and nitro reductases are different and the nitro reductases can be divided into those that are oxygen sensitive and those that

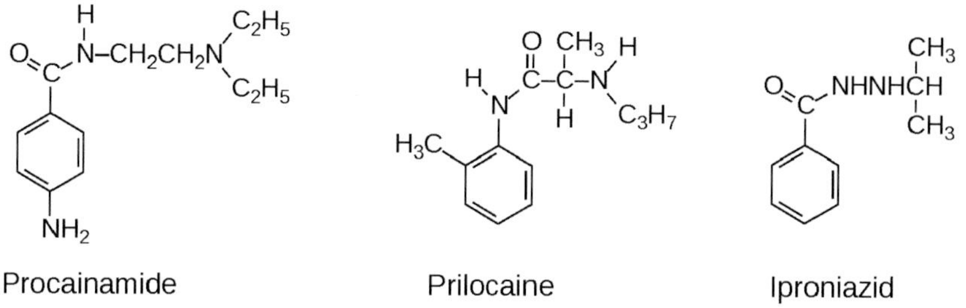

Procainamide Prilocaine Iproniazid

Figure 3.4 Examples of amides and hydrazides that are hydrolysed.

are not. The nitro groups in nitrazepam, clonazepam and chloramphenicol are reduced to primary aromatic amines. The azo dye prontosil, which is inactive *in vivo*, is reduced to an active metabolite, sulfanilamide (Figure 3.5), from which the antimicrobial sulfonamides were developed.

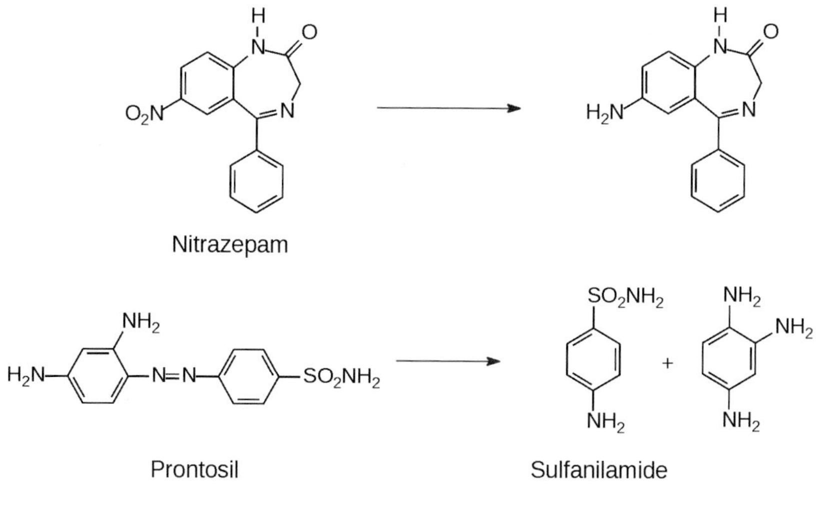

Nitrazepam

Prontosil Sulfanilamide

Figure 3.5 Examples of nitro and azo reduction.

3.2.2 Examples of phase 1 oxidation

3.2.2.1 Oxidation of alcohols

Alcohols may be oxidized via various routes, including microsomal CYP2E1 and catalase, but in mammals the major route is catalysed by cytosolistic alcohol dehydrogenase (ADH) to the corresponding aldehyde or ketone. Ethanol and methanol are oxidized to acetaldehyde and formaldehyde, respectively. Further oxidation by aldehyde dehydrogenase (ALDH) gives either acetic acid or formic acid (Figure 3.6). Formaldehyde and formic acid are much more toxic than acetaldehyde and acetic acid, so in methanol poisoning ethanol may be given to compete with methanol for ADH and so reduce the rate of production of the toxic metabolites. Disulfiram inhibits ALDH so that after consumption of ethanol, acetaldehyde concentrations increase causing flushing, nausea and vomiting. Disulfiram may be used in aversion therapy for alcoholics.

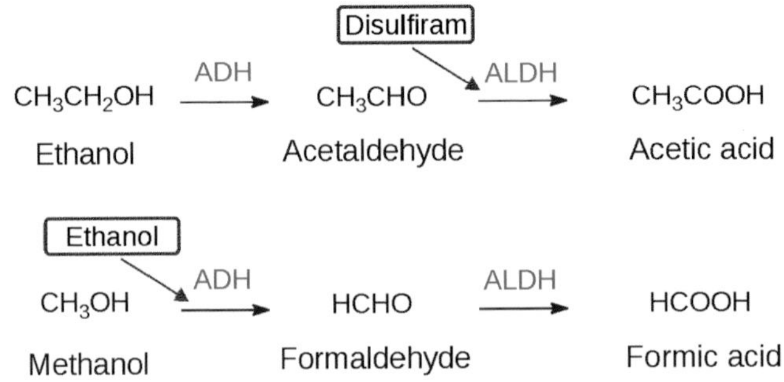

Disulfiram
CH_3CH_2OH —ADH→ CH_3CHO —ALDH→ CH_3COOH
Ethanol Acetaldehyde Acetic acid

Ethanol
CH_3OH —ADH→ HCHO —ALDH→ HCOOH
Methanol Formaldehyde Formic acid

Figure 3.6 Oxidation of ethanol and methanol by alcohol dehydrogenase (ADH) and aldehyde dehydrogenase (ALDH). Inhibitors are shown in purple-edged boxes.

3.2.2.2 Aliphatic and aromatic hydroxylations

Microsomal hydroxylations are commonly catalysed by MFOs and occur with both aromatic and aliphatic moieties. Aromatic hydroxylations are exemplified by oxidation of phenobarbital to *p*-hydroxyphenobarbital, which is devoid of pharmacology activity, and hydroxylation of phenothiazines and tricyclic antidepressants.

Examples of aliphatic hydroxylation include 4-hydroxylation of debrisoquine (Section 10.5.1) and 3-hydroxylation of 1,4-benzodiazepines (Figure 3.7).

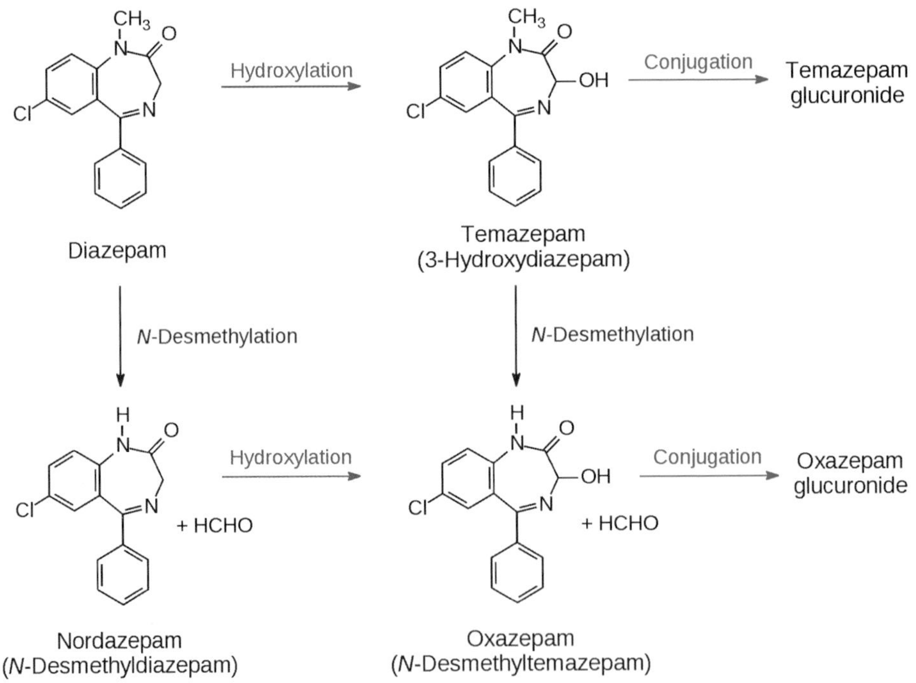

Figure 3.7 Metabolism of diazepam. Aliphatic hydroxylation gives temazepam, whilst N-desmethylation produces nordazepam and formaldehyde. Desmethylation of temazepam, or hydroxylation of nordazepam, gives oxazepam. The 3-hydroxy metabolites are conjugated with glucuronic acid and excreted in the urine.

3.2.2.3 *Oxidative desalkylation*

Although cytochrome P450 catalysed oxidations insert an atom of oxygen into the drug, the initial product may be unstable and rearrange. This is the case with oxidative *N*- and *O*-desalkylation reactions. In the case of desmethylations, the methyl group is lost as formaldehyde, which contains the oxygen atom from the initial oxidation. Loss of larger alkyl groups results in the production of the corresponding aldehyde. With *in vitro* metabolism studies, the aldehyde may be trapped and measured as an indicator of the degree of desalkylation. *N*-Desalkylation is common with tertiary and secondary amines (Figure 3.7). Examples of drugs undergoing *O*-desalkylation include phenacetin (described earlier) and codeine (to give morphine).

3.2.2.4 *N- and S-oxidation*

Tertiary amines and sulfides are oxidized to *N*-oxides and sulfoxides, respectively. Sulfoxides may be further oxidized to sulfones. Thioridazine (Figure 3.8) is an interesting example of a drug that forms an *N*-oxide, a sulfone and two sulfoxides (more if the stereoisomers are taken into consideration). All phenothiazines form 5-sulfoxides, which are considered to be pharmacologically inactive. However, oxidation of the 2-thiomethyl group in thioridazine gives the 2-sufoxide (mesoridazine) and 2-sulfone (sulforidazine), which are pharmacologically active and marketed as drugs in some countries.

N-Oxides are labile and usually easily reduced to the parent amine or may spontaneously desalkylate to the secondary amine. A few drugs are marketed as *N*-oxides, chlordiazepoxide and minoxidil (Figure 3.13), for example. Because of their labile nature, it is difficult to demonstrate whether the activity resides in the parent compound or one or more of its metabolites.

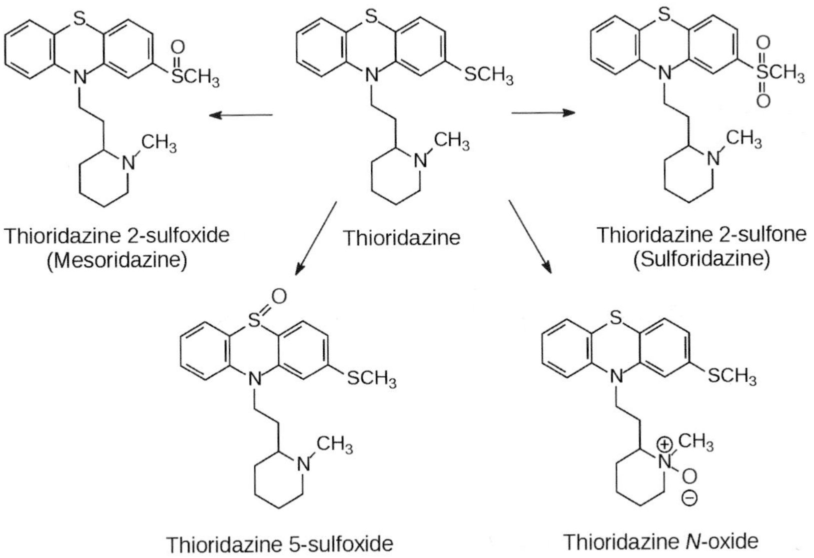

Thioridazine 2-sulfoxide (Mesoridazine)
Thioridazine
Thioridazine 2-sulfone (Sulforidazine)
Thioridazine 5-sulfoxide
Thioridazine *N*-oxide

Figure 3.8 *Examples of S- and N-oxidation metabolites of thioridazine.*

3.2.2.5 Oxidative deamination

Monoamine oxidase was mentioned earlier and is responsible for deamination of the endogenous amine transmitters norepinephrine, dopamine and serotonin (5-HT). Compounds with similar structures, for example ephedrine and tyramine, are deaminated to the corresponding aldehydes, which may be oxidized to carboxylic acids or reduced to alcohols. Other enzymes capable of deamination include diamine oxidase and cytochromes, for example CYP2C3, which has been shown to deaminate amfetamine to phenylacetone (Figure 3.9).

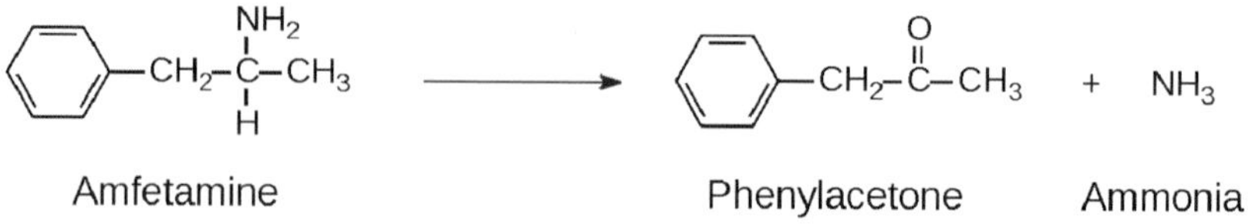

Amfetamine → Phenylacetone + Ammonia (NH_3)

Figure 3.9 Deamination of amfetamine.

3.2.2.6 Desulfuration

Sulfur can be exchanged for oxygen. This exchange in the general anaesthetic thiopental produces the anxiolytic barbiturate pentobarbital. In insects, malathion is converted to the much more toxic malaoxon whereas in mammals the major pathway is hydrolysis to the considerably less toxic dicarboxylic acid (Figure 3.10). This example of selective toxicity may also be referred to as lethal synthesis – lethal to insects that is.

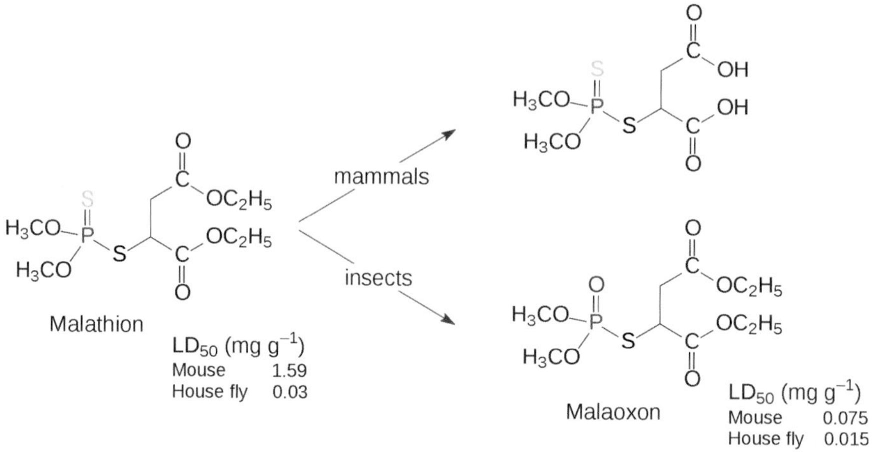

Malathion → (mammals) dicarboxylic acid; Malathion → (insects) Malaoxon

Malathion LD50 (mg g^{-1})	
Mouse	1.59
House fly	0.03

Malaoxon LD50 (mg g^{-1})	
Mouse	0.075
House fly	0.015

Figure 3.10 Comparison of metabolism of malathion and comparative toxicity of malathion and malaoxon in insects and mammals.

3.2.2.7 Combination of reactions

For those compounds with suitable groups, several of the reactions described above may occur. Figure 3.11 shows some of the metabolic pathways of chlorpromazine. Additionally,

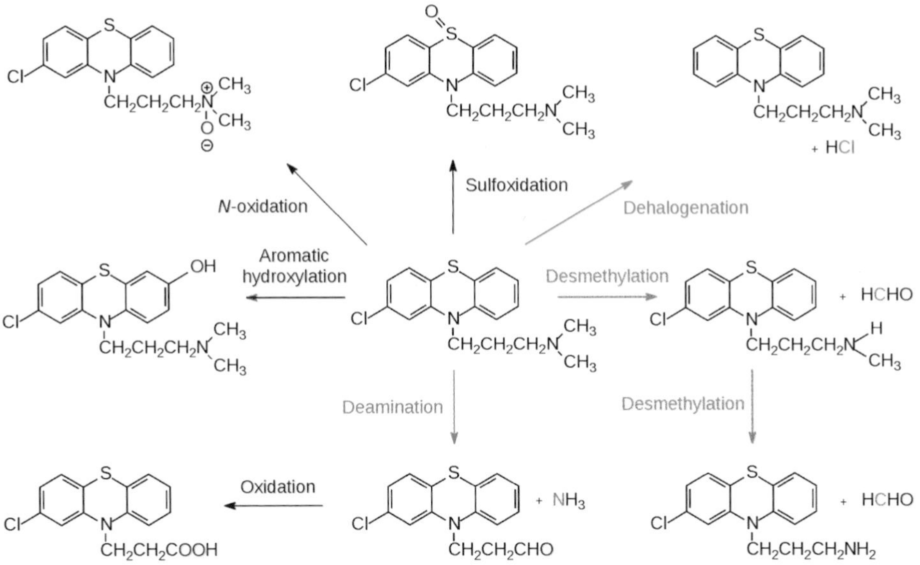

Figure 3.11 *Examples of some of the pathways by which chlorpromazine is metabolized. Red: introduced oxygen atoms; blue: deamination; green: dehalogenation; red/brown: desmethylation.*

hydroxylation may occur at other sites and that, plus conjugation reactions (Section 3.2.4), results in there being over 160 potential metabolites – and chlorpromazine is not the most chemically complex of the phenothiazine drugs.

3.2.3 Miscellaneous reactions

In this class are examples such as ring-opening, for example phenytoin to diphenylureidoacetic acid, and ring-formation, for example as occurs with proguanil. Transesterification can occur when ethanol has been ingested. This has been shown to occur with methylphenidate (to ethylphenidate) and cocaine, when cocaethylene is formed. Cocaethylene is active but has a longer half-life than cocaine and it has been suggested that this is why users take their cocaine with alcohol.

3.2.4 Phase 2 metabolism

Phase 2 reactions are those where molecules with suitably reactive groups are conjugated with endogenous substances such as glucuronate, sulfate, acetate, amino acids and reduced glutathione (GSH). Usually the products are considerably less pharmacologically active, more water-soluble and more amenable to excretion than the original substance, but there are some notable exceptions.

3.2.4.1 D-glucuronidation

This conjugation is a two-stage process, the first stage being the synthesis of the donor molecule, uridine diphosphate glucuronic acid (UDPGA), in the cytosol.

Glucose-1-phophate is combined with uridine triphosphate and then oxidized to the carboxylic acid:

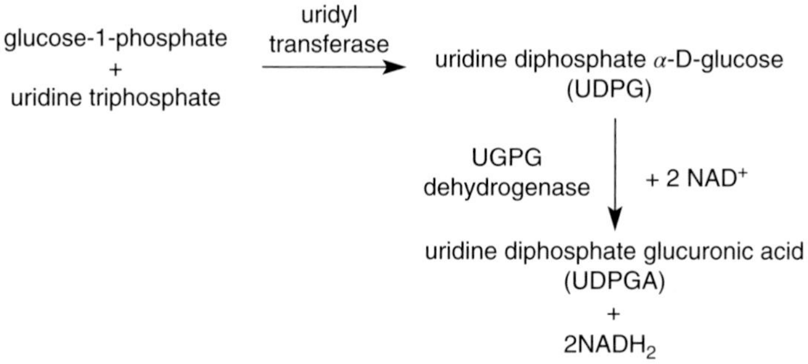

glucose-1-phosphate + uridine triphosphate —(uridyl transferase)→ uridine diphosphate α-D-glucose (UDPG)

UDPG —(UGPG dehydrogenase, + 2 NAD^+)→ uridine diphosphate glucuronic acid (UDPGA) + $2NADH_2$

The substrate–donor interaction is catalysed by UDP-glucuronosyltransferases, which are membrane-bound enzymes. The reaction always produces the β-glucuronide (Figure 3.12).

Large numbers of different glucuronides are possible, including *ether* glucuronides, formed with phenolic or alcoholic hydroxyl groups, *ester* glucuronides, formed with carboxylic acids, *N-glucuronides*, formed with aromatic amines, alicyclic amines and sulfonamides, and *S-glucuronides*, formed with various thiol compounds.

Generally, glucuronides are very water soluble, amenable to excretion via the kidney and bile (Sections 3.3.1 and 3.3.2), and pharmacologically inactive. However, morphine glucuronides are remarkable exceptions. There are three morphine glucuronides: 3-glucuronide, 6-glucuronide and 3,6-diglucuronide. The 3-glucuronide accumulates in plasma to give higher concentrations than the parent compound and the 6-glucuronide is considered to be at least as active as morphine at μ-opioid receptors. It is possible that other glucuronide conjugates possess pharmacological activity, as has been suggested for 1-hydroxymidazolam glucuronide, which accumulates in patients with renal failure.

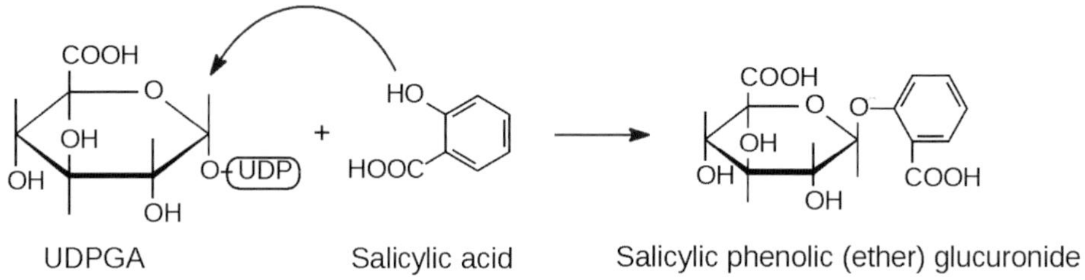

Figure 3.12 *Glucuronidation of salicylic acid produces the β-glucuronide.*

3.2.4.2 *Sulfation*

Ethereal sulfation is catalysed by several cytoplasmic sulfotransferases, depending on the substrate. The sulfate donor, 3′-phosphoadenosine-5′-phosphosulfate (PAPS), is formed from sulfate and ATP. Sulfate conjugates are usually very water soluble and readily excreted, and probably less pharmacologically active than the parent drug. The antihypertensive minoxidil is interesting. First, it is an *N*-oxide, and, second, sulfation occurs via this group (Figure 3.13). Minoxidil is also used to reduce or to prevent hair loss, and the

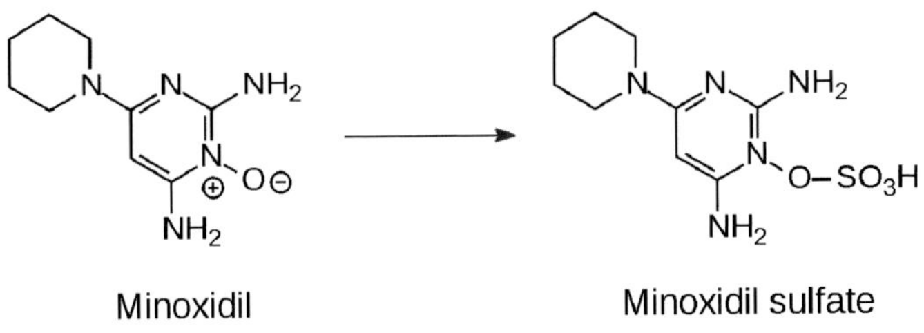

Minoxidil Minoxidil sulfate

Figure 3.13 Sulfation of minoxidil.

sulfate conjugate has been shown to be 14 times more potent than minoxidil in stimulating cysteine incorporation in hair follicles.

3.2.4.3 N-acetylation

Aromatic primary amines and hydrazines are *N*-acetylated by *N*-acetyltransferase type 2 (NAT2), which is chiefly expressed in the liver, colon and intestinal epithelium. The donor molecule is acetylCo-A. NAT2 acetylates isoniazid, hydralazine, phenelzine, procainamide, dapsone and several antibacterial sulfonamides, including sulfadimidine (sulfamethazine) as shown in Figure 3.14.

Most acetylated metabolites have reduced pharmacological activity and are considered to be inactive, this is particularly true of the sulfonamide antimicrobials. An exception is *N*-acetylprocainamide (acecainide), an antiarrhythmic agent that can be given either intravenously or orally.

It would be expected that acetylation would result in reduced aqueous solubility. This is only partially true; the acetyl metabolites of the older sulfonamides were less soluble in urine and caused crystalluria, but not the newer, more acidic, ones. Acetylation reduces the pK_a sufficiently to increase the degree of ionization at physiological pH values, which may compensate for the increase in lipophilicity.

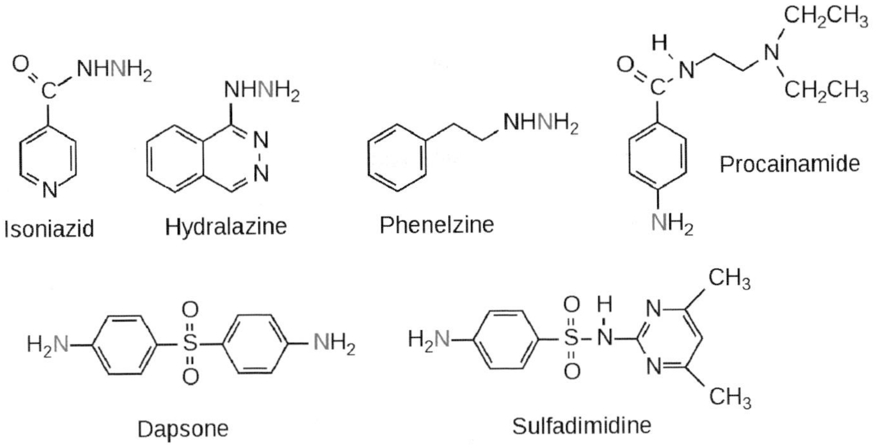

Isoniazid Hydralazine Phenelzine Procainamide

Dapsone Sulfadimidine

Figure 3.14 Examples of arylhydrazines and arylamines acetylated by human NAT2. Acetylated nitrogen are atoms highlighted in blue.

3.2.4.4 Methylation

N-, *O*- and *S*-methylations are known and are common biochemical reactions, occurring with amines, phenols and thiols. The methyl donor is *S*-adenosylmethionine (SAM). Norepinephrine and related phenylethanolamines are *N*-methylated by phenylethanolamine-*N*-methyltransferase, which is found in the soluble fraction of adrenal homogenates. Histamine is *N*-methylated by imidazole *N*-methyltransferase.

Several enzymes are involved in *O*-methylation. Catecholol-*O*-methyltransferase (COMT) methylates both endogenous and foreign catecholamines, including isoprenaline (isoproterenol). Phenolic metabolites from phase 1 oxidations may be methylated, for example 7-hydroxychlorpromazine (Figure 3.11).

A number of exogenous thiols are methylated by *S*-methyltransferase. Endogenous substrates include the thiol-containing amino acids cysteine and homocysteine, and GSH. Inorganic sulfur, selenium and tellurium compounds are methylated to volatile derivatives.

3.2.4.5 Glycine conjugates

Aromatic amino acids, such as benzoic and salicylic acids, are conjugated with glycine to produce hippuric and salicyluric acids, respectively. The acids react with coenzyme A to form an acyl donor that reacts with glycine, and possibly other amino acids (Figure 3.15).

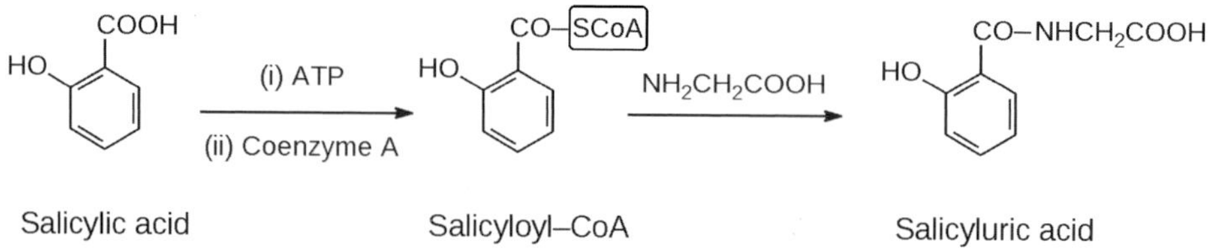

COOH
HO
(i) ATP
(ii) Coenzyme A
CO–SCoA
HO
NH_2CH_2COOH
$CO–NHCH_2COOH$
HO
Salicylic acid
Salicyloyl–CoA
Salicyluric acid

Figure 3.15 Formation of salicyluric acid.

3.2.4.6 Conjugation with glutathione

The reduced form of glutathione is an important reducing agent and one of its roles is to reduce methaemoglobin to haemoglobin. GSH usually acts as a nucleophile and can react chemically or enzymatically via a family of glutathione transferases. Substrates include aromatic nitro and halogen compounds, and oxidized products of phase 1 metabolism. Probably the best-known reaction of GSH is that with the reactive paracetamol metabolite, NAPQI (Figure 3.16). The GSH conjugate undergoes further (phase 3) metabolism to give the mercapurate derivative, which is excreted in urine.

3.3 Excretion

The organs most involved in excretion are the kidney and liver. Although the liver is an important site of drug metabolism it is also an important site of excretion, drugs and their metabolites being excreted via the bile. Other sites might not be so important with regards to the amounts excreted but may be of toxicological significance. Excretion of drugs in

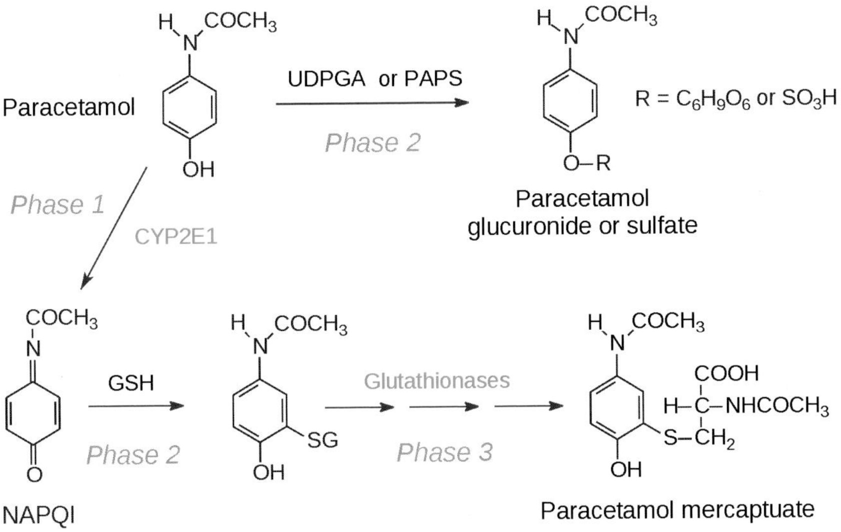

Figure 3.16 Metabolism of paracetamol (acetaminophen).

milk may have consequences for a feeding infant. Expired air is used for medico-legal estimates of blood alcohol whilst measurement of drugs in saliva, sweat, hair and nails are often of forensic interest.

3.3.1 Urine

The functioning unit of the kidney is the nephron, there being over 1.2 million in a human kidney (Figure 3.17).

3.3.1.1 Glomerular filtration

The Bowman's capsule filters small molecules up to the size $M_r = \sim 66{,}000$. Approximately 90% of the peptide hormone insulin ($M_r = \sim 6{,}000$) in plasma water appears in the ultrafiltrate whereas the figure for serum albumin ($M_r = \sim 69{,}000$) is normally <0.1%. Approximately 180 L of plasma is filtered per day, equivalent to 13–15 times the volume of ECF. The plasma filtrate has the composition of plasma water because small, unbound drugs are filtered, but bound drug is retained.

3.3.1.2 Tubular secretion

The proximal convoluted tubular cells have transport proteins for acids, bases and neutral substances. Organic anion transport proteins (OAT1 and OAT3) and organic cation transporters (OCT1) are located in the basolateral membrane. OATP1B3 transports neutral compounds such as digoxin and steroids as well as some anions. Efflux pumps in the apical membrane include P-gp and MRP4 (Figure 3.18). OCT4 secretes cations from the cell and takes up uric acid from the filtrate. MATE1 excretes anions, a process which is driven by the uptake of protons. Many acidic drugs and their glucuronide and glycine conjugates are transported. Examples include penicillins, some sulfonamides and thiazide

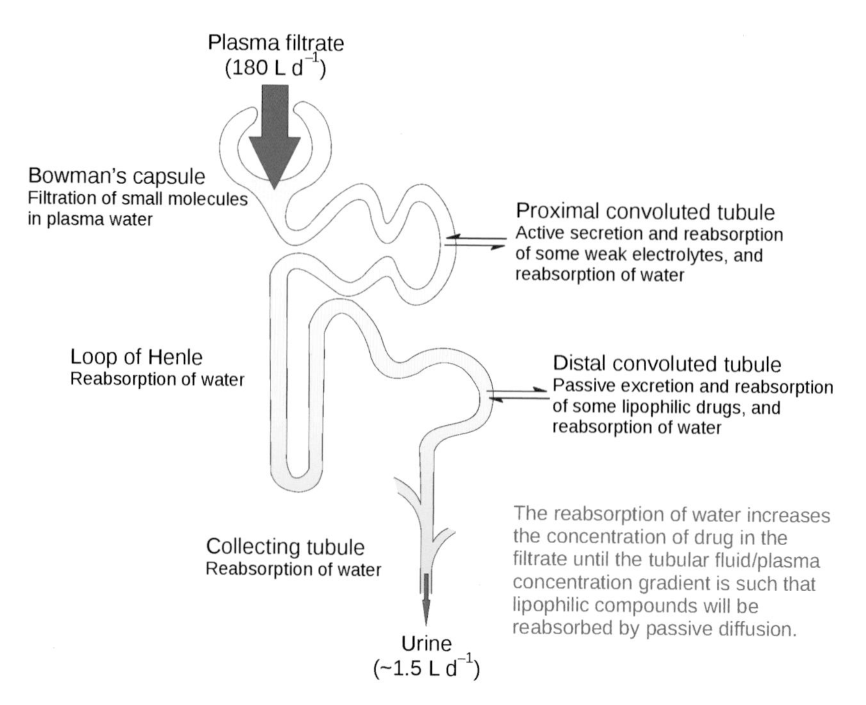

Figure 3.17 Diagrammatic representation of the principal processes involved in the renal excretion of drugs.

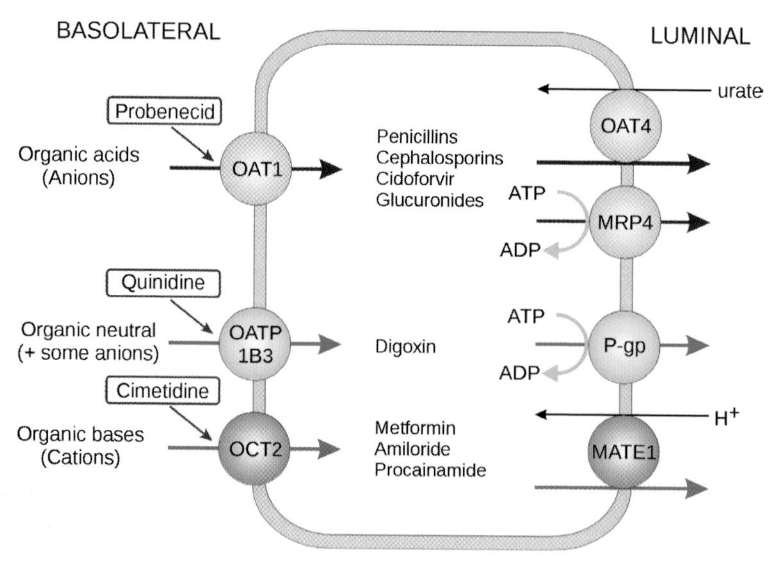

Figure 3.18 Simplified diagram of transporters of kidney proximal tubular cells. Secretion into the cell is largely by facilitated diffusion whilst secretion into the tubular fluid is mediated by active transport. Inhibitors are shown in purple-edged boxes.

diuretics. Examples of basic drugs secreted include quinine, pethidine (meperidine), amiloride and triamterene.

Penicillins have short elimination half-lives (~0.5–1 h), which can be increased by co-administration of probenecid which competes for OAT1/3 so that the renal excretion of the penicillin is reduced. Probenecid is no longer used for this purpose but it is sometimes used as a uricosuric drug because it inhibits reabsorption of uric acid in the PCT, thereby increasing its excretion.

3.3.1.3 *Passive reabsorption*

To some extent the kidney tubule behaves like a typical lipid barrier. Lipophilic molecules may diffuse in either direction and, if it were a static system, the concentrations in tubular fluid and plasma water would equilibrate. However, as a very large proportion of the filtered water is reabsorbed (Figure 3.17), the concentration of the tubular fluid markedly increases so that the net movement of drug or metabolite is from tubular fluid to plasma, that is, lipophilic molecules are reabsorbed. The reabsorption of weak electrolytes will be affected by the degree of ionization and the pH-partition hypothesis applies (Section 3.3.1.5).

Lipophilic drugs tend to be highly protein bound; they are hydrophobic and tend whenever possible to move from aqueous to non-aqueous environments. Protein binding reduces glomerular filtration and that, coupled with tubular reabsorption, means that little parent drug appears in the urine. This nicely illustrates the need for prior metabolism of such drugs as an aid to excretion – albeit as metabolites.

3.3.1.4 *Renal clearance*

The functioning of the kidney and how it handles substances has been investigated and described in terms of the physiological concept, renal clearance:

$$CL_{\mathrm{R}} = \frac{U}{P} \times \text{urine flow rate} \tag{3.1}$$

where U is the concentration of substance in urine and P the concentration of substance in plasma. The ratio U/P is dimensionless so the units of clearance are those of urine flow, usually mL min^{-1}. Renal clearance can be defined as the volume of plasma flowing through the kidneys from which substance is removed per unit time, for example per minute. The two kidneys receive ~20–25% of the cardiac output ~1200 mL min^{-1} of blood, or ~650 mL min^{-1} plasma. If 10% of a substance was removed from the plasma as it passed through the kidney, then the clearance would be 65 mL min^{-1}, or if it were 20% it would be 130 mL min^{-1}. The proportion removed is known as the extraction ratio, E, and so

$$CL_{\mathrm{R}} = EQ \tag{3.2}$$

where Q is the plasma flow rate. It follows from Equation 3.2 that the maximum value of renal clearance cannot exceed the plasma flow, that is when $E = 1$.

The renal clearance of a substance that is freely filtered, not actively secreted by renal tubular cells and not reabsorbed, is a measure of the glomerular filtration rate (GFR). Inulin is a highly water-soluble polysaccharide that is not bound to plasma proteins; it is not secreted, nor reabsorbed, and is used to assess GFR. Its value is maximal in males of about 20 years (125 mL min^{-1}). It is less in women and declines with age. Creatinine clearance is sometimes used to assess GFR or the concentration of serum creatinine is taken as an indicator of renal function, the premise being that reduced filtration is reflected in elevated serum creatinine. However, creatinine is a metabolite of creatine, which is chiefly found in skeletal muscle in varying amounts, so exogenously administered inulin is a more accurate method of measuring GFR. The interest in GFR is that it is an indicator of kidney function and doses of some drugs are adjusted on the basis of GFR.

p-Aminohippuric acid (PAH) is actively secreted and totally cleared as it passes through the kidneys, that is, its extraction ratio is 1. This means that renal clearance of PAH equals the plasma flow through the kidneys, and PAH has been used to measure this. Kidney *blood* flow can be derived from the haematocrit (H) as it is the plasma flow divided by $(1 - H)$.

Because of tubular reabsorption, it is not always obvious from the renal clearance whether a drug undergoes active tubular secretion. Penicillin G has a renal clearance of approximately 350 mL min^{-1} and as this far exceeds GFR, penicillin must be actively secreted into tubular fluid. However, the converse is not true. A drug may be actively secreted but reabsorption may be such that $CL_R < \text{GFR}$. Some indication may be obtained by comparing CL_R with the filtration rate of the unbound fraction. Competition between the protein carriers and plasma proteins probably means that plasma protein binding has little, if any, effect on the active tubular secretion of drugs.

3.3.1.5 *Effect of urine pH and flow rate*

The pH of urine may vary, depending on diet and other factors, but pH 6.3–6.6 is a reasonable range under normal physiological conditions. Changing urine pH, say by administering sodium bicarbonate (sodium hydrogen carbonate, $NaHCO_3$) to make it alkaline, can have a major effect on the excretion of some weak acids and bases. The ionization of salicylic acid (pK_a = 3.0) at pH 6.3 is 99.95% whereas at pH 7.3 it is 99.995%. This extremely small difference in the degree of ionization represents a 10-fold difference in the proportion non-ionized (0.05–0.005%). The difference at pH 8.3, a value that can be easily obtained by giving bicarbonate, is 100-fold. Alkalinization of urine may be used to treat overdoses with drugs such as aspirin and phenobarbital but the old practice of increasing urine flow by giving large volumes of water and possibly concomitant administration of a diuretic (forced alkaline diuresis, FAD) is no longer recommended. Whilst it is true that the renal clearance of phenobarbital increases with increasing urine flow rate, the risks of electrolyte imbalance and pulmonary oedema do not justify the use of this approach. Simply adjusting urine pH is sufficient.

Similar calculations can be performed for a base like amfetamine (pK_a = 9.8). At pH 7.8 the percentage non-ionized is 1% but at pH 4.8 the percentage falls to 0.001%, which makes a major change to the amount not reabsorbed and hence renal clearance. This is reflected in the amounts excreted at different pH values (Figure 3.19).

3.3.2 Biliary excretion

Drugs excreted in the bile enter the duodenum via the hepatic duct. The rate of release of bile is not uniform but is increased in response to food. Those drugs and metabolites that remain in the GI tract will be removed from the body in the faeces, but there may be some

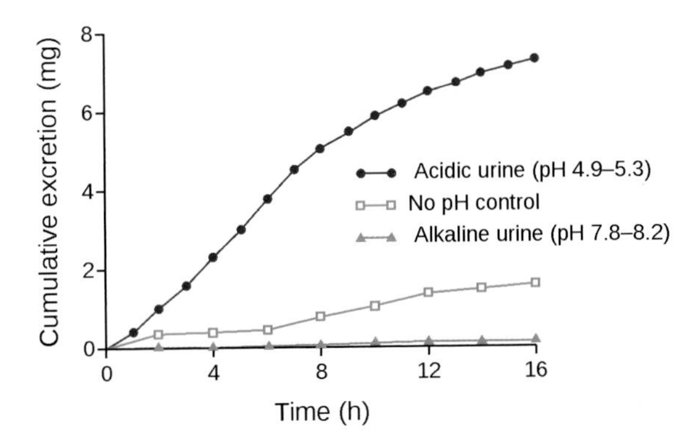

Figure 3.19 Effect of urine pH on amount of metamfetamine (methamphetamine) excreted in a single subject. Redrawn with permission from Beckett & Rowland (1965).

reabsorption back into the body via the hepatic portal system (Section 2.3.1). When temoporfin, an anti-cancer drug, was injected intravenously over 99.9% of the dose was recovered in the faeces, demonstrating how important biliary excretion can be. It also illustrates that finding an orally administered drug in faeces does not necessarily indicate that it has not been absorbed.

Bile canaliculi run between adjacent rows of hepatocytes to join the branching bile duct. Drugs may diffuse passively into the bile but this will result in relatively low concentrations similar to those of the unbound drug. Thus one would expect most drugs to be present in bile to some extent. Because the bile flow is only ~0.5–0.8 mL min^{-1}, the biliary clearance, CL_{Bile}, of such drugs will be low:

$$CL_{Bile} = \frac{\text{concentration in bile}}{\text{concentration in plasma}} \times \text{bile flow rate} \qquad (3.3)$$

However, some drugs and, particularly, water-soluble metabolites such as glucuronides, are actively secreted into bile and can attain bile/plasma ratios in excess of 500 and the biliary clearance of these molecules may be several hundred millilitres per minute. Drugs enter the hepatocytes by passive diffusion or via SLC proteins, including OATP1B1, OATP1B3, OAT2 and OCT1. Proteins of the ATP-binding cassette family (P-gp, MRP2 and BCRP) and MATE1 located on the luminal membrane actively secrete their substrates into bile (see Figure 10.5). There appears to be a molecular weight cut-off for transport with substrates having to have M_r values > 500. In the rat the figure is ~325 and there are species variations. Above the minimum, the degree of biliary excretion increases with increasing M_r. Another feature is that polar acidic molecules make good substrates. This is supported by the fact that high concentrations of glucuronide metabolites are often found in bile. Glucuronides are polar and acidic, and glucuronidation increases the molecular weight of the parent drug by 193 g mol^{-1}. It is likely that active transport has evolved to transport bile acids and compounds such as bilirubin glucuronide.

Suitable molecules may be actively transported without prior metabolism. The water-soluble cardiac glycoside ouabain (M_r = 728.8) can have a bile:plasma ratio as high as 500. It is likely that temoporfin (M_r = 680.7) is actively transported into bile. Bromosulfophthalein (M_r = 705.6) is used to test liver function.

3.3.3 Expired air

The lungs are an important site for the excretion of volatile drugs and metabolites. The majority of gaseous and volatile general anaesthetics are excreted in expired air and because they are administered by the same route, the dose and hence the level of anaesthesia can be controlled by adjusting the partial pressures of these agents in the breathing apparatus. Deep alveolar air concentrations tend to equilibrate with those in blood, as would be predicted by Henry's law. This is the basis of breath alcohol measurements. At 34 °C the distribution between blood and alveolar air is 2100:1, that is, 2100L of air contain the same amount of ethanol as 1 L of blood. Other volatile liquids are excreted via the lungs. For example, it has been shown that ~40% of an oral dose of benzene is excreted in expired air. Inorganic selenium salts are metabolized to dimethyl selenide (Section 3.2.4.4), which is exhaled, giving a garlic smell to the breath. Extensive metabolism of an organic compound will eventually produce carbon dioxide, which may be excreted in expired air. Study protocols for metabolism studies with ^{14}C-labelled drug may include trapping any exhaled $^{14}CO_2$.

3.3.4 Saliva

Excretion of drugs in saliva appears to occur largely by passive diffusion. The pH of saliva can vary from 5.5, the lower limit for parotid saliva, to 8.4 in certain ruminants. The average pH of 'mixed' saliva is ~6.5, about 1 unit lower than plasma pH. Thus the unbound concentrations of basic drugs are usually higher in saliva than in plasma water whist the converse is true of weak acids. The saliva to plasma equilibrium ratio for an acid will be:

$$\frac{S}{P} = \frac{\left(1+10^{\left[\mathrm{pH(s)}-\mathrm{p}K_{\mathrm{a}}\right]}\right) f_{u(p)}}{\left(1+10^{\left[\mathrm{pH(p)}-\mathrm{p}K_{\mathrm{a}}\right]}\right) f_{u(s)}} \tag{3.4}$$

where pH(s) is the saliva pH, pH(p) is the plasma pH and $f_{\mathrm{u(p)}}$ and $f_{\mathrm{u(s)}}$ are the unbound fractions in plasma and saliva, respectively. There is less protein in saliva than in plasma, with what there is being chiefly antimicrobial and digestive enzymes secreted by the salivary glands. Consequently, the amount of drug-binding protein in saliva is usually considered to be insignificant, so $f_{\mathrm{u(s)}}$ is taken to be one. For salicylic acid, pK_a = 3.0, which is 50–90% protein bound in plasma, depending on the dose, the S/P ratio would be expected to be 0.06–0.01 for a saliva pH = 6.5. The equivalent equation for bases is:

$$\frac{S}{P} = \frac{\left(1+10^{\left[\mathrm{p}K_{\mathrm{a}}-\mathrm{pH(s)}\right]}\right) f_{u(p)}}{\left(1+10^{\left[\mathrm{p}K_{\mathrm{a}}-\mathrm{pH(p)}\right]}\right) f_{u(s)}} \tag{3.5}$$

Pethidine, a weak base, pK_a = 8.7, is 40–50% bound to plasma proteins so, for at a saliva pH of 6.5, the S/P ratio calculates to be 4.6–3.8. This is consistent with reports that pethidine concentrations in saliva are higher than those in plasma.

There is little evidence that active transport is important for secretion of drugs into saliva. Active transport of ions results in saliva being hypotonic, having less sodium but higher concentrations of potassium than plasma. S/P ratios for lithium are 2–3 and vary with plasma concentration, indicating that there is carrier-mediated transport, probably via the potassium pump.

The interest in salivary excretion has arisen (i) because it was seen as a non-invasive method of sampling for pharmacokinetic studies, for example calculating phenazone (antipyrine) half-lives, (ii) as a way of monitoring drugs, both therapeutic and drugs of abuse, and (iii) as a method of estimating the plasma concentrations of the unbound drug. However, there are several caveats and considerations. Three pairs of glands are responsible for most of the saliva that comprises oral fluid. Stimulation, either mechanical or with citric acid, changes the percentage produced by the different glands and so changes the composition of the fluid. Salivary flow is under autonomic control. Stimulation of β-receptors increases the amount of protein and mucin, and results in a more viscous fluid whereas α-adrenoceptor excitation increases protein. Muscarinic stimulation increases saliva flow, which is accompanied by increases in sodium and bicarbonate concentrations, leading to increased pH values. Clearly drugs acting at these receptors will influence the composition of the oral fluid.

Despite the problems, it is possible to use saliva for kinetic investigations when it is usual to validate against blood samples (Figure 3.20). However, it is drug monitoring that is driving the increasing interest in oral fluid analysis. For drugs of abuse, many of which are basic, sampling oral fluid is non-invasive and can be done under supervision to reduce the risk of adulteration of the sample. Oral fluid is useful for situations where venepuncture may be difficult, such as the nervous patient or, more importantly, for therapeutic drug monitoring in children, who are often less embarrassed than adults about spitting into a container. It has been suggested that phenytoin concentrations in saliva are a good approximation to the unbound concentration in serum and that this is a better indicator for dose adjustment than total serum concentrations.

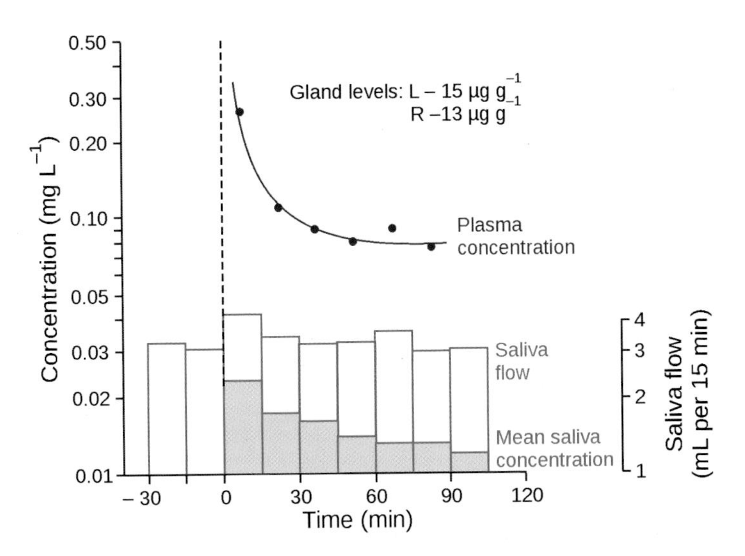

Figure 3.20 Concentrations of clonidine in plasma, submaxillary gland tissue (L = left, R = right) and submaxillary gland saliva in a cat after 250 mg kg^{-1}, intravenously. Redrawn with permission from Cho & Curry (1969).

3.3.5 Stomach and intestine

A number of basic drugs following i.v. doses are excreted into gastric juice in the stomach. This is predictable from the pH-partition hypothesis. Amongst the drugs studied this has been demonstrated for quinine and nicotine. Water-soluble drugs and ionized forms of weak electrolytes are likely to be excreted into the intestine in a way analogous to the excretion of drugs into the stomach. With both the stomach and the intestine diffusion from blood into the GI lumen is an essential part of the reversible diffusion reaction involved in the absorption of drugs. When the concentration gradient is favourable, for example after i.v. administration, the net movement will be into the GI lumen.

3.3.6 Breast milk

Excretion in milk is a minor pathway in terms of the amount of drug eliminated from the mother. It is of interest because of the potential effects of the drug on the suckling infant and the environmental exposure of populations to veterinary drugs in cows' milk.

A number of drugs appear in milk, usually in small concentrations, suggesting that the mechanism of transfer is passive diffusion, that is, the breast behaves as a lipid membrane. Breast milk is slightly more acidic on average (pH 7.2) than plasma and contains some protein, chiefly whey proteins, lactoferrin and α-lactalbumin. Albumin is present at ~0.4 g L^{-1}, approximately 1/10th of the concentration in plasma. Assuming drugs enter milk by passive diffusion, a method for estimating the amount of a drug that will pass into in milk has been proposed (Atkinson & Begg, 1990). Factors taken into consideration are the pK_a, the degree of protein binding in plasma and the lipophilicity, assessed as log $D_{7.2}$ (octanol partition coefficient at pH 7.2). The method predicted a milk:plasma ratio of 2.33 for venlafaxine, which was in good agreement with an average observed value of 2.5. Despite the high concentration of venlafaxine in milk (Figure 3.21), the proportion of the mothers' dose consumed by the infants ($n=6$) was calculated to be <6% (mean = 3.2%).

Interestingly, as the recommendations to breast feed increase, so the number of studies showing that the risk from maternal drug use is minimal is increasing. However, it is wise

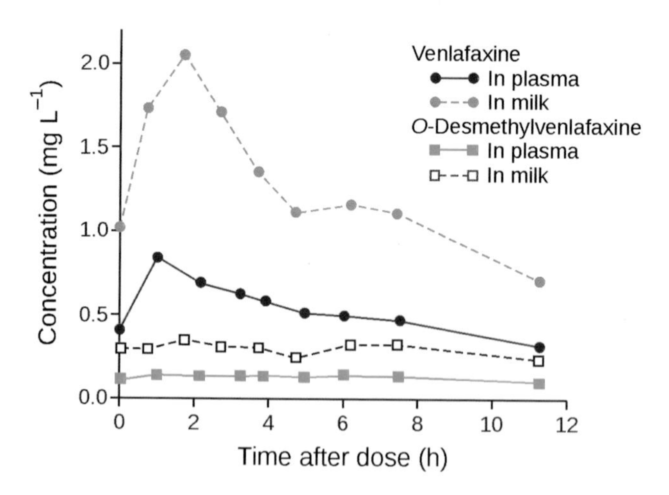

Figure 3.21 *Steady-state venlafaxine and O-desmethylvenlafaxine concentrations in plasma and milk in a single subject. Redrawn with permission from the data of Ilett* et al. *(2002).*

to treat each case individually and to monitor the wellbeing of the infant. The death of a 13-day-old breast-fed baby has been ascribed to the codeine his mother was taking for post-partum pain (Madadi *et al.*, 2007). This was an unusual case because the mother was an ultrafast metabolizer of codeine to morphine (Section 10.5.1) and also rapidly converted morphine to morphine 6-glucuronide (Section 10.8.1).

3.3.7 Other routes of excretion

Amongst the remaining minor routes of excretion are sweat, keratinous tissues and genital secretions. The increasing interest in these routes of excretion is because the samples provide a way of demonstrating exposure to drugs, by detecting either the drug directly or one or more of its metabolites.

3.3.7.1 Sweat

There are two kinds of sweat gland, eccrine and apocrine. The latter, which form during puberty, are associated with hair in the pubic region and under the armpits, and secrete a thick liquid into the hair follicle. Eccrine glands produce a watery sweat that is ~99% H_2O. It has long been known that drugs are excreted in sweat. This was demonstrated for quinine as early as 1844 and for morphine in 1942. There appear to have been few systematic studies as to the mechanisms involved, but a simple diffusion model probably applies as some basic drugs have sweat:plasma ratios >1 whilst weak acids have ratios <1, as predicted from the pH-partition hypothesis for distribution drugs between plasma water and sweat, pH 5.8 (Johnson & Maibach, 1971).

3.3.7.2 Hair and nails

Drugs and their metabolites enter hair cells, probably by passive diffusion, while the hair is growing. Because the cells subsequently die, the drugs not only remain in the hair but they are not metabolized and may be measured months, years and even centuries later. This explains the forensic interest in measuring drugs in hair as a means of assessing previous exposure. For example, by examining segments of hair it is possible to determine periodic exposure to drugs and poisons.

3.3.7.3 Genital secretions

Obviously, excretion of drugs in vaginal secretions or seminal fluid is not going to play a major role in the removal of a drug from the body. Interest in these fluids arises because of the potential for drugs to interfere with sperm motility, have teratogenic effects or provide forensic evidence in a case of sexual assault. The excretion of drugs in semen has been reviewed by Pichini *et al.* (1994). In men approximately 30% of seminal plasma originates from the prostate and ~60% from seminal vessels. The mechanism by which drugs enter seminal fluid is probably by passive diffusion and the equilibrium plasma:seminal fluid ratio of weak electrolytes will be given by modified forms of the Henderson–Hasselbalch equation, analogous to Equations 3.4 and 3.5. However, the situation is complicated by the fact that prostatic fluid is acidic (pH ~6.6) while that of vesicular fluid is alkaline (pH ~7.8). Some antibiotics have been shown to have fluid:plasma ratios >1, for example ciprofloxacin (5.8) and norfloxacin (5.4), but for many other drugs the ratio is

<1, for example carbamazepine (0.4–0.7), phenytoin (0.17) and valproic acid (0.07), possibly reflecting the effect of plasma protein binding.

3.3.8 Cycling processes

Drugs that are excreted into saliva or the stomach are likely to be reabsorbed from the GI tract, particularly those that are readily absorbed when given orally. Consequently, excretion via these routes will make little contribution to the overall removal of such drugs from the body. Instead these drugs will be 'trapped' in a cycling process: blood → saliva → intestine → blood or blood → stomach → intestine → blood.

3.3.8.1 Enterohepatic cycling

This is a further, and possibly more significant, form of cycling. Drugs that are excreted in bile may be reabsorbed from the intestine, in much the same way as described above. However, as explained previously (Section 3.3.2), a number of drugs are excreted in bile as glucuronide metabolites. Water-soluble glucuronides are not normally absorbed from the GI tract. However, if the conjugate is hydrolysed back to the original drug by β-glucuronidases in intestinal bacteria then the drug can be reabsorbed, conjugated and secreted again, leading to cycling of drug and drug-conjugate (Figure 3.22). Obviously only drugs that undergo phase 2 glucuronidation directly without a prior phase 1 reaction can do this. Examples of such drugs include chloramphenicol, phenolphthalein and morphine.

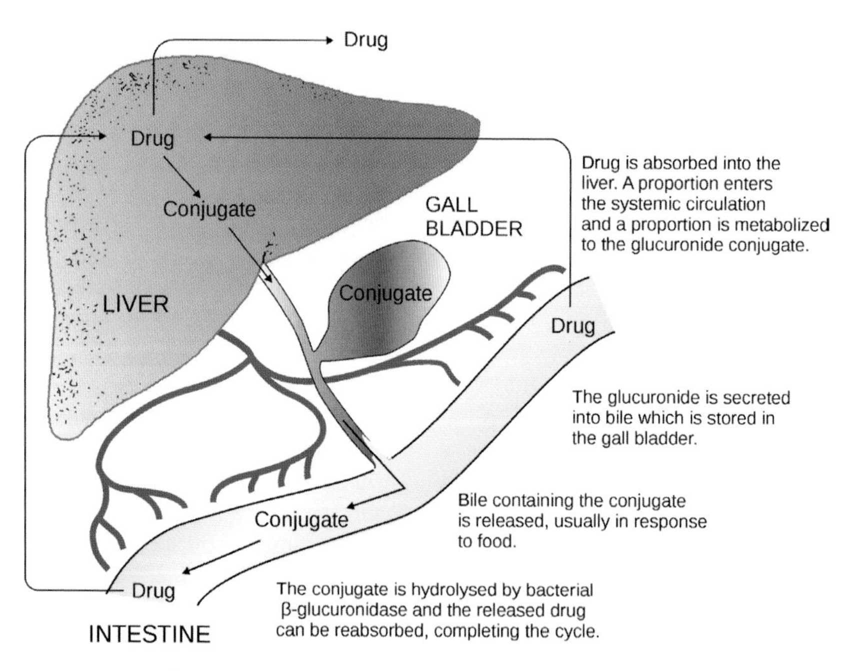

Figure 3.22 *Enterohepatic cycling. The orally administered drug enters the intestine from the stomach and is absorbed into the hepatic portal vein, carried to the liver and converted to the glucuronide, which is secreted into the bile. Intestinal bacteria hydrolyse the glucuronide, liberating the parent drug, which can be reabsorbed.*

3.3.8.2 Significance of cyclic processes

A drug trapped in a cycle persists in the body for longer than would otherwise occur and this reduces the rate of decline in plasma concentrations to some extent. Because the excretion of drugs via saliva and the stomach usually represents a steady release of small proportions of the dose it is most unlikely that the cycling will lead to any discernible fluctuations in plasma concentrations. On the other hand, the release of bile is pulsatile and greater quantities are released in response to food, particularly fatty meals. Some studies have reported increases in plasma concentrations after meals, which have been ascribed to enterohepatic cycling. Rises in plasma concentrations of β-blocking drugs, benzodiazepines and phenothiazines have been observed at late times following intravenous injection.

Enterohepatic cycling will increase the half-life of a drug but the effect on the activity may be small if only a small proportion of the dose is being cycled. Furthermore, for the majority of drugs, drug molecules in the cycle are not available to the receptors, apart possibly from the laxative phenolphthalein and the anti-diarrhoeal morphine.

Summary

This chapter has examined the commonly encountered metabolic reactions and the various enzymes and isoforms involved. The examples were chosen because, in the main, they are relevant to consideration of topics such as pharmacogenetics, toxicity and drug interactions. Knowledge of the routes and mechanisms of metabolism and excretion is paramount to understanding how these influence pharmacokinetics, as discussed in the next two chapters.

3.4 Further reading

Aps JK, Martens LC. Review: The physiology of saliva and transfer of drugs into saliva. *Forensic Sci Int* 2005; 150: 119–31.

Gibson GG, Skett P. *Introduction to Drug Metabolism.* 3rd edn. Cheltenham: Nelson Thornes, 2001. A classic text on drug metabolism.

International Society for the Study of Xenobiotics. History of xenobiotic metabolism. http://www.issx.org/?History. Accessed 9 June 2015. Interesting insight into early discoveries of drug metabolism.

Nelson DR. Cytochrome P450 Home page: http://drnelson.utmem.edu/CytochromeP450.html. Accessed 5 June 2015.

3.5 References

Atkinson HC, Begg EJ. Prediction of drug distribution into human milk from physicochemical characteristics. *Clin Pharmacokinet* 1990; 18: 151–67.

Beckett AH, Rowland M. Urinary excretion of methylamphetamine in man. *Nature* 1965; 206: 1260–1.

Cho AK, Curry SH. The physiological disposition of 2(2,6-dichloroanilino)-2-imidazoline (St-155). *Biochem Pharmacol* 1969; 18: 511–20.

Ilett KF, Kristensen JH, Hackett LP, Paech M, Kohan R, Rampono J. Distribution of venlafaxine and its *O*-desmethyl metabolite in human milk and their effects in breastfed infants. *Br J Clin Pharmacol* 2002; 53: 17–22.

Johnson HL, Maibach HI. Drug excretion in human eccrine sweat. *J Invest Dermatol* 1971; 56: 182–8.

Madadi P, Koren G, Cairns J, Chitayat D, Gaedigk A, Leeder JS, Teitelbaum R, Karaskov T, Aleksa K. Safety of codeine during breastfeeding: fatal morphine poisoning in the breastfed neonate of a mother prescribed codeine. *Can Fam Physician* 2007; 53: 33–5.

Pichini S, Zuccaro P, Pacifici R. Drugs in semen. *Clin Pharmacokinet* 1994; 26: 356–73.

4

Single-compartment Pharmacokinetic Models

Learning objectives

By the end of the chapter the reader should be able to:

- demonstrate that the elimination half-life is directly proportional to apparent volume of distribution and inversely proportional to systemic clearance
- explain why, for first-order elimination, plasma concentrations during a constant rate infusion of drug must asymptote to a steady-state value
- discuss the effect elimination half-life has on attainment of steady-state and fluctuations in plasma concentrations during multiple dosing
- plot ln C versus t data following i.v. injection, derive parameters including systemic clearance and comment on the values obtained
- plot ln C versus t data following an oral dose and estimate λ and k_a
- describe the principle by which sustained release is obtained
- explain why some drugs exhibit non-linear elimination and describe how individual values of V_{max} and Km can be derived.

Introduction to Drug Disposition and Pharmacokinetics, First Edition. Stephen H. Curry and Robin Whelpton.

Companion website: www.wiley.com/go/curryandwhelpton/IDDP

4.1 Introduction

Pharmacokinetics is the science of using models (especially mathematical equations) to describe the time course of a drug and/or its metabolites in the body, usually in terms of the concentrations in plasma, although other fluids, such as whole blood or urine, may be used. Further information may be gained from the analysis of tissues or faeces. Probably the most important applications of the subject are in new drug development and in understanding and controlling optimum drug use. It is generally assumed that there is a relationship between the concentration of drug in plasma and the effects that are elicited. This and the concept of a therapeutic window were introduced in Chapter 1. If a suitably sized single oral dose is taken (Figure 4.1) there will be a lag-time while the tablet or capsule disintegrates and the drug begins to appear in the plasma (t_{lag}, Section 4.4.1) and until sufficient drug has been absorbed for the plasma concentration to reach the threshold for the effect to begin.

The effect will cease when elimination has reduced the concentration to below the minimum concentration for the desired effect to be apparent. The duration of effect will be the time between these two points. A larger dose may be taken to increase the intensity, or even the duration of the effect, but it is clear from Figures 1.1 and 4.1 that there is a limit to the size of a single dose if toxic concentrations are to be avoided. Doubling the dose generally doubles the peak concentration and may increase the intensity of the effect, but it will not necessarily double the duration of action. To obtain a longer duration of action, it is necessary to give smaller doses at regular intervals and if the dose and dosage intervals are chosen appropriately the concentrations can be maintained within the therapeutic window (Figure 4.2). The curves of this figure were generated using a simple pharmacokinetic model, which will be explained later in this chapter.

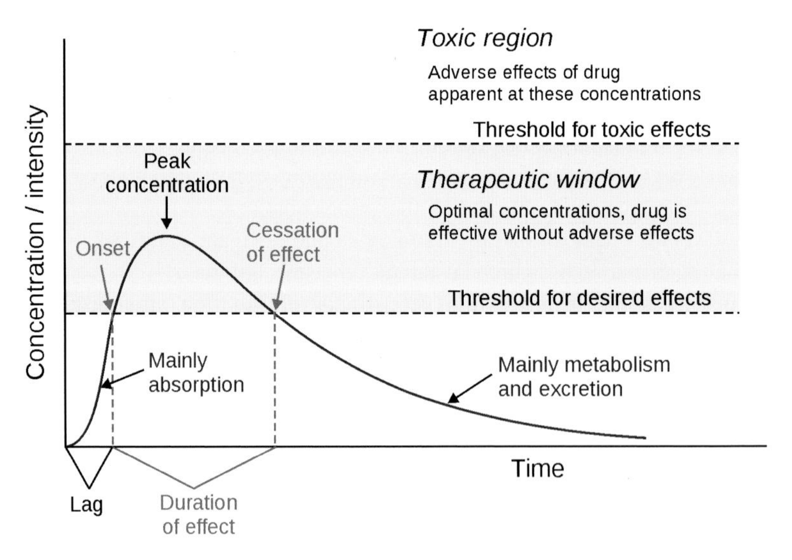

Figure 4.1 Model curve showing a theoretical perfect relationship between drug effect (intensity) and drug concentration following a single dose by any route other than intravenous injection.

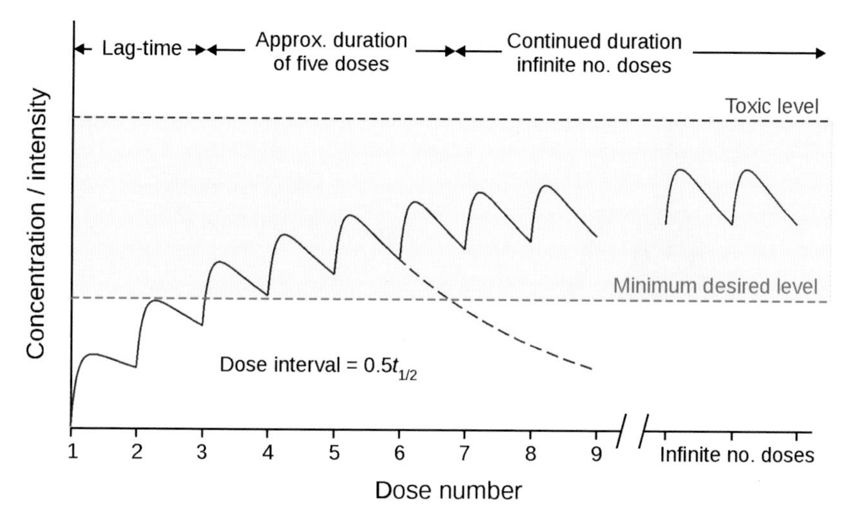

Figure 4.2 Model curves showing a theoretical perfect relationship between drug effect (intensity) and drug concentration during a series of eight doses, with a dose interval 0.5 × $t_{½}$ *of the drug. The broken line following the fifth dose represents the decay if that had been the last dose. Also shown are the concentrations after an infinite number of doses.*

The crucial point about deriving a model is that it can be used to predict what will happen when the parameters are changed. Pharmacokinetics can be used to:

- assess the rate and extent of absorption
- predict the effects of changing doses and routes of administration
- model the effect of enzyme induction/inhibition and changes in excretion
- model relationships between concentrations and effect (PK-PD).

Various types of model may be used, but probably the best known, and most widely used, are *compartmental* models. The simplest model, the single-compartment model, assumes that the drug in the body can be treated as if it were in a homogeneous solution in a 'pool' or 'compartment'. Multiple-compartment models assume that the drug is distributed between two or more pools (Chapter 5). The concept of compartments is discussed in more detail in Section 5.1.6.

A single-compartment model is one where the drug is considered to be present as a well-stirred, homogeneous solution, rather like the solution of a dye in the flask depicted in Figure 4.3(a). The volume of the flask is analogous to the apparent volume of distribution of the drug, as discussed in Section 2.4.1.1. A drug is eliminated by metabolism and/or excretion as discussed earlier in Chapter 3, and in the figure these processes are represented by the water flowing through the flask, removing a proportion of dye as it does so. In pharmacokinetics the model is usually illustrated as a box or circle with an input and an output (Figure 4.3(b)). The amount (A) of drug at any time will be the volume of the compartment (V) multiplied by the concentration (C) at that time.

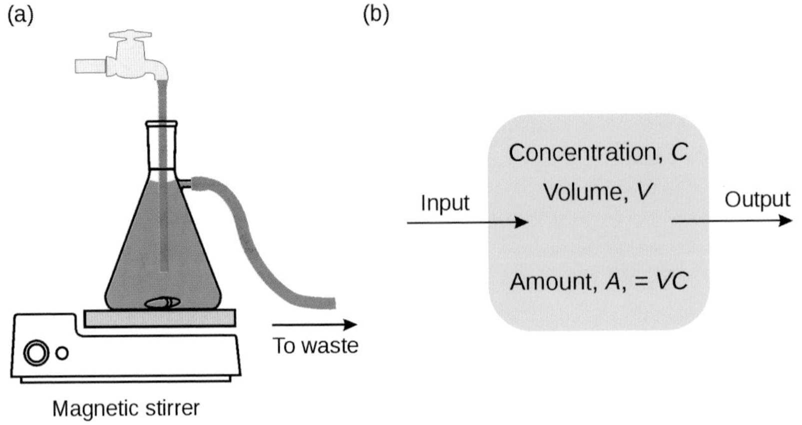

(a) (b)

Magnetic stirrer

To waste

Input

Concentration, C

Volume, V

Amount, A, = VC

Output

Figure 4.3 *(a) Single-compartment models are based on the concept of a well-stirred solution of drug. (b) Typical representation of a single-compartment model.*

Without any recourse to mathematics it should already be apparent from Figure 4.3(a) that if dye (representing the drug) is rapidly introduced into the flask (representing a bolus i.v. injection) the flow of water will gradually wash out the dye and, furthermore:

- the faster the flow of water, the faster the dye will be removed, and
- the larger the flask, the longer it will take to remove the dye, although of course the initial concentration will be lower in a larger flask.

Intuitively, dye is being removed according to first-order elimination kinetics, observable in the flask as a gradual reduction in the colour of the solution, and this can be confirmed by writing a differential equation of the form of Equation 1.7 described in Section 1.3.3.1. The concentration C is declining exponentially, and so is the amount A. Because the volume is constant, $A = V \times C$, and the rate of elimination can be written:

$$-\frac{\mathrm{d}A}{\mathrm{d}t} = \lambda A = \lambda VC \tag{4.1}$$

where λ is a first-order elimination rate constant.

4.2 Systemic clearance

Systemic clearance, CL, also known as plasma clearance or whole body clearance, can be defined as the volume of plasma from which drug is removed per unit time. It is usually expressed as mL min^{-1} or L h^{-1}. CL is the sum of all the *organ* clearances, so if a drug is eliminated by the kidney and the liver and by no other routes:

$$CL = CL_{\mathrm{R}} + CL_{\mathrm{H}} \tag{4.2}$$

where CL_{R} and CL_{H} are the renal and hepatic clearances, respectively. Note that because of the *additivity of clearance*, in the example above, the hepatic clearance can be calculated if the renal clearance is obtained via Equation 3.1. Organ clearance is the volume of

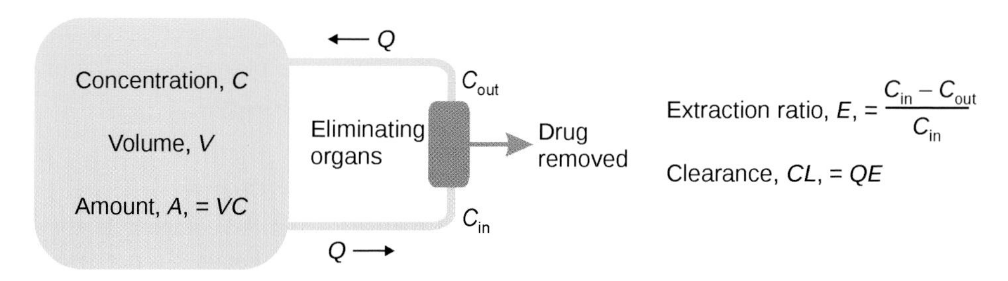

Figure 4.4 The concept of systemic clearance. The extraction ratio, E, is the proportion of drug removed as it flows through the eliminating organs and clearance equals the flow rate, Q, multiplied by the extraction ratio.

plasma flowing through the organ from which the drug is removed per unit time. This is represented in Figure 4.4.

The organ clearance cannot exceed the plasma flow rate through the organ (Equation 3.2, Section 3.3.1.4). Clearance can be used to calculate the rate of elimination of the amount of drug from the body if the plasma concentration is known:

$$\text{rate of elimination} = C \times CL \tag{4.3}$$

and this is true under all circumstances, even when *CL* varies with concentration (Section 4.7). When *CL* is constant, Equation 4.3 is a first-order equation because the concentration is declining exponentially and therefore Equations 4.1 and 4.3 are equal, so that:

$$C \times CL = \lambda VC \tag{4.4}$$

and cancelling *C* from either side gives:

$$CL = \lambda V \tag{4.5}$$

Because the elimination kinetics are first-order, the half-life is given by Equation 1.10, which can be rearranged to:

$$\lambda = \frac{0.693}{t_{1/2}} \tag{4.6}$$

Substituting Equation 4.6 into Equation 4.5 and rearranging gives:

$$t_{1/2} = \frac{0.693\ V}{CL} \tag{4.7}$$

According to Equation 4.7, $t_{1/2}$ is a *dependent* variable that increases when the apparent volume of distribution increases and decreases when the clearance increases. This is in agreement with the model of Figure 4.3(a), where the volume of the flask is analogous to *V* and the flow rate of water represents *CL*. Because no dye is returned to the flask, the extraction ratio $E = 1$ and so, in this example, the clearance equals the flow, *Q*. The concept of extraction can also be applied to uptake of drug by non-eliminating organs (Section 7.3). A more general method for obtaining *CL* using the area under the plasma concentration–time curve (*AUC*) is described later (Equation 4.16 and Section 5.1.2.4).

4.2.1 Why drugs have different elimination half-lives

The Büchner flask in Figure 4.3(a) may seem far removed from the complex anatomy and physiology of the body, but the single-compartment model, although it is not always applicable, is a useful introduction to pharmacokinetic modelling. First, Equation 4.7 explains why drugs have different half-lives; it is because they have different volumes of distribution and/or systemic clearance. The apparent volume of distribution will depend upon the nature of the drug (Table 2.1) and the individual. Lipophilic drugs tend to have large apparent volumes of distribution and, if they enter adipose tissue, will be more widely distributed in someone who is obese, resulting in a longer elimination half-life. Also, changes in binding may result in changes in the distribution of drugs that are bound to plasma protein (Section 2.5.1). Clearance is a measure of how well the eliminating organs remove the drug. Hepatic clearance will increase when enzymes are induced and decrease in the presence of an enzyme inhibitor. Similarly, making urine alkaline with $NaHCO_3$ will increase the renal clearance of salicylate (Section 3.3.1.5) and hence decrease the half-life.

4.3 Intravenous administration

When a drug is injected as a rapid intravenous dose (often referred to as a bolus dose), mixing within its volume of distribution is considered to be instantaneous – compartments are assumed to be well-stirred homogeneous solutions of drug. Clearly this must be an approximation because it takes a finite time to give the injection and a finite time for the drug to be distributed, even within blood. However, this may take less than a minute (Figure 4.5(a)), which is a negligible amount of time compared with the half-lives of the majority of drugs. There is no absorption involved and, after the initial mixing, the plasma concentrations show a monophasic exponential decline:

$$C = C_0 \exp(-\lambda t) \tag{4.8}$$

If serial blood samples are taken and the plasma concentrations, *C*, of the drug are measured, then the rate constant λ can be obtained from the slope of the plot of ln *C* versus time:

$$\ln C = \ln C_0 - \lambda t \tag{4.9}$$

as seen in Figure 4.5(b). Back-extrapolation of the line to $t = 0$ gives a value for ln C_0 and hence C_0. This concentration is the *theoretical* plasma concentration of the drug at $t = 0$, assuming instantaneous equilibration within its volume of distribution. In the dye model of Figure 4.3(a) it would be possible to stop the flow of water, add the dye, allow it to equilibrate so that the initial concentration is C_0 and then to restart the flow of water. Clearly this is impossible in a living creature and so the value of C_0 has to be obtained by extrapolation. At $t = 0$, the *entire* dose of the drug is in the body, as there has been no

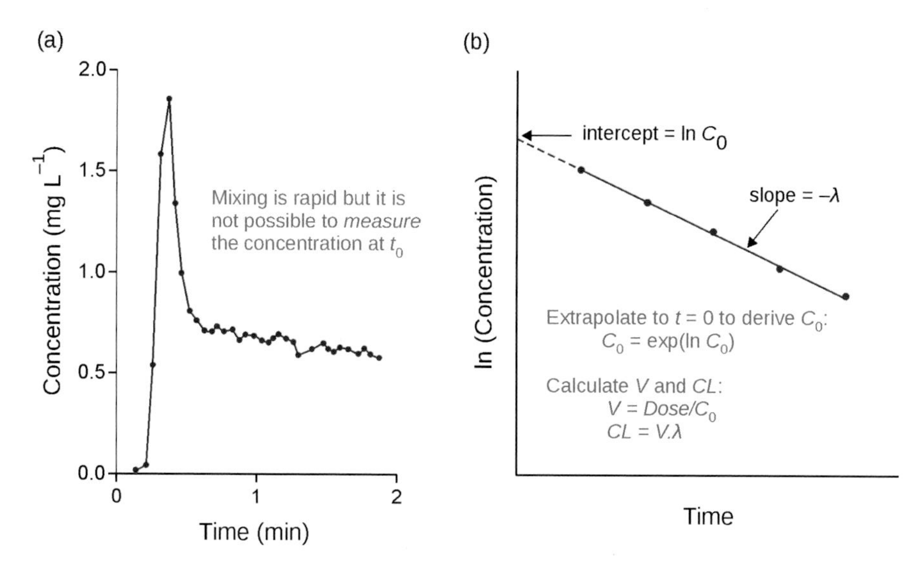

Figure 4.5 *(a) Concentrations of phenylbutazone in one jugular vein of an anaesthetized sheep given a rapid intravenous dose in the other jugular vein. Redrawn with permission from McQueen & Wardell (1971). (b) Back-extrapolation of a* ln *C versus t plot to obtain a value of* ln C_0 *and hence* C_0.

elimination, and the apparent volume of distribution, V, can be calculated from Equation 2.4 as described in Section 2.4.1.1.

4.3.1 Half-life

The half-life, as defined earlier, is the time taken for the initial concentration to fall by one half. It would be better to refer to it as 'half-time', as there is no 'life' involved. However, it is probably futile to rail against the use of half-life, as the term is used almost universally. There is, of course, a direct analogy with the half-life concept used with the decay of radioactive nuclei. Sometimes $t_{½}$ is qualified, for example 'biological' or 'metabolic' half-life, adjectives that can be misleading or, worse, incorrect. Because it is measured in plasma, 'plasma half-life' has been suggested, but it is possible that $t_{½}$ refers to rising plasma concentrations, say after an oral dose. The important thing is to ensure that if a description is used, it is unambiguous.

Half-life is a useful concept when applied to first-order reactions because it is constant. Thus, if 50% is lost in $1 \times t_{½}$ a further 50% of what is remaining will be lost in the next $t_{½}$, that is, after $2 \times t_{½}$ 75% will have been lost and 25% of the original amount will remain. Note that the amount remaining, A, can be readily calculated from the number of half-lives that have elapsed, n:

$$A = \frac{A_0}{2^n} \tag{4.10}$$

where A_0 is the original amount. Note that n need not be an integer (Table 4.1). After five half-lives nearly 97% of the dose has been eliminated and after seven half-lives >99% has

Table 4.1 Proportion of the dose eliminated or remaining as a function of the number of half-lives elapsed after an i.v. bolus dose of drug

Time elapsed (number of half-lives, *n*)	Amount eliminated (%)	Amount remaining (%)
0.5	29.3	70.7
1	50	50
1.5	64.6	35.4
2	75	25
3	87.5	12.5
5	96.87	3.13
7	99.22	0.78

been eliminated. Because the rate of elimination is higher when the concentrations are high shortly after the injection, almost 30% of the dose is lost in a time equivalent to $t_{½}$ divided by 2.

Sometimes it is more convenient to obtain an estimate of the elimination half-life and to use Equation 4.6 to calculate the elimination rate constant, particularly when the concentration data are plotted on a logarithmic scale (Figure 4.6). Note how the half-life is constant and can be obtained by choosing any convenient concentration that is easily divisible by 2. It is worth noting that the 'slope' of the line is not, as is frequently claimed, equal to $-\lambda$. Any attempt to derive a slope between two time points will give the average slope of the C versus t curve between those times.

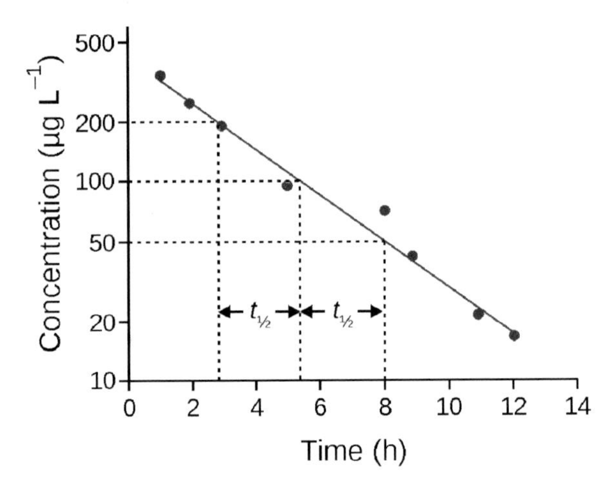

Figure 4.6 Prednisolone concentrations in a kidney transplant patient after an i.v. dose of 20 mg. Redrawn with permission from Gambertoglio et al. *(1980).*

4.3.2 Area under the curve

The area under the plasma concentration–time curve (AUC) can be computed from the integral of the equation defining the decay (Appendix 1):

$$AUC = \int_0^\infty C\,dt \tag{4.11}$$

C is given by Equation 4.8, so:

$$AUC = \int_0^\infty C_0 \exp(-\lambda t)\,\mathrm{d}t \tag{4.12}$$

$$AUC = \frac{C_0}{\lambda} \tag{4.13}$$

but from Equation 2.4, $C_0 = D/V$, so:

$$AUC = \frac{D}{\lambda V} \tag{4.14}$$

Note that the AUC is directly proportional to the dose injected and inversely proportional to the volume of distribution and the elimination rate constant. This is to be expected as a larger volume will lead to lower plasma concentrations and the rate of elimination will be higher when λ is larger, again reducing the plasma concentrations. Furthermore, $CL = \lambda V$ (Equation 4.5) so:

$$AUC = \frac{D}{CL} \tag{4.15}$$

and again, as might be expected, the AUC is reduced when CL is large. Equation 4.15 provides a way of calculating CL without the use of Equation 1.8, that is, without defining the model,

$$CL = \frac{D}{AUC} \tag{4.16}$$

because the AUC can be derived using the trapezoidal method (Appendix 1).

4.4 Absorption

During drug absorption, the concentration gradient is very high and so the contribution of diffusion back from the pool of unabsorbed drug can be ignored and absorption can be considered as a non-reversible phenomenon. The model now requires an equation to describe the input (Figure 4.1). Absorption from the GI tract can be very variable for some drugs and the input function can be complex. The kinetics of absorption after i.m. injections are often first order, and absorption after oral dosing may approximate to first-order kinetics. Thus, the rate of change of the amount of drug in the body is given by the rate at which it is being absorbed and the rate at which it is being eliminated:

$$\frac{\mathrm{d}A}{\mathrm{d}t} = k_\mathrm{a} A_\mathrm{a} - \lambda A \tag{4.17}$$

where k_a is the first-order rate constant of absorption and A_a is the amount of drug remaining to be absorbed. Provided that all of the drug leaving the site of administration appears in the plasma, the rate of disappearance from the site of administration, which is basically inaccessible, must equal the rate of appearance in plasma. Integration of Equation 4.17 produces:

$$A = \frac{A_0 k_a}{k_a - \lambda}\left[\exp(-\lambda t) - \exp(-k_a t)\right] \tag{4.18}$$

where A_0 is the dose, D. Dividing by V converts the amounts into plasma concentrations, and because only a fraction, F, of the dose may reach the systemic circulation, Equation 4.18 can be rewritten:

$$C = \frac{FD}{V}\frac{k_a}{k_a - \lambda}\left[\exp(-\lambda t) - \exp(-k_a t)\right] \tag{4.19}$$

Equation 4.19 can be written as:

$$C = C_0' \exp(-\lambda t) - C_0' \exp(-k_a t) \tag{4.20}$$

where

$$C_0' = \frac{FD}{V}\frac{k_a}{k_a - \lambda}.$$

It is therefore clear, via Equation 4.20, that Equation 4.19 represents a concentration–time curve that is the *difference* of two exponential terms that have a common intercept on the y-axis (Figure 4.7(a)). Both exponential terms asymptote to zero as t increases but the one with the largest rate constant decays the fastest, so that at later times Equation 4.20 approximates to a single exponential, as shown by the almost straight line in the terminal

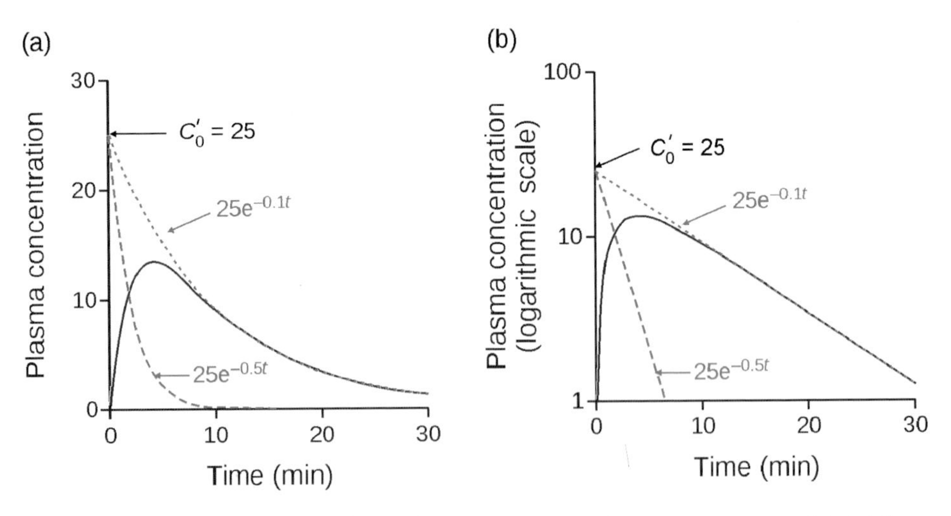

Figure 4.7 Simulated curves for first-order input into a single-compartment model: (a) linear scales and (b) semilogarithmic scales. $C_0' = 25$ arbitrary units, $k_a = 0.5\,min^{-1}$, $\lambda = 0.1\,min^{-1}$.

portion of Figure 4.7(b). Because Equation 4.20 is the difference between two exponentials it follows that the area under the curve, $AUC_{(0\text{-}\infty)}$, is the difference in the areas under the two exponentials.

The maximum concentration, C_{max}, occurs when the rates of input and output are equal, and $dC/dt=0$. The time of the peak, t_{max}, is given by differentiating Equation 4.19 with respect to t and rearranging:

$$t_{max} = \frac{1}{k_a - \lambda} \ln \frac{k_a}{\lambda} \tag{4.21}$$

Substitution into Equation 4.19 gives C_{max}, or from a simplified equation:

$$C_{max} = \frac{FD}{V} \exp\left(-\lambda t_{max}\right) \tag{4.22}$$

Note that, according to the above, t_{max} is independent of the dose, but C_{max} is directly proportional to the dose. This relationship does not hold for non-linear examples (Section 4.7). Also note that the peak concentration is *not* the time that absorption ceases and elimination starts. According to the model both start at $t=0$ and continue to $t=\infty$. Absorption continues to make a significant contribution to the concentration–time curve at times beyond t_{max}, as can be seen in Figure 4.7. The fraction of the dose absorbed, f_{A_a} at t_{max}, is a function of the relative sizes of the rate constants of absorption and elimination (Kaltenbach *et al.*, 1990):

$$f_{A_a} = 1 - \left(\frac{k_a}{\lambda}\right)^{-k_a/(k_a-\lambda)} \tag{4.23}$$

When $k_a=2\lambda$, 75% of the dose will be absorbed by t_{max}, whereas for 90% of the dose to be absorbed when $t=t_{max}$, then k_a must equal 8λ. This is to be expected because as $k_a \rightarrow \infty$, Equation 4.19 approximates to the intravenous case (Equation 4.8) when f_{A_a} must equal 100%. It is therefore important when performing pharmacokinetic experiments to ensure that data are collected for sufficient time for accurate assessment of the elimination rate constant, that is, when absorption is making a negligible contribution to the concentration–time data (Section 5.1.3).

4.4.1 Lag time

The rise in plasma drug concentrations does not, in fact, commence immediately after an oral dose. Inevitably there is a lag phase because, even after administration of an oral solution, it takes a finite time for the drug to reach its site of absorption. The lag time will be longer for tablets and capsules that have to disintegrate, and delayed gastric emptying may delay absorption even further. The lag time after consumption of enteric-coated tablets may be several hours. The lag time, t_{lag}, moves the plasma concentration–time curve to the right so that Equation 4.19 becomes:

$$C = \frac{FD}{V}\frac{k_a}{k_a - \lambda}\left\{\exp\left[-\lambda\left(t - t_{lag}\right)\right] - \exp\left[-k_a\left(t - t_{lag}\right)\right]\right\} \tag{4.24}$$

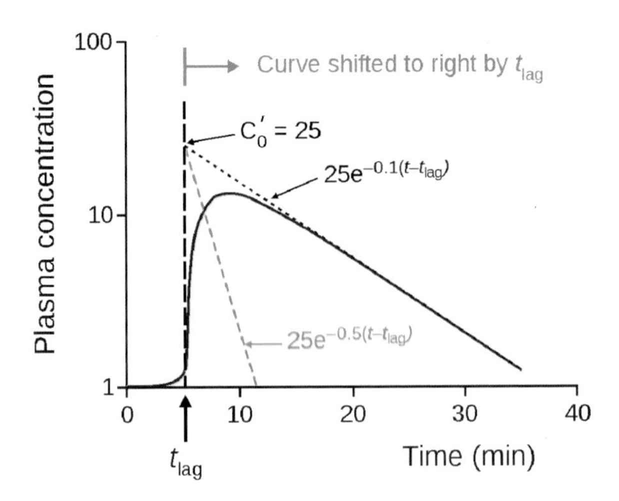

Figure 4.8 Effect of lag time. The curve is moved to the right but the other parameters remain unchanged so the curves intercept on a line representing t_{lag}.

An estimate of t_{lag} can be obtained by iterative curve-fitting of Equation 4.24 by adding t_{lag} as an extra variable or, if the method of residuals is used (Appendix 3), the construction lines will intersect at $t=t_{lag}$, rather than intersecting at the y-axis, where $t=t_0$ (Figure 4.8).

4.4.2 Area under the curve

The *AUC* is obtained in the same way as for the intravenous case, that is, integration of the plasma concentration–time curve (Equation 4.19):

$$AUC_{po} = \int_0^{\infty} \frac{FD}{V} \frac{k_a}{k_a - \lambda} \left[\exp(-\lambda t) - \exp(-k_a t)\right] dt \tag{4.25}$$

$$AUC_{po} = \frac{FD}{\lambda V} \tag{4.26}$$

and because the intercept following an i.v. injection, $C_0 = D/V$:

$$AUC_{po} = F\frac{C_0}{\lambda} \tag{4.27}$$

Comparison of *AUC* after i.v. (Equation 4.13) and oral doses (Equation 4.27) gives:

$$F = \frac{AUC_{po}}{AUC_{iv}} \tag{4.28}$$

Note that although the example considers an *oral* dose, the equation and those which follow apply equally to other sites of administration from which absorption is first order. Equation 4.28 is used to calculate the value of F, the respective areas usually being obtained by the trapezoidal method (Appendix 1). However, because the area under a single exponential curve is the intercept at $t=0$ divided by the rate constant (see Figure 4.7(a)), the AUC_{po} can be calculated from:

$$AUC_{po} = \frac{C'_0}{\lambda} - \frac{C'_0}{k_a} \tag{4.29}$$

Obviously this approach requires the data to be analysed to estimate the value of k_a. The equation for clearance (c.f. Equation 4.16) is:

$$CL = F\frac{D}{AUC} \tag{4.30}$$

This means that CL can *only be obtained after intravascular injection*, when F is known to be 1.

4.4.2.1 Apparent oral clearance

Sometimes Equation 4.30 is rearranged to:

$$\frac{CL}{F} = \frac{D}{AUC} = CL_{oral} \tag{4.31}$$

and the result referred to as 'oral clearance' or '*apparent* oral clearance' or sometimes even just 'clearance', a potential cause of confusion. Obviously, a value for CL cannot be derived without knowing the proportion of the dose which reaches the systemic circulation, F. Because an accurate value of F cannot be obtained without the use of i.v. doses, it would seem to be better to use the data from i.v. studies to obtain systemic clearance. To avoid ambiguity, any value obtained for D/AUC from extravascular doses should be referred to as CL/F or CL_{oral} to ensure that there is no confusion with systemic clearance, CL. In fact, all one is doing when comparing oral clearances is comparing the areas under the curves, and attempting to explain the results in terms of clearance simply leads a circular argument, as in Section 11.3.1.

4.4.3 Bioavailability and bioequivalence

The word 'bioavailability' may be used to describe the extent to which a drug is released from its pharmaceutical dosage form to be available to exert an effect. Regulatory authorities have defined bioavailability as:

> 'The rate and extent to which the therapeutic moiety is absorbed and becomes available to the site of drug action.'

However, what is normally studied is *systemic availability*. The area under the blood concentration curve–time curve from $t=0$ to $t=\infty$ is directly proportional to the amount of drug that enters the systemic circulation. However, plasma concentrations are usually measured on the assumption that there is also a direct relationship between the blood and plasma concentrations of the drug. When a drug is injected intravenously, the entire dose enters the circulation and so the AUC_{iv} can be used to estimate F using Equation 4.28, or if equal-sized doses cannot be given:

$$F = \frac{AUC_{po}}{AUC_{iv}} \times \frac{dose_{iv}}{dose_{po}} \tag{4.32}$$

In sequential experimental designs, the doses of drugs to be compared are given to the same subjects with a suitable time interval between the doses to ensure that all of the first dose has been removed before the second dose is given.

4.4.3.1 *Effect of systemic availability on plasma concentration–time curves*

Low systemic availability will reduce the plasma concentrations relative to those of an equal size intravenous dose of equal size. Figure 4.9(a) shows typical curves for an intravenous injection and an oral administration for which $F=1$. Note that the concentration following the i.v. injection declines very rapidly, as would be expected for a drug eliminated according to first-order kinetics. The concentrations after the oral dose increase from zero to the maximum concentration, t_{max}, the point at which the rate of absorption equals the rate of elimination, and then declines because from this point the *rate* of elimination is greater than the *rate* of absorption. Because $F=1$, $AUC_{(0-\infty)}$ is the same for each route of administration (although of course, infinite time cannot be shown on the figure). Part of the area under the curves, shaded green, is common to both routes of administration but the area shaded blue represents an area that is only under the i.v. route. Similarly, the yellow area is only under the oral curve. When $F=1$ the total areas under the curves are equal and so it follows that the blue area must equal the yellow area. Attempting to determine F by measuring the concentration at a single time point would be completely misleading, for example

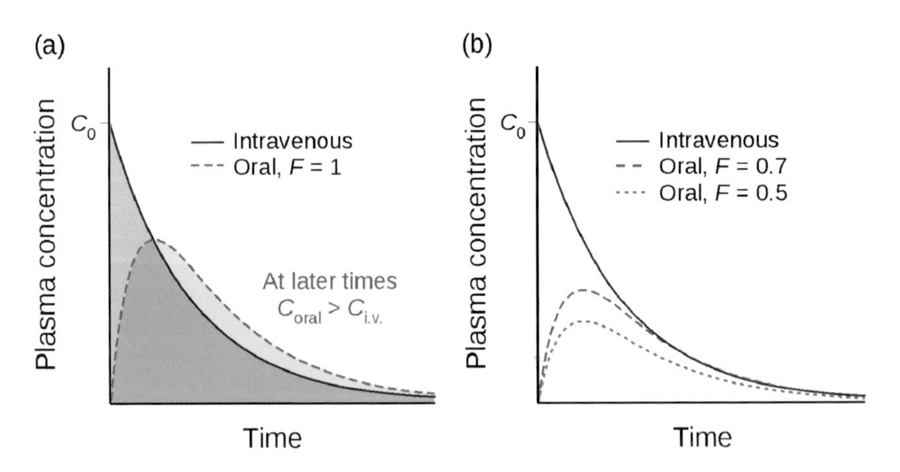

Figure 4.9 Comparison of plasma concentration–time curves for various values of F. (a) F=1. At later times the concentrations after the oral dose must be higher than after i.v. (b) For F=0.7 concentrations at later times are superimposed, whereas for F=0.5 all the oral concentrations are less than the i.v. ones. C_0 is the initial plasma concentration after the i.v. injection.

there can be a *time* when the oral concentrations are higher than those after the i.v. dose (Figure 4.9(a)). It is, at least in theory, possible for the concentrations to be superimposable at later times. This occurs when $F \sim 0.7$ (Figure 4.9(b)) whereas for lower values of F *all* the oral concentrations are less than those after the i.v. injection.

4.4.3.2 Factors affecting bioavailability

The many reasons why bioavailability may be limited include those discussed previously: poor absorption because of the inherent nature of the drug, pre-systemic metabolism, interaction with food and other drugs. A further factor is tablet formulation. Enteric-coated tablets are designed not to release their contents before reaching the intestine, which leads to an inevitable delay before absorption can occur. This is not an issue when repeated doses are given at regular intervals because the delay is only observed following the first dose. Examples of enteric-coated preparations include aspirin and ferrous sulfate, which may cause gastric irritation, and pancreatic enzymes that are coated to prevent digestion in gastric contents. If rapid absorption is required then soft gelatin capsules may be used or an oral solution may be given.

A particular issue is when the bioavailability of a particular drug is highly variable and this has led to therapeutic failures in the past. The variability may be between preparations from one manufacturer or may exist between different formulations from different manufacturers (Section 11.2).

4.4.3.3 Bioequivalence

Bioequivalence studies are initiated to investigate differences between products; usually so-called 'generics' (an unfortunate term) are compared with established preparations. For example, a generic diazepam might be compared with a proprietary brand such as Valium. The aim is not to show that the test compound is better than the established one but to show equivalence to it. Thus, if the innovator product (the proprietary brand) has low bioavailability then the new generic product must also have low bioavailability. In the US the Food and Drug Administration (FDA) has defined bioequivalence as 'the absence of a significant difference in the rate and extent to which the active ingredient or active moiety in pharmaceutical equivalents or pharmaceutical alternatives becomes available at the site of drug action when administered at the same molar dose under similar conditions in an appropriately designed study.'

Typically a bioequivalence study will involve *in vivo* testing of the generic drug against the standard drug in a cross-over design using 24–36 healthy, normal volunteers. Sometimes the study will call for the experiments to be conducted after meals but usually the subjects are fasted before they are given the drugs. Sufficient blood samples must be collected so that the C_{max}, t_{max} and AUC can be measured. Protocols for bioequivalence evaluation will be designed with strict statistical control so that they adequately test for, say, ±20% differences between pairs of products.

4.4.4 Sustained-release preparations

Drugs are usually formulated to be absorbed rapidly to give a rapid onset of effect. Consequently, one would expect k_a to be greater than λ. However, k_a can be less than λ and this forms the principle of sustained release. When the rate constant of release and, hence,

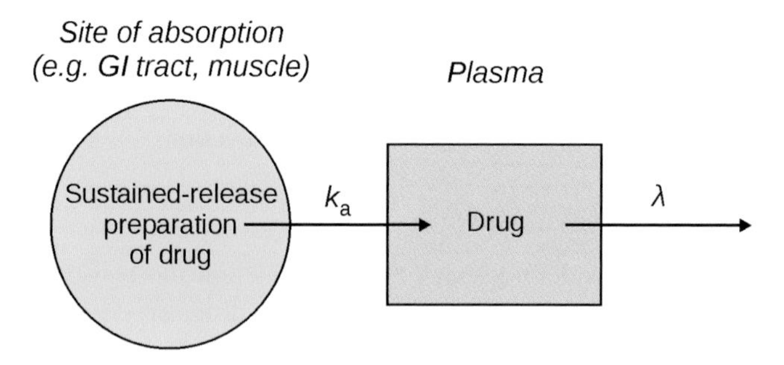

Figure 4.10 *Principle of sustained release:* $k_a << \lambda$ *so the overall rate is determined by* k_a.

absorption (k_a) is less than the elimination rate constant (λ) then, as with any sequential reaction, the rate constant of the slowest step is rate determining (Figure 4.10).

Sustained release results in a longer duration of action, which in turn means less frequent dosing, more convenient dosing and better patient compliance. Sustained-release preparations are available for most routes of administration, including sublingual, oral, subcutaneous, intramuscular, transdermal and rectal. Oral preparations may make use of wax matrices or tablets with different layers disintegrating at different rates, and capsules containing hundreds of pellets of different types (and often of different colours), each type disintegrating at a different rate. Transdermal preparations include glyceryl trinitrate, hyoscine and nicotine patches where the patch sometimes includes a rate-determining membrane to ensure a steady release of drug. Several preparations of insulin are available for subcutaneous injection, each giving different rates of release. Sustained-release preparations for intramuscular use may be referred to as depot injections. Long-acting i.m. preparations of penicillin G are microcrystalline salts; procaine penicillin G acts for approximately 3 days, whereas benzathine penicillin G acts for up to 7 days.

When the hepanoate ester of fluphenazine, fluphenazine enanthate, was injected as a depot injection, fluphenazine was slowly released, resulting in low but sustained plasma concentrations of fluphenazine with a half-life of ~3.5 days. However, when non-esterified fluphenazine was injected the plasma half-life of fluphenazine was ~12 h (Figure 4.11(a)). Thus fluphenazine appeared to have different elimination half-lives depending on which preparation was injected. Subjecting the enanthate data to pharmacokinetic analysis, that is, fitting the data to Equation 4.19, gave estimates of the rate constants for the rising and declining phases, from which the half-life of the steeper phase was found to be approximately 11 h, in good agreement with that obtained for the elimination half-life following injection of the rapidly absorbed preparation.

A growth and decay still occurred as would be expected from Equation 4.19. Initially there was a large amount of drug to be absorbed, and so the *rate* of absorption, $k_a A_a$, exceeded the *rate* of elimination, λA (which initially was zero because A was zero). In these circumstances, the half-life estimated from the decay phase is actually that for absorption and the rate constant of elimination is obtained from the steeper exponential curve. When the rate constants are substituted into Equation 4.19 the concentration curve is of the same form as that of Figure 4.7, but the rate constants *appear* to have been exchanged. The term 'flip-flop' has been coined to describe this situation. The difference in the exponential terms is negative, but so is the term representing the intercept at the y-axis (Equation 4.19). Similarly, t_{max} (Equation 4.21) is the same as in the more usual

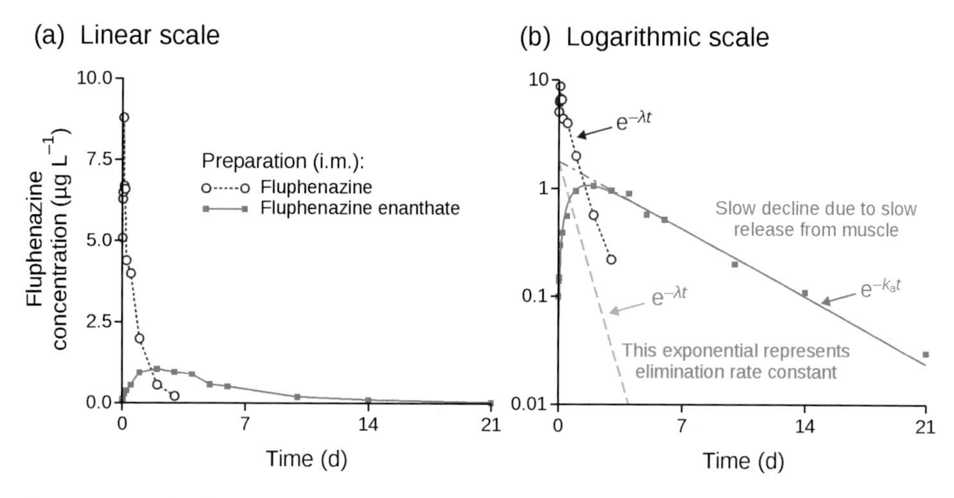

Figure 4.11 Fluphenazine enanthate was developed so that the rate of release from muscle is rate determining. (a) Linear scales: the areas under the curve are similar after the two preparations. (b) Logarithmic scales: resolution of the data into its two exponentials makes λ appear to be the rate constant of absorption, but the decline after injection of fluphenazine (red, open circles) confirms it is the elimination rate constant. Redrawn with permission from the data of Curry et al. *(1979).*

situation of $k_a > \lambda$. Thus, it is not possible from the concentration–time data to assign the rate constants to elimination or absorption. The only way to know how to ascribe the rate constants is from unambiguous data, such as that following an independent i.v. dose. In the example of Figure 4.11, the rate constant of elimination was assigned by giving an intramuscular injection of a preparation that was not sustained release because an i.v. preparation was not available.

4.5 Infusions

Equations exist for the oral single-dose situation in which absorption is at a constant rate, that is, with zero-order kinetics, and elimination occurring with first-order kinetics. This can be important with controlled-release oral dosage forms. However, the principles are easily demonstrated by considering constant rate intravenous infusions.

For drugs that are given intravenously, an infusion over minutes or hours, rather than a bolus dose, may be appropriate, especially for safety reasons. Also, orally administered drugs are often taken for extended periods of time, sometimes for the rest of a patient's life. Second and later doses of oral drugs given repeatedly are usually given before the previous dose has been completely eliminated, so that the drug accumulates in the body.

4.5.1 Zero-order input

Drugs may be infused using a motorized syringe pump and miniaturized versions are available for ambulatory patients. A constant rate infusion represents zero-order input, R_0. At the start of the infusion the plasma concentration will rise as the infusion proceeds, but

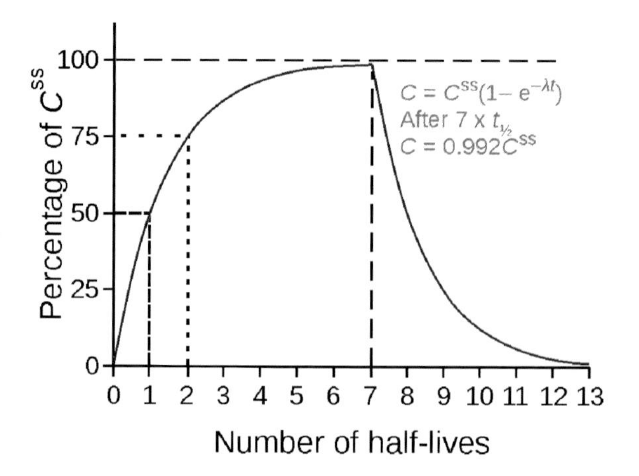

Figure 4.12 Attainment of steady-state concentrations is a function of the elimination half-life. Over 99% of steady state is achieved in a time equivalent to $7 \times t_{½}$.

for a drug that is eliminated according to first-order kinetics, it will not continue to rise indefinitely but 'level off' to what is referred to as the steady-state concentration, C^{ss}. This must be the case because as the plasma concentration increases the rate of elimination increases until the two rates become equal. The equation defining the plasma concentration during an infusion is:

$$C = C^{ss}\left[1 - \exp(-\lambda t)\right] \tag{4.33}$$

This is the equation of an exponential curve that asymptotes to C^{ss}. It is the same shape as the decay curve at the end of the infusion, which of course asymptotes to zero, but it is inverted. Because the curves are the same shape, it can be appreciated that during the infusion the concentration will be 50% of C^{ss} at $t = 1 \times t_{½}$, 75% of C^{ss} at $t = 2 \times t_{½}$ and so on. In fact, the values are the same as those for the percentage eliminated in Table 4.1, therefore after $7 \times t_{½}$ the concentration will be $>0.99C^{ss}$ even though, according to Equation 4.33, C^{ss} is only achieved at infinite time (Figure 4.12).

Post infusion the plasma concentration, C_{postinf}, decays according to an equation analogous to Equation 1.8, but rather than C_0 the highest concentration is the concentration at the time (T) that the infusion was stopped:

$$C_{\text{end of inf}} = C^{ss}\left[1 - \exp(-\lambda T)\right] \tag{4.34}$$

Post infusion the time is the time elapsed ($t - T$) so the declining concentrations are given by:

$$C_{\text{postinf}} = C^{ss}\left[1 - \exp(-\lambda T)\right]\exp\left[-\lambda(t - T)\right] \tag{4.35}$$

Figure 4.13 illustrates the principles of infusion using a simulation for the antibiotic amikacin. In this type of simulation we ask 'what if' questions, in a sense the opposite of data analysis. The amikacin simulation shows the concentrations in plasma to be expected

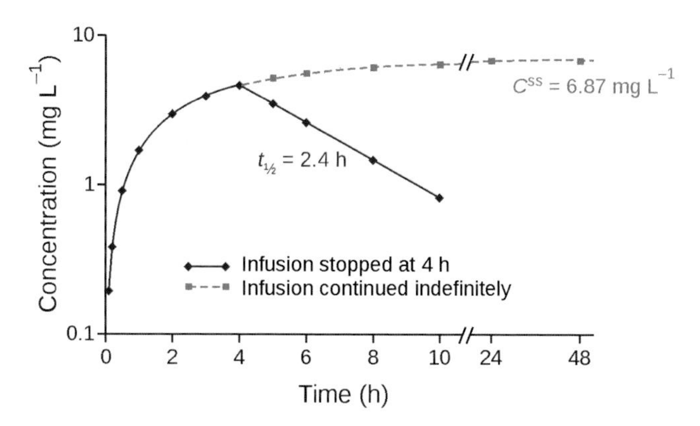

Figure 4.13 *Amikacin infusion: simulated curves derived using Equations 4.33, 4.34 and 4.35.*

if amikacin is infused at 0.625 mg min^{-1} (or 0.0089 mg min kg^{-1}) into a 70 kg subject in whom the clearance is 1.3 mL min^{-1} kg^{-1}. The apparent volume of distribution is 0.27 L kg^{-1}. The C^{ss} that results is approximately 7 mg L^{-1}. When the infusion is stopped, the drug concentration falls with a half-life of 2.4 h. This simulation shows two 'what if' questions being asked, in that the effect of infusion to steady state is compared with infusion for 4 h, after which the concentrations decline.

4.5.2 Loading dose

Because drugs with long elimination half-lives will approach steady state slowly this may result in an unacceptable delay before therapeutic concentrations are achieved. This delay can be overcome by injecting a large bolus dose, known as a 'loading dose', at the start of the infusion. The aim is to reach steady-state conditions immediately, and so the loading dose must be such that the injection gives a plasma concentration of C^{ss}. This amount is, as always, the concentration multiplied by the apparent volume of distribution, so:

$$\text{loading dose} = VC^{ss} \tag{4.36}$$

The rate of elimination is the *amount* of drug multiplied by the elimination rate constant, and at steady state this equals the rate of infusion, and combining with Equation 4.5 gives:

$$R_0 = \lambda VC^{ss} = C^{ss}CL \tag{4.37}$$

and so doubling the rate of infusion would double C^{ss}. Equation 4.37 provides a convenient way of calculating the infusion rate required to achieve a particular steady-state concentration, provided that the systemic clearance is known, $C^{ss} = R_0/CL$.

Combining Equations 4.36 and 4.37 gives:

$$\text{loading dose} = R_0/\lambda \tag{4.38}$$

This equation was used to calculate a loading dose for the data of Figure 4.14.

Loading doses, both by injection and orally, are widely used with antibacterial drugs, such as sulfonamides, and in cancer therapy, such as with trastuzumab. They can also be

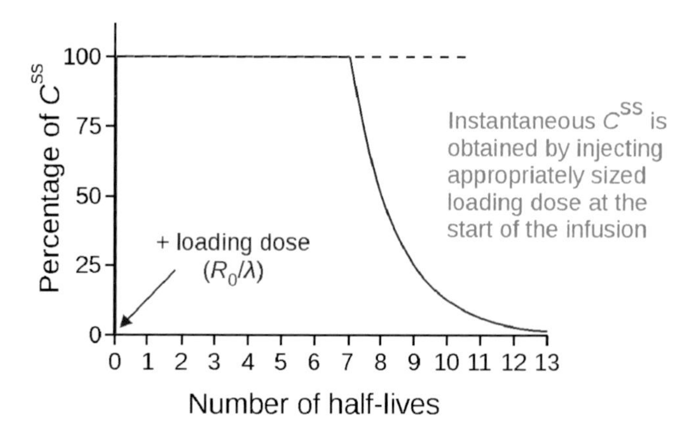

Figure 4.14 Simulation of an intravenous infusion when an appropriately sized loading dose is injected simultaneously at the start of the infusion.

used with non-steroidal anti-inflammatory drugs such as naproxen. However, there are situations in which pharmacokinetic considerations imply that a loading dose might make sense, but pharmacological considerations dictate otherwise. This occurs with the tricyclic antidepressants, in that with these drugs the patient must initially develop a tolerance to the acute cardiovascular effects, occurring at relatively low concentrations in plasma, so that the antidepressant effect, which occurs at relatively high concentrations, can be utilized. Thus the desired effect cannot be obtained after the first dose, in part for pharmacokinetic reasons.

4.6 Multiple doses

When a sustained effect is required from oral dosing, drugs are usually prescribed as divided doses, taken over a period of time. Steady-state conditions will apply for drugs that are eliminated according to first-order kinetics, as they do for infusions, but now the plasma concentrations will fluctuate between doses. Because of the fluctuation this situation is sometimes referred to as 'pseudo-steady-state'. The rate at which steady state is achieved is the same as for the infusion, >99% after $7 \times t_{1/2}$. This is demonstrated schematically for repeated i.v. doses in Figure 4.15. An initial injection gives a concentration of 100 units, which falls to 50 units in $1 \times t_{1/2}$, when the dose is repeated. This gives a new maximum of 150 units, which falls to 75 in the next half-life, a further injection takes the concentration to 175 units, and so on, until at steady-state the peak and trough concentrations are 200 and 100 units, respectively.

At steady state, the maximum (peak) and the minimum (trough) concentrations after i.v. bolus doses can be calculated:

$$C_{max}^{ss} = \frac{D}{V}\left(\frac{1}{1-\exp(-\lambda\tau)}\right) \tag{4.39}$$

$$C_{min}^{ss} = \frac{D}{V}\left(\frac{1}{1-\exp(-\lambda\tau)}\right)\exp(-\lambda\tau) \tag{4.40}$$

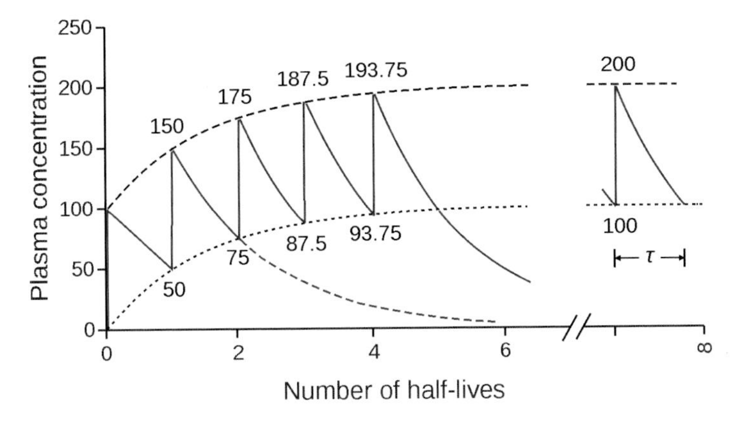

Figure 4.15 Modelled curve showing repeated i.v. injections at intervals equivalent to the half-life. Note the steeper decline in the curves as the concentration increases.

where τ is the dosing interval. A loading dose can be calculated:

$$\text{loading dose} = D\left(\frac{1}{1-\exp(-\lambda\tau)}\right) \tag{4.41}$$

The 'average' steady-state concentration is not the mean of the peak and trough concentrations but is based on the area under the curve as this is considered to give an estimate of the patient's exposure to the drug. It is the area under the curve divided by the dosage interval, τ, so from Equations 4.26 and 4.27:

$$C_{av}^{ss} = \frac{AUC}{\tau} = F\frac{C_0}{\lambda\tau} = \frac{FD}{V\lambda\tau} = \frac{FD}{CL\tau} \tag{4.42}$$

The equation actually applies to both intravenous doses (when $F=1$) and oral doses. Note how clearance can be used to calculate the maintenance dose to produce a required average steady-state concentration. With oral doses, because the average concentration is proportional to the dose and inversely proportional to τ, giving half the dose twice as often will give the same average concentration, but the peak to trough fluctuations will be less (see Figure 4.16).

Systemic clearance can be derived from a constant rate infusion to steady state to determine C^{ss} (Equation 4.37), but the same is not true of Equation 4.42 because C_{av}^{ss} has to be calculated from the *AUC* after a single i.v. dose.

Equations for multiple-dosing with first-order input (oral, i.m.) have been derived. The situation is complex if the doses are given during the absorptive phase, but the equations that assume doses are given post absorption are simpler and relative easy to use. For example, the concentration at any time t in the nth dose is given by:

$$C_n = \frac{FDk_a}{V(k_a-\lambda)}\left[\left(\frac{1-\exp(-n\lambda\tau}{1-\exp(-\lambda\tau)}\right)\exp(-\lambda t)-\left(\frac{1-\exp(-nk_a\tau}{1-\exp(-k_a\tau)}\right)\exp(-k_a t)\right] \tag{4.43}$$

where τ is the dosing interval. Equation 4.43 may look formidable, but with the aid of a spreadsheet and breaking it into to its component parts, it is relatively easy to simulate the concentration–time curves, as was done for multiple dosing in Figure 4.2. A loading dose can be calculated:

$$\text{loading dose} = D\left(\frac{1}{\left[1-\exp\left(-\lambda\tau\right)\right]\left[1-\exp\left(-k_{a}\tau\right)\right]}\right) \tag{4.44}$$

In this situation, D, the amount of drug injected at each second and later dose interval, is often referred to as the 'maintenance' dose to distinguish it from the loading dose. The concentration–time curves of Figures 4.16 and 4.17 were simulated using Equations 4.42 and 4.44.

4.6.1 Effect of half-life and dosage interval

Drugs with a short elimination half-life may prove difficult to use clinically, particularly when the therapeutic window is narrow. These drugs approach steady state quickly, as would be expected, but if the half-life is much less than the dosage interval (τ), the peak–trough fluctuations will be greater than those for a drug with half-life greater than τ. This is because a short half-life results in a greater amount of drug being eliminated between the doses than would be the case for a drug with a long half-life, in the same period. As a consequence it may prove difficult to maintain the concentrations within a narrow therapeutic window (Figure 4.16).

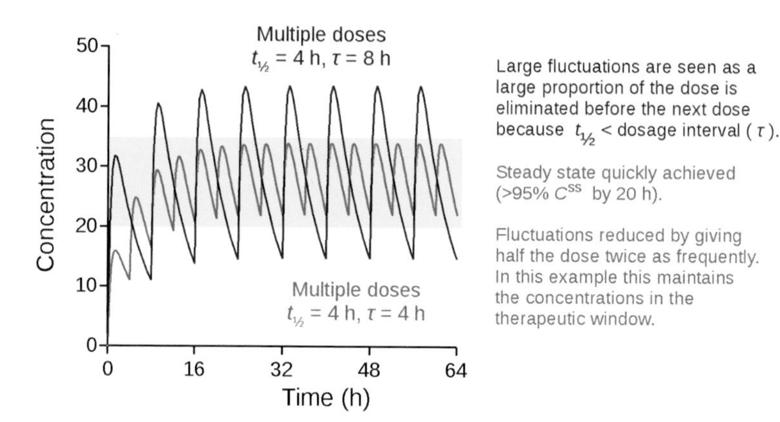

Figure 4.16 Multiple dosing at regular intervals with a drug with a short elimination half-life.

This situation can arise with morphine ($t_{½}$ ~2 h), when the patient may suffer pain at pre-dose (trough) concentrations and show signs of respiratory depression at peak concentrations. One way of dealing with such a situation is to give smaller doses more frequently as this reduces the difference between peak and trough levels whilst maintaining the same average concentration (Equation 4.42, Figure 4.16). However, there is a practical limit to how frequently a drug can be administered and other approaches may be required, including the use of portable infusion pumps. An alternative approach is to use sustained-release preparations (Section 4.4.4) so that the 'effective' half-life is longer and the time course

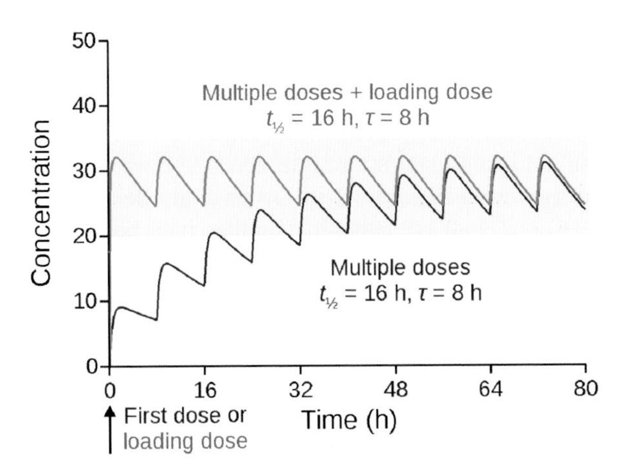

Figure 4.17 *Multiple dosing at regular intervals with a drug with a long elimination half-life (red), also showing the effect of a loading dose (blue). The shaded area represents the therapeutic window.*

resembles that of Figure 4.17, when it takes longer to reach steady state but the fluctuations are markedly reduced. A loading dose may be used to overcome any problems due to the delay in onset of effects.

Of course, there is not such a problem when a drug has a large therapeutic window. However, this is not the case for lithium for which serum concentrations should be maintained within the range 0.5–1.25 mmol L^{-1}. Figure 4.18 illustrates how pharmacokinetic parameters derived from the serum concentrations after the first dose can be used to predict the steady-state values for a patient to be given 900 mg daily over a prolonged period of time.

The half-life of lithium in this patient was 18 h, the apparent volume of distribution 0.66 L kg^{-1} and the clearance 0.35 mL min^{-1} kg^{-1}. The bioavailability (F) was unity. Figure 4.18 shows the concentrations after the first dose (red), then the predicted rise

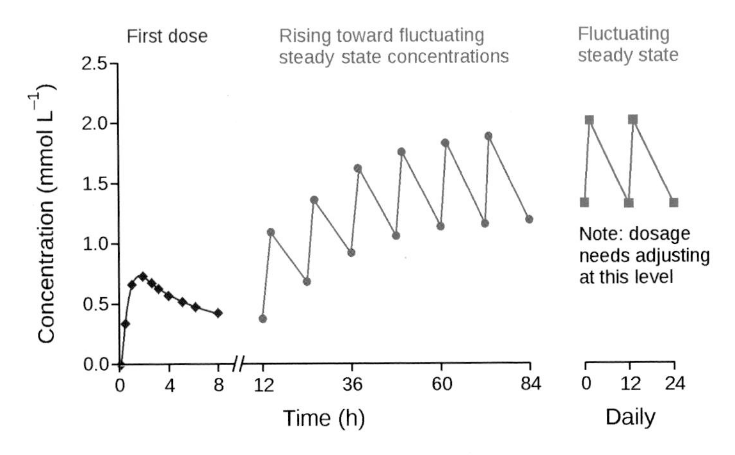

Figure 4.18 *Calculation of lithium steady-state serum concentrations from the data for the first oral dose.*

towards the pharmacokinetic steady state (blue), which shows that at this dosage in this patient the concentrations (green) would overshoot the therapeutic range and so a lower dose should be given to avoid the risk of toxicity.

Whether or not fluctuation is desired often depends on the clinical objective. For example, a sleeping aid taken night after night needs to have the maximum fluctuation achievable in order to prevent the occurrence of residual effects during the day. Hence, temazepam ($t_{½}$ ~12 h) and with no active metabolites may be considered to be a more suitable hypnotic than diazepam ($t_{½}$ ~48 h), which also has an active metabolite that accumulates in plasma (Section 6.2.6). On the other hand, for an anticonvulsant it would be better if there were little or no fluctuation.

4.7 Non-linear kinetics

A feature of first-order kinetics is that C^{ss} is proportional to the dose administered, but there are situations when this is not the case and the kinetics are referred to as non-linear. The plasma concentrations may increase disproportionately or may not increase as much as would be predicted from linear models when the dose is increased (Figure 4.19). Antiepileptic drugs are examples of drugs exhibiting non-linear kinetics. Phenytoin metabolism tends to saturation at doses in the therapeutic range, whist gabapentin shows saturable absorption. Valproic acid concentrations may be less than predicted because of dose-dependent binding to plasma proteins, although the clinical effects may be greater (Section 2.4.3). Carbamazepine concentrations decline on repeated dosing because of auto-induction of enzymes. Phenobarbital shows linear kinetics at therapeutic doses but non-linear kinetics in overdose.

Usually it is possible to apply the Michaelis–Menten equation to non-linear elimination:

$$v = \frac{V_{max} C}{Km + C} \tag{4.45}$$

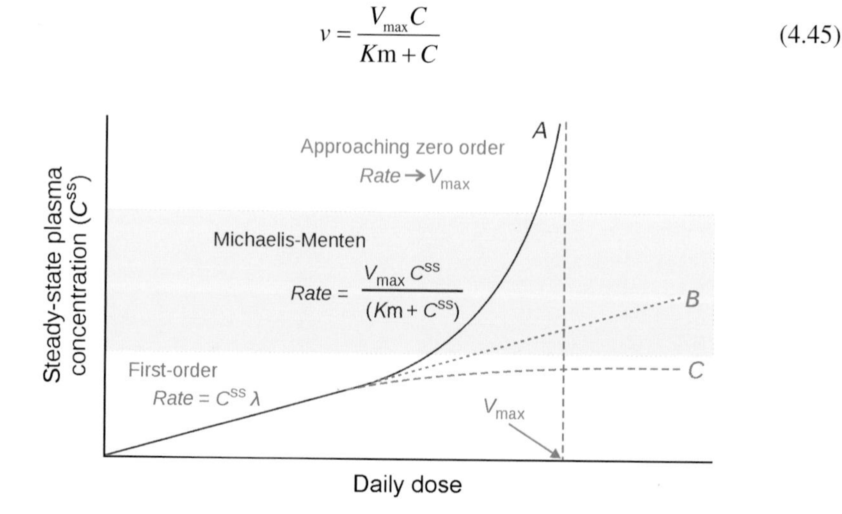

Figure 4.19 Comparison of non-linear and linear kinetics. A: saturation of elimination; B: linear kinetics; C: saturation of absorption when it may be necessary to increase the dose to reach therapeutic concentrations. With increasing doses curve A asymptotes to V_{max}. The shaded area represents the therapeutic window.

As the substrate concentration is increased, the enzyme becomes saturated with substrate so that the velocity, v, of the reaction cannot increase anymore. Thus, for a given concentration of enzyme, there is a maximum velocity, or rate, V_{max}. The Michaelis constant, Km, is numerically equal to the substrate concentrations at one half V_{max}. This is easily demonstrated by substituting $V_{max}/2$ for v in Equation 4.45 and rearranging.

The majority of drugs exhibit first-order elimination kinetics, even those that are extensively metabolized. This is because the effective drug concentration is very low compared to that of the enzyme, therefore at any time little enzyme is bound by drug and the concentration of non-bound enzyme is more or less constant. At low substrate concentrations $Km >> C$, and so C makes a negligible contribution to the denominator of the Michaelis–Menten equation, that is $Km + C \rightarrow Km$ and Equation 4.45 becomes:

$$v \approx \frac{V_{max}}{Km} C \tag{4.46}$$

This is the equation of a first-order reaction because the rate is directly proportional to the substrate concentration. Applying a similar argument to the situation when $C >> Km$, the denominator approximates to C because it is now Km that makes a negligible contribution to $Km + C$ and:

$$v = \frac{V_{max} C}{C} = V_{max} \tag{4.47}$$

which is the situation described earlier when the enzyme is saturated. The reaction rate, V_{max}, is constant and so the kinetics are zero order. Equations 4.46 and 4.47 represent the extreme values of rate in the Michaelis–Menten equation, which is said to 'limit' to these values at very low and very high substrate concentrations. It follows then that the order of a Michaelis–Menten reaction must be somewhere between 0 and 1, and increases towards 1 (first order) as the substrate concentration declines.

Because half-life is dependent on apparent volume of distribution and clearance, non-linearity must be a result of non-linearity in one or both of these variables. The rate of elimination is $C \times CL$ (Equation 4.3), but for a drug eliminated according to Michaelis–Menten kinetics the rate is also given by Equation 4.45, so putting the equations equal to each other and dividing each side by C gives:

$$CL = \frac{V_{max}}{Km + C} \tag{4.48}$$

As concentration increases to the point where $C >> Km$ (zero-order case) then:

$$CL = \frac{V_{max}}{C} \tag{4.49}$$

Thus, for non-linear kinetics, clearance will decrease with increasing drug concentrations. In the case of valproic acid, concentration-dependent changes in protein binding result in concentration-dependant changes in the apparent volume of distribution and hence non-linear kinetics.

4.7.1 Phenytoin

Non-linear kinetics, such as those exhibited by phenytoin, can be modelled using the Michaelis–Menten equation. At low doses the kinetics approximate to first order but the kinetics at higher doses can only be adequately described by Equation 4.45. As the dose is increased further the kinetics approach zero order. This can result in C^{ss} increasing very rapidly for small increases in dose, making it difficult to maintain the serum concentrations in the therapeutic window. Individual variations in apparent Km and V_{max} complicate things further (Figure 4.20(a)). Note the large individual variations in V_{max}. If it is necessary to individualize the dose a rearrangement of Equation 4.45 can be used. At steady state, the daily dose, R, can be used as an estimate of the rate of elimination (mg d^{-1}) so:

$$R = V_{max} - \frac{R}{C^{ss}} Km \tag{4.50}$$

Because there are two unknowns, C^{ss} values must be obtained for at least two dosage rates, when Equation 4.50 can be solved for V_{max} and Km. Alternatively, a plot of R/C^{ss} versus R is a straight line of slope –Km and y-intercept V_{max}. The values can then be used to calculate a daily dose that will give a concentration in the therapeutic range for that particular patient.

Non-linear kinetics are apparent from curvature of the ln C verses t plot (Figure 4.20(b)). At high concentrations, the rate of elimination may be approaching V_{max} but because the *proportionate* change is less than that at lower concentrations, the slope of the semilogarthmic plot is less. As the concentration falls, the rate is less but the proportionate change tends to become constant as the kinetics approximate to first order, when the slope of the lines will be the same, irrespective of the initial concentration (see also Figure 4.21).

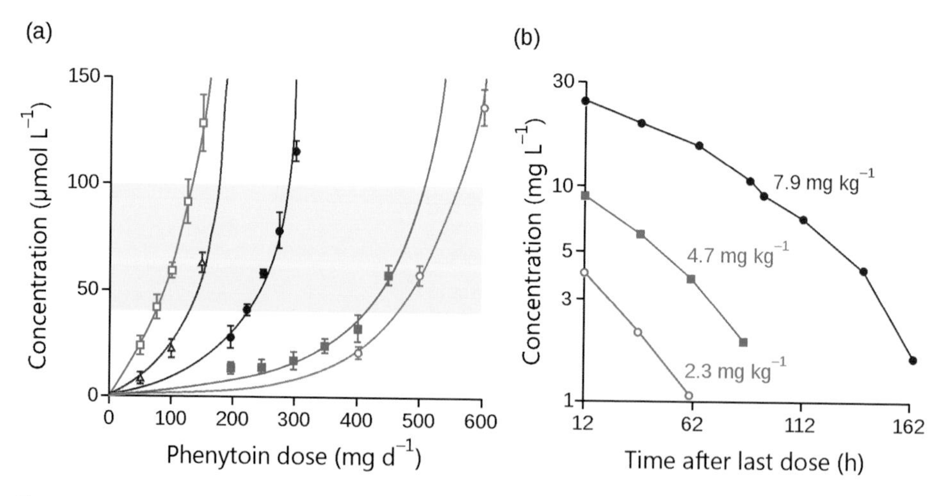

Figure 4.20 (a) Steady-state serum concentrations of phenytoin in five patients. The shaded area is the therapeutic window. Redrawn with permission from Richens & Dunlop (1975). (b) Phenytoin plasma concentrations in a single subject given three dosage regimens, each lasting 3 days (Gerber & Wagner, 1972).

4.7.2 Ethanol

Ethanol is metabolized primarily by liver ADH, although other routes of elimination exist, including metabolism by other enzymes and excretion of unchanged drug in urine and expired air. Because the *K*m of human ADH is very low, and the 'dose' of ethanol, by pharmacological standards, is high, the enzyme becomes saturated even with moderate drinking. In the 1930s Erik Widmark used a single-compartment model and zero-order kinetics to model blood ethanol concentrations. However, with advances in analytical methods it became clear that at low concentrations the kinetics were more akin to first-order decay. This led to the suggestion that at concentrations above $0.2\,g\,L^{-1}$ the kinetics are zero order whilst below this concentration the kinetics are first order. Inspection of Equation 4.45 clearly shows that there cannot be an abrupt switch from first-order to zero-order kinetics and this can be demonstrated by substituting values into the equation (Table 4.2). If first-order kinetics applied, then the values calculated below $0.2\,g\,L^{-1}$ would be directly proportional to the concentration. Clearly they are not. Similarly, if the kinetics were zero order above $0.2\,g\,L^{-1}$ then the rate would be constant, in this example V_{max}, which was $8\,g\,h^{-1}$. Again this is not the case, although the rate does asymptote to V_{max} at very high concentrations. It is generally agreed that a subject would have imbibed their last drink if they ever reached the point of metabolizing ethanol at V_{max}.

Table 4.2 Calculated rates of ethanol metabolism as a function of concentration

C ($g\,L^{-1}$)	0.01	0.02	0.05	0.1	0.2	0.04	0.8	1.6	4.0	8.0
Rate ($g\,h^{-1}$)	0.73	1.33	2.67	4.0	5.33	6.4	7.11	7.54	7.8	7.9

*K*m = $0.1\,gL^{-1}$, $V_{max} = 8\,g\,h^{-1}$

In a theoretical decay curve for ethanol (Figure 4.21(a)), the initial part of the curve is almost linear, reflecting approximately zero-order elimination, but the line becomes progressively more curved. Plotting the data with a logarithmic *y*-axis (Figure 4.21(b))

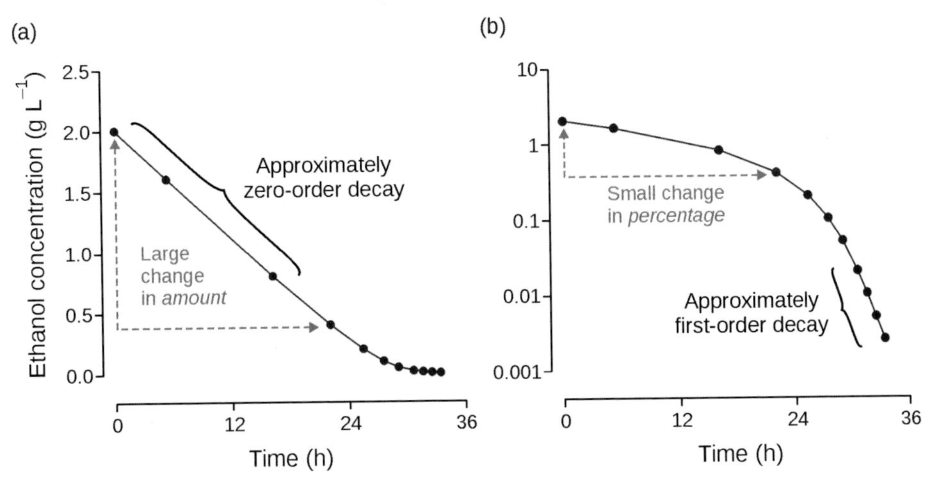

Figure 4.21 Simulated decay curve for ethanol ($C_0 = 2\,g\,L^{-1}$, $V_{max} = 8\,g\,h^{-1}$, $Km = 0.1\,g\,L^{-1}$). (a) Concentration plotted on a linear y-axis. (b) Concentration plotted on a logarithmic y-axis.

gives the typical shape expected for Michaelis–Menten elimination kinetics. At high concentrations the elimination rate approaches V_{max}, but the *proportionate* change is small, hence the shallow slope of the log-transformed curve. At later times (lower plasma concentrations) the rate of elimination is much lower, but the proportionate change is greater and will become constant when the elimination is first order.

4.8 Relationship between dose, and onset and duration of effect

For a drug exhibiting single-compartment kinetics there are some simple theoretical relationships between the dose given and the duration and intensity of effect. Assuming that after an intravenous dose the effect ceases at a minimum effective concentration, C_{eff}, then the duration of effect increases by $1 \times t_{½}$ for every doubling of the dose (Figure 4.22(a)). If the *y*-axis were dose and a horizontal line was drawn representing the duration of action, then the duration of action of the lowest dose would be less than that at higher doses, which would double with each doubling of the dose.

Similarly, if at some concentration or time after the injection the effect wears off and the dose is repeated then the time for the second dose to wear off will be longer because there is now a greater amount of drug in the body. However, if subsequent doses are given the time to reach the concentration at which the effect ceases is now constant between doses (Figure 4.22(b)). If a horizontal line were drawn on Figure 4.22(a) then the duration of action of the lowest dose would be less than that at higher doses. In the discussion above it has been assumed that the onset of effect was more or less instantaneous.

The above only applies to a bolus intravenous injection and assuming a single-compartment model applies. Even for a simple single-compartment model the duration of action will

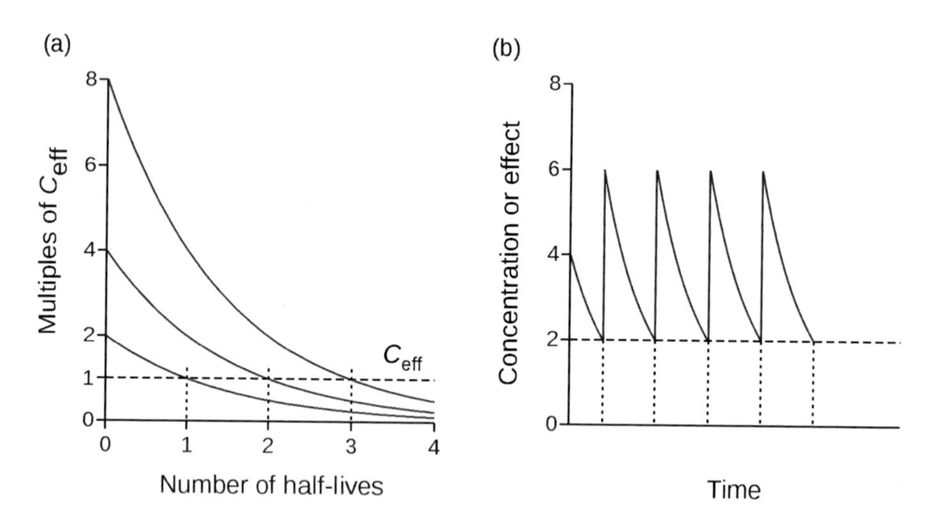

Figure 4.22 (a) Doubling the dose increases the time to reach the minimum effective concentration, C_{eff} by one elimination half-life. (b) Multiple doses given when the plasma concentration or effect has fallen to some arbitrary level. The duration of effect after the first dose is less than that after the second and subsequent doses. After the second dose the maximum concentration/effect is the same.

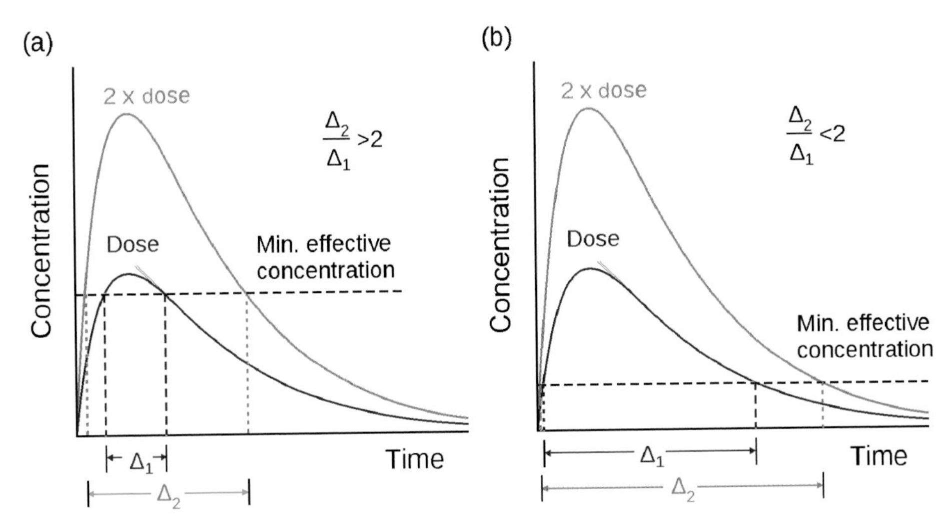

Figure 4.23 *The proportionate change in the duration of effect (Δ) on doubling the dose depends on where the minimum effective concentration line is initially. (a) When the line is close to the maximum concentration, doubling the dose more than doubles the duration. (b) When the line is much lower, doubling the dose may have little effect on either the onset or duration of action.*

not necessary double for a doubling of an oral dose. The change in duration is dependent on where the minimum effective concentration is relative to $C_{max.}$ (Figure 4.23). Thus, when the simplest models for oral doses apply, an increase in dose causes a proportional increase in C_{max}, no difference in t_{max} and a variable increase in duration of effect (Figure 4.23). However, if the effect were proportional to the *AUC*, as is often the case for antimicrobial drugs, a doubling of dose would lead to a doubling of effect.

4.9 Limitations of single-compartment models

The astute reader will have noticed that most of the figures in this chapter use modelled data. This reflects the difficulty in finding convincing biological examples of data which fit a single-compartment model, particularly after an i.v. bolus injection. In Figure 4.6, which has been presented as representing decay from a single compartment, the first sample was not collected until 1 h after the injection. Thus, an earlier decay phase may have been present but missed. This is supported by the fact that the apparent volume of distribution, calculated using Equation 2.4, is ~50 L. This is larger than the volume of plasma or extracellular fluid, with which drug may be expected to equilibrate rapidly. In fact, it approximates to total body water, including water in tissues, into which a drug will penetrate relatively slowly, so in all probability analysis of earlier samples would have revealed another exponential decay representing distribution of the drug into tissues. Even ethanol concentrations, which are frequently treated as if they conform to single-compartment kinetics, show multi-exponential decay after i.v. infusion. Because it is now recognized that the timing of samples may influence conclusions about the most appropriate compartmental model, more recent studies are usually designed with sampling at sufficiently early collection times to define any distributional phase(s).

There are examples of what appear to be single-compartment models when some drugs are given orally or by i.m. injection. This is because the absorption phase masks more complex kinetics, as with fluphenazine enanthate described above (Figure 4.11(a)). In such cases there is no alternative but to analyse the data as if it were from a single compartment, even if other routes of administration (e.g. i.v.) show otherwise. These situations may be referred to as apparent or pseudo single-compartment cases. Consequently, it is not unusual to find examples in the literature, for individual drugs, in which biexponential decay equations (two-compartment model, Chapter 5) are used for i.v. doses and single-compartment equations, as described in this chapter, are used for oral doses.

Summary

The fundamental concepts of pharmacokinetics have been described, including half-life, clearance, zero-order and first-order input into a single-compartment model, multiple dosing and attainment of steady state. However, because single-compartment models are of limited use in describing the plasma concentration–time relationships for the majority of drugs more complex models are required. These are the subject of the next chapter.

4.10 Further reading

Curry SH, Whelpton R. *Disposition and Pharmacokinetics: from Principles to Applications.* Chichester: Wiley-Blackwell, 2011. See Chapter 8 for further details of bioequivalence and the effects of tablet formulation.

Nelson E. Kinetics of drug absorption, distribution, metabolism, and excretion. *J Pharm Sci* 1961; 50: 181–92.

4.11 References

Curry SH, Whelpton R, de Schepper PJ, Vranckx S, Schiff AA. Kinetics of fluphenazine after fluphenazine dihydrochloride, enanthate and decanoate administration to man. *Br J Clin Pharmacol* 1979; 7: 325–31.

Gambertoglio JG, Amend WJ, Jr., Benet LZ. Pharmacokinetics and bioavailability of prednisone and prednisolone in healthy volunteers and patients: a review. *J Pharmacokinet Biopharm* 1980; 8: 1–52.

Gerber N, Wagner JG. Explanation of dose-dependent decline of diphenylhydantoin plasma levels by fitting to the integrated form of the Michaelis–Menten equation. *Res Commun Chem Pathol Pharmacol* 1972; 3: 455–66.

Kaltenbach ML, Curry SH, Derendorf H. Extent of drug absorption at the time of peak plasma concentration in an open one-compartment body model with first-order absorption. *J Pharm Sci* 1990; 79: 462.

McQueen EG, Wardell WM. Drug displacement from protein binding: isolation of a redistributional drug interaction in vivo. *Br J Pharmacol* 1971; 43: 312–24.

Richens A, Dunlop A. Serum-phenytoin levels in management of epilepsy. *Lancet* 1975; 2: 247–8.

5

Multiple-compartment and Non-compartment Pharmacokinetic Models

Learning objectives

By the end of this chapter the reader should be able to:

- explain the bi-phasic decay in concentrations seen with drugs that impart the characteristics of a two-compartment model
- describe the relationship between the drug concentrations in central and peripheral compartments
- discuss which tissues might comprise the central compartment and which the peripheral compartment(s)
- discuss the role of redistribution as a determinant of the duration of drug action
- calculate CL, V_{area} and terminal rate constant using a non-compartmental approach
- explain how the most appropriate model should be chosen.

5.1 Multiple-compartment models

5.1.1 Introduction

In the previous chapter several aspects of pharmacokinetics were discussed using the simplest of compartmental models. However, it is often the case that the plasma concentration–time data cannot be explained adequately using a single-compartment model. Indeed, it is

Introduction to Drug Disposition and Pharmacokinetics, First Edition. Stephen H. Curry and Robin Whelpton.

Companion website: www.wiley.com/go/curryandwhelpton/IDDP

difficult to find convincing examples of drugs which conform to single-compartment pharmacokinetics after intravenous bolus doses. Generally, more complex, multiple-compartment models are required, or it may not be possible to define a compartment model at all and other approaches have to be adopted.

5.1.2 Intravenous injections

If, when the concentration data are plotted on logarithmic scales, the plot is curvilinear and may be described as the sum of two or more exponential decays:

$$C = C_1 \exp(-\lambda_1 t) + C_2 \exp(-\lambda_2 t) + C_3 \exp(-\lambda_3 t) + \ldots \tag{5.1}$$

then a multiple-compartment model can be used This may be described as the drug conferring on the body the characteristics of a two- (or multiple) compartment model. This is exemplified by Figure 5.1, which shows data for methylene blue fitted to a three-compartment model. Methylene blue is used in the treatment of both congenital and drug-induced methaemoglobinaemia, which may be life threatening.

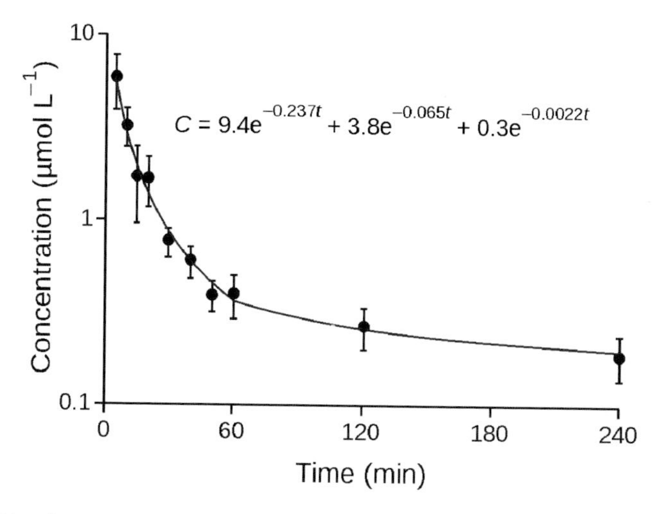

Figure 5.1 Blood concentrations of methylene blue (mean ± SEM, n = 7) after 100 mg intravenously. Data fitted using sum of three exponential terms (solid line). Reproduced with permission from Peter et al. *(2000).*

Experimental data are rarely good enough to fit more than three exponential terms and two-compartment models are more commonly encountered. Figure 5.2 shows data simulated using a dye model (Appendix 2) to represent drug elimination from a two-compartment model.

For a two-compartment model the concentration, C, is given by:

$$C = C_1 \exp(-\lambda_1 t) + C_2 \exp(-\lambda_2 t) \tag{5.2}$$

Sometimes (particularly in American literature) Equation 5.2 is written:

$$C = A \exp(-\alpha t) + B \exp(-\beta t) \tag{5.3}$$

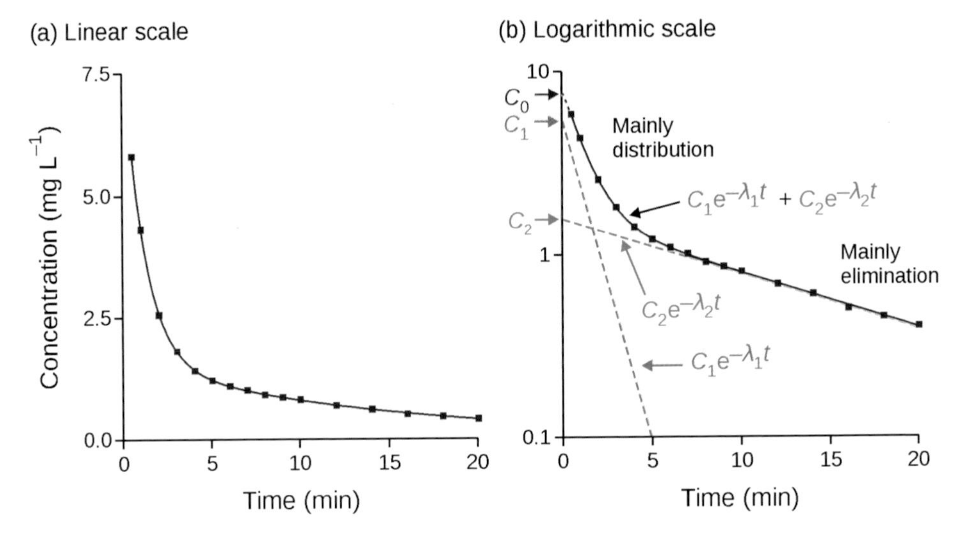

Figure 5.2 Example of data that fits a two-compartment model using (a) linear coordinates and (b) a semi-logarithmic plot.

Note that λ_1 is greater than λ_2 ($\alpha > \beta$) because λ_1 (or α) is the rate constant of the steeper part of the decay curve. Indeed this initial steep part may be referred to as the 'α phase'. The last exponential defines the *terminal* phase, and by convention the rate constant of this phase is λ_z. Thus, for a two-compartment model, $\lambda_2 = \beta = \lambda_z$. A representation of a two-compartment model is depicted in Figure 5.3. Rather than the one homogenous solution of drug of the single-compartment model, this model has two solutions, with reversible transfer of drug between them. By definition, plasma is always a component of the central compartment, which is sometimes referred to as the plasma compartment. The other compartment is the peripheral or tissue compartment. The volumes of these compartments are V_1 and V_2, respectively. The use of other terms, for example V_p, can lead to untold confusion – is that p for plasma or p for peripheral? The rate constants are labelled according to the direction of movement to which they refer, so the constant relating to transfer from the

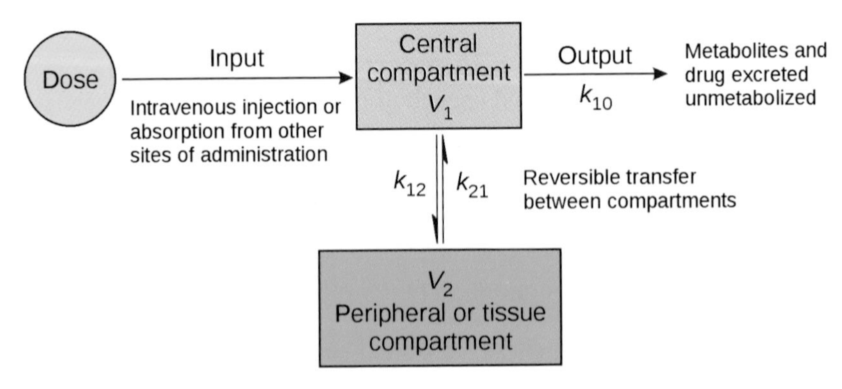

Figure 5.3 Diagrammatic representation of a two-compartment model.

central compartment (1) to the peripheral compartment (2) is k_{12}. The rate constant for loss from the central compartment by metabolism or excretion of unchanged drug is k_{10} (Figure 5.3), where 0 depicts compartment zero, that is, outside the body. Sometimes it may be written k_{10}, that is, 'O' for outside.

Other two-compartment models are possible where drug is metabolized or excreted from the peripheral compartment, or from both central and peripheral compartments. However, the liver and kidneys are often part of the central compartment so the model of Figure 5.3 is usually appropriate.

The shape of the curves of Figure 5.2 can be explained as follows:

1. Shortly after injection there is rapid transfer of drug from plasma, in the central compartment, to the tissues of the peripheral compartment, with only a small proportion of the dose being eliminated during this time.
2. At later times the concentrations of drug in the compartments equilibrate and the decay from plasma is chiefly a result of elimination of drug from the body, by metabolism and/or excretion of unchanged drug.

5.1.2.1 Concentrations in the peripheral compartment

The concentrations in the peripheral compartment can be calculated:

$$C_{\text{periph}} = \frac{Dk_{21}}{V_2(\lambda_1 - \lambda_2)}\left[\exp(-\lambda_2 t) - \exp(-\lambda_1 t)\right] \tag{5.4}$$

after the plasma concentration–time data have been analysed to calculate the required constants (Sections 5.1.2.2 and 5.1.2.3). However, unlike the biological situation, it is possible to demonstrate this relationship using the dye model (Appendix 2) and to *measure* the concentrations in V_2 (Figure 5.4). When the drug is injected at $t=0$, instantaneous mixing in the central compartment is assumed and there is no drug in the peripheral compartment. As drug diffuses into the peripheral compartment the concentration increases until the concentrations become equal (Figure 5.4(a), photo C). At this time the forward and backward rates of transfer are equal and, for an instant, there is no net movement of drug. This condition is referred to as steady state after which net transfer is from the peripheral to the central compartment because elimination is reducing the concentration in the central compartment. Steady state occurs for an instant when the peripheral compartment concentration is maximal:

$$t_{\max} = \frac{1}{(\lambda_1 - \lambda_2)}\ln\left(\frac{\lambda_1}{\lambda_2}\right) \tag{5.5}$$

At later times (in theory only at infinite time) the compartments equilibrate and the *ratio* of concentrations remains constant, as shown by a parallel decline on a semilogarithmic plot (Figure 5.4(b)). Note that post-equilibrium concentrations in the peripheral compartment are higher than those in the central compartment (Figure 5.4(a), photo D). This is important because the rate of return of drug from tissues can have a major influence on the terminal elimination half-life.

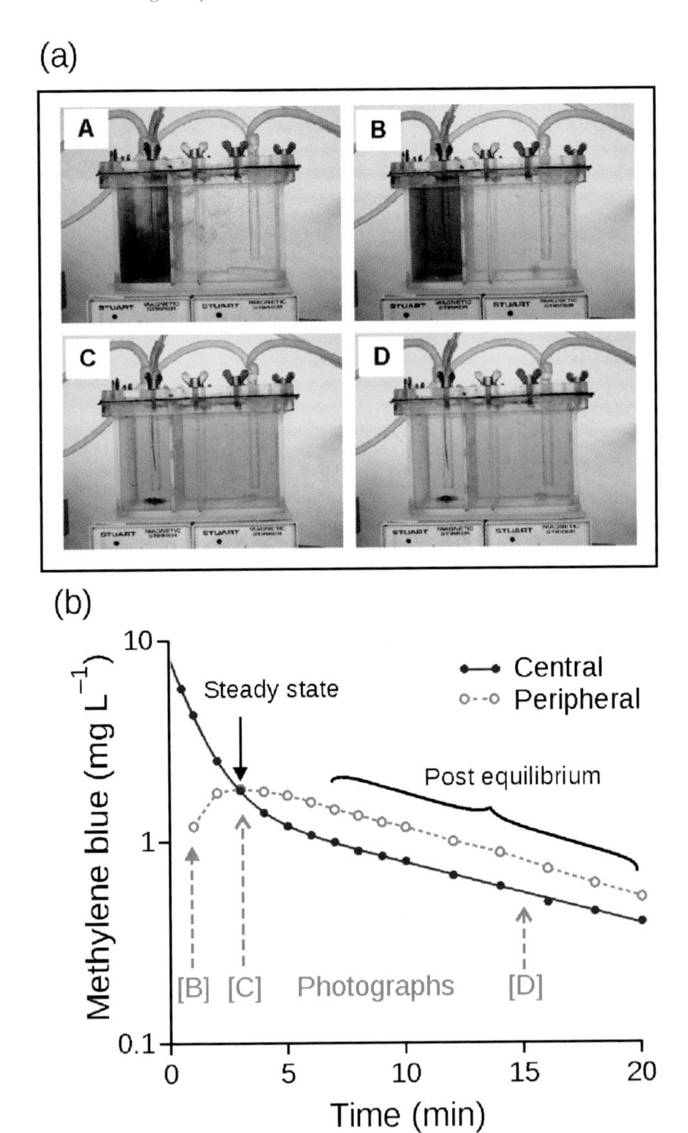

Figure 5.4 (a) Photographs taken following i.v. bolus injection of dye into the apparatus described in Appendix 2: A: immediately after the injection; B: 1 min after the injection; C: when the concentrations in the two compartments were equal; D: post equilibrium. (b) Dye concentrations measured in the central and peripheral compartments.

5.1.2.2 Microconstants

The rate constants k_{10}, k_{12} and k_{21} are known as microconstants and have to be derived from C_1, C_2, λ_1 and λ_2, which are obtained by resolving the plasma concentration–time data into its two exponential terms (Section 5.2). The sum of the microconstants equals the sum of λ_1 and λ_2:

$$\lambda_1 + \lambda_2 = k_{10} + k_{12} + k_{21} \tag{5.6}$$

and

$$k_{21} = \frac{C_1\lambda_2 + C_2\lambda_1}{C_1 + C_2} \tag{5.7}$$

$$k_{10} = \frac{\lambda_1\lambda_2}{k_{21}} \tag{5.8}$$

$$k_{12} = \lambda_1 + \lambda_2 - k_{10} - k_{21} \tag{5.9}$$

Confusion sometimes arises over the relationship between k_{10} and λ_2. The former is the rate constant for loss of drug from the central compartment and is greater than λ_2, which is the rate constant for loss of drug from the body and is the derived from the terminal phase of the plasma concentration–time curve once equilibrium has been fully established. The value of λ_2 may be influenced by the other rate constants, particularly k_{21}, which controls the rate of return of drug from tissues to the central compartment.

The importance of the microconstants is that they are required to calculate the apparent volumes of distribution as well as allowing the peripheral compartment concentrations to be calculated.

5.1.2.3 Apparent volumes of distribution

Initially, the concept of apparent volume of distribution may be difficult for some. It can be thought of as an estimation of the extent to which a drug is distributed in the body and this distribution is a determinant of the elimination half-life as described earlier. It can also be considered as a mathematical tool, a constant of proportionality which allows the amount of drug in the body at any time to be calculated from knowledge of the plasma concentration at that time, and as a means of calculating systemic clearance. However, its influence is not always obvious as many of the pharmacokinetic calculations can be performed using rate constants. Of course, the values of those rate constants are determined by V and CL. The situation is further complicated by the fact that a two- compartment model requires an additional volume of distribution, V_{area}, as explained below.

Adopting the same approach as that for the single-compartment model, the volume of the central compartment is:

$$V_1 = \frac{D}{C_0} = \frac{D}{C_1 + C_2} \tag{5.10}$$

because at $t=0$ none of the dose has been transferred to the peripheral compartment, and C_0 is the sum of the constants (C_1 and C_2) of the two exponential terms. At steady state, the concentrations in each compartment are equal (C^{ss}), and the forward and backward rates are equal and there is (instantaneously) no net movement of drug:

$$V_1C^{ss}k_{12} = V_2C^{ss}k_{21} \tag{5.11}$$

Cancelling C^{ss} and rearranging gives:

$$V_2 = V_1\frac{k_{12}}{k_{21}} \tag{5.12}$$

The sum of apparent volumes of the individual compartments is known as the volume of distribution at steady state:

$$V_{ss} = V_1 + V_2 \tag{5.13}$$

The volume of distribution at steady state indicates the extent to which the drug is distributed in the body. This is illustrated by the dye simulation because the calculated V_{ss} is in good agreement with the notional total volume of the box, 1500 mL (Table 5.1). However, it is clear from Figure 5.4 that the only time that multiplying V_{ss} by the plasma concentration will give the true amount of drug in the body is that instant at which steady-state conditions occur. Post equilibrium, using V_{ss} will give an underestimate of the amount of drug in the body because the peripheral concentrations are higher than those in plasma (Figure 5.4(a), photo D). Consequently a further calculation of apparent volume of distribution is required. This known as V_{area}:

$$V_{area} = \frac{D}{AUC\,\lambda_2} \tag{5.14}$$

This apparent volume of distribution is best thought of as a constant of proportionally that allows the amount of drug in the body to be calculated post equilibrium. In the simulation the calculated volume was 1902 mL, larger than the physical volume of the box, as predicted. Furthermore, V_{area} changes with changes in clearance (which changes λ_2). Consider the effect of increasing the flow of water on Figure 5.4(a), photo D. As the clearance increases (depicted by the flow) the concentration in the central compartment will fall but the concentrations in the peripheral compartment do not fall to the same extent so the differences in post-equilibrium concentrations are greater and so a larger value of V_{area} is required (Table 5.1).

If there were no flow of water the dye would not be removed at all *and the concentrations in the central and peripheral compartments would be equal*, that is, $V_{area} = V_{ss}$. The effect of changing the flow rate, representing a change in clearance, is shown in Figure 5.5.

The term V_{extrap}, which uses a construction line through the terminal phase of the plasma concentration data extrapolated to $t = 0$, may be encountered, particularly in older literature. In the example of Figure 5.2, the intercept is C_2 and $V_{extrap} = D/C_2$, which is analogous to the single-compartment case. However, when applied to multiple-compartment models, this approach overestimates the apparent volume of distribution, particularly when V_2 is large relative to V_1. For example, using the data of Table 5.1, D/C_2 calculates to be 2550 mL, larger than both V_{ss} (1475 mL) and V_{area} (1902 mL).

Table 5.1 Results from i.v. injection of dye into a two-compartment model

Parameter	Found	Parameter	Found	Parameter	Found	Notional
Dose (mg)	4*	C_0 (mg L^{-1})	8.13	V_1 (mL)	492	500
C_1 (mg L^{-1})	6.56	k_{21} (min^{-1})	0.219	V_2 (mL)	965	1000
λ_1 (min^{-1})	0.847	k_{10} (min^{-1})	0.267	V_{ss} (mL)	1457	1500
C_2 (mg L^{-1})	1.57	k_{12} (min^{-1})	0.430	V_{area} (mL)	1902	n.a.
λ_2 (min^{-1})	0.0691			CL (mL min^{-1})	131.5	130

*Not calculated.
n.a.: not applicable.

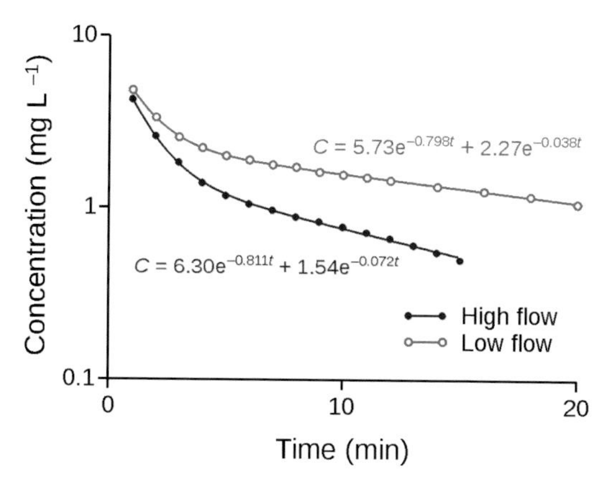

	Flow rate	
Parameter	High	Low
C_0 (mg L^{-1})	7.84	8
CL (mL min^{-1})	137	59
AUC (mg min L^{-1})	29.2	67.6
V_{ss} (mL)	1442	1411
V_{area} (mL)	1900	1572

Figure 5.5 Dye simulation demonstrating the effect that changing *CL* has on *V*. Note how reducing *CL* increases *AUC* and reduces V_{area}. Data provided by medical students of the London Hospital Medical College.

5.1.2.4 Clearance

The area under the curve is the sum of the areas under the two exponential phases so, by analogy with Equation 4.13, for a two-compartment model:

$$AUC = \frac{C_1}{\lambda_1} + \frac{C_2}{\lambda_2} \tag{5.15}$$

Several approaches may be used to calculate clearance. Equation 4.16 is applicable to single- and multiple-compartment models, and if the *AUC* is obtained using the trapezoidal method it is not even necessary to define the number of compartments (Section 5.2):

$$CL = \frac{D}{AUC} \tag{4.16}$$

Because the rate of elimination from the body $= C \times CL$, where C is the plasma concentration (or the concentration in V_1) then the rate of elimination from the central compartment/plasma is the amount ($V_1 \times C$) multiplied by the elimination rate constant, k_{10}:

$$C \times CL = V_1 \times C \times k_{10} \tag{5.16}$$

Cancelling C from each side gives:

$$CL = V_1 k_{10} \tag{5.17}$$

Combining Equations 5.14 and 4.16 and rearranging:

$$CL = V_{area} \lambda_2 \tag{5.18}$$

and because Equations 5.17 and 5.18 both are equal to CL:

$$V_{area}\lambda_2 = V_1 k_{10} \tag{5.19}$$

The *amount* of drug in the body post equilibrium is V_{area} multiplied by the plasma concentration so the rate of elimination post equilibrium is $(V_{area} \times C)\lambda_2$.

5.1.3 Absorption

The addition of an absorption phase to a two-compartment model leads to an extremely complex situation. A working equation for first-order absorption into a two-compartment model is:

$$C = C_1' \exp(-\lambda_1 t) + C_2' \exp(-\lambda_2 t) - (C_1' + C_2')\exp(-k_a t) \tag{5.20}$$

The values C'_1 and C'_2 are not the same as C_1 and C_2 as they are affected by both the rate and extent of absorption. The situation is further complicated by the relative sizes of the rate constants. If the rate constant of absorption is large then it may be possible to resolve the data to obtain estimates of k_a, λ_1 and λ_2. However, it is often not possible to ascribe values to k_a and λ_1 without investigating the kinetics using an intravenous dose. Provided that $k_a > k_{21}$ (Figure 5.6, red line) it is usually possible to resolve the plasma concentration–time data to solve for the variables of Equation 5.20. However, when $k_a < k_{21}$ the distributional phase is not apparent and the model *appears* to be that of a single compartment, as shown by the blue line of Figure 5.6. In this situation one would not be able to describe the drug as conferring the characteristics of a two-compartment model without also observing the plasma concentrations following an intravenous dose. This was the case with fluphenazine enanthate (Section 4.44), and

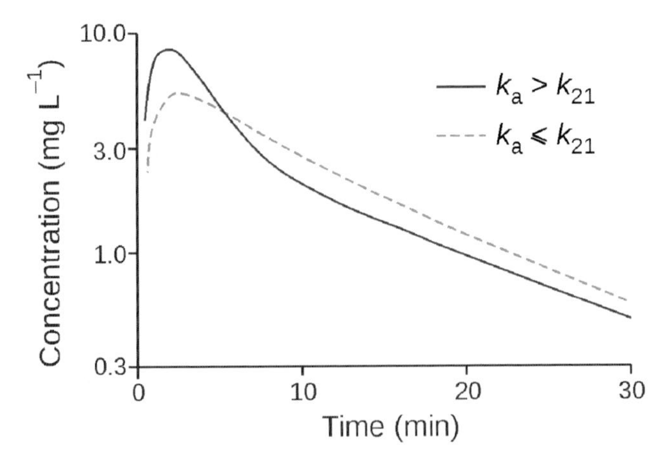

Figure 5.6 Empirical demonstration of how the relative size of the absorption rate constant affects the shape of the plasma concentration–time curve. These curves were obtained using the dye model, with different sized Büchner flasks connected in series to provide first-order input in the two-compartment model (Appendix 2, Figure A2.2) so there can be no doubt that these curves are from the same two-compartment model.

is likely to be the case with most sustained-release preparations which are, of course, designed to have low values of k_a.

5.1.4 Infusions

When a drug is infused into a two-compartment model the shape of the rising phase is the inverse of the decay curve at the end of the infusion, just as it was in the single-compartment case (Section 4.5). Thus, the rising phase is composed of two exponential terms, with an initial steep rise. When the infusion is stopped the decay is biphasic and it may be possible to resolve the decay curve into two exponential terms (Figure 5.7). The initial steep phase is not as obvious as that after an intravenous bolus injection because during the infusion drug has been transferred to the peripheral compartment and the concentration gradient between the two compartments is not as great, particularly as the concentration approaches steady state (Figure 5.7(b)). When the infusion is stopped the post-infusion concentration curves can be extrapolated to give intercepts, C'_1 and C'_2 on a line representing the time that the infusion was stopped (the *y*-axis of Figure 5.7(b)). These values can be used to calculate the constants after an i.v. bolus injection, C_1 and C_2:

$$C_1 = \frac{DC'_1\lambda_1}{R_0\left[1-\exp(-\lambda_1\tau)\right]} \tag{5.21}$$

$$C_2 = \frac{DC'_2\lambda_2}{R_0\left[1-\exp(-\lambda_2\tau)\right]} \tag{5.22}$$

where D is the dose of the intravenous bolus injection.

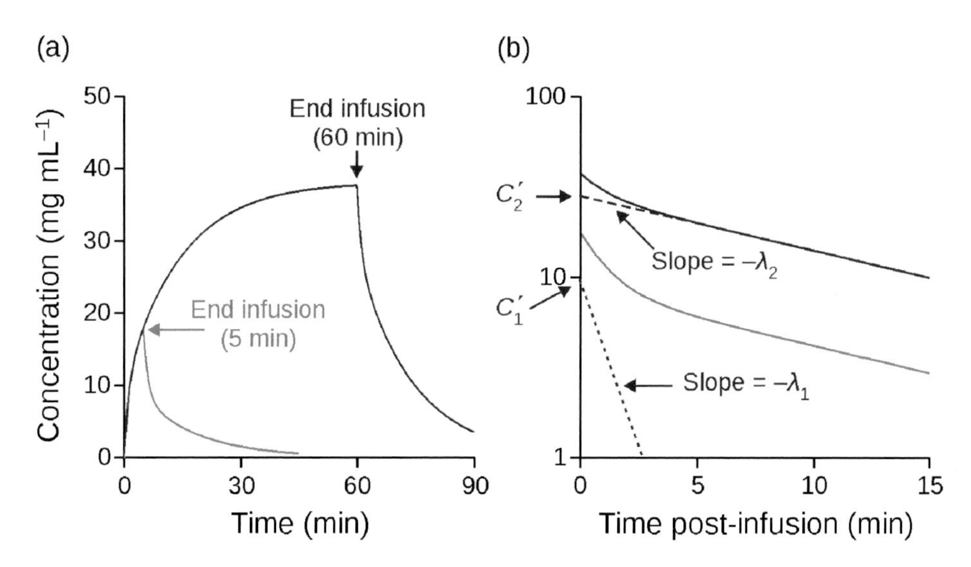

Figure 5.7 (a) Constant rate infusions into a two-compartment model. (b) Semilogarithmic plot showing how the post-infusion decay is influenced by the duration of the infusion.

The infusion curves of Figure 5.7 were modelled using the dye model data of Figure 5.2. When the values of intercepts from the post-infusion concentrations of Figure 5.7(b) were substituted into Equations 5.21 and 5.22, the values of C_1 and C_2 were 6.56 and 1.57 mg L^{-1}, respectively (cf. values in Table 5.1). The same values were obtained for the 5-min or the 60-min infusion.

Using the model data of Figure 5.7(b) it was possible to demonstrate the validity of Equations 5.21 and 5.22, and there is no doubt that infusions can be a useful way of determining kinetic parameters, particularly when it is considered that it might be unsafe to inject an i.v. bolus dose. However, in practice there may be problems. The longer the infusion is maintained, the more difficult it becomes to discern the distribution phase and this results in increased probability of both experimental and calculation errors. It is usually better to use shorter infusion times if that is possible.

5.1.4.1 *Loading dose*

It is possible to calculate a loading dose, R_0/k_{10}, to give an instantaneous steady-state concentration, C^{ss}, in plasma. However, this ignores the transfer of drug from the central to the peripheral compartment and so initially the plasma concentration falls and then rises and asymptotes to C^{ss}. Again this can be demonstrated empirically using the dye model (Figure 5.8, red line). To avoid the plasma concentrations during the infusion falling below C^{ss}, a larger loading dose, R_0/λ_2, has to be used, but this may result in the initial concentrations being unacceptably high (Figure 5.8, blue line). This problem is of practical importance for some drugs, for example when lidocaine is used to control ventricular arrhythmias. One solution is to use a 'loading infusion', that is, to infuse at a higher rate initially and then reduce the rate to that required for the desired steady-state concentration. An alternative scheme, using intravenous administration of an initial bolus loading dose in conjunction with a constant rate and an exponential intravenous drug infusion, has been proposed (Vaughan & Tucker, 1976).

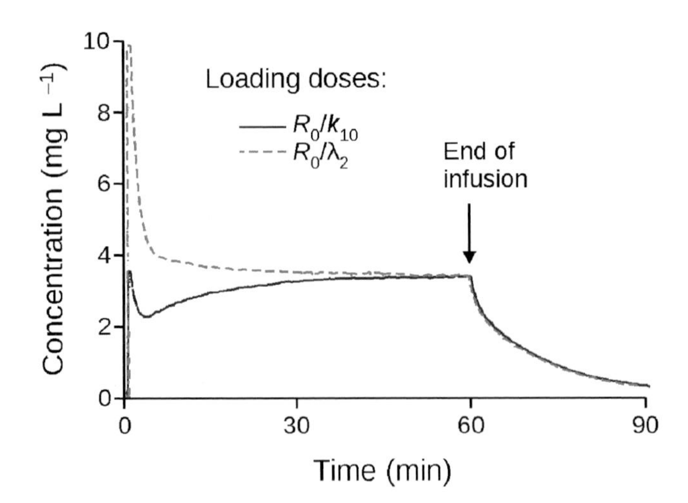

Figure 5.8 Infusion into a two-compartment model with different loading doses. (a) R_0/k_{10} gives instantaneous steady-state concentration, but this falls as drug is transferred to the peripheral compartment. (b) A loading dose R_0/λ_2 ensures the concentrations do not fall below C^{ss}, but the peak concentration may be unacceptably high.

5.1.5 Multiple dosing

The principles outlined in Chapter 4 apply. On repeated dosing the plasma concentrations will increase and tend towards steady-state conditions. As before the situation is simpler at steady state when following repeated intravenous doses:

$$C = C_1' \exp(-\lambda_1 t) + C_2' \exp(-\lambda_2 t) \tag{5.23}$$

where t is any time during the dosage interval and

$$C_1' = C_1 \left(\frac{1}{1 - \exp(-\lambda_1 \tau)} \right) \tag{5.24}$$

and

$$C_2' = C_2 \left(\frac{1}{1 - \exp(-\lambda_2 \tau)} \right) \tag{5.25}$$

where τ is the dosing interval. The average plasma concentration at steady state is given by:

$$C_{\text{av}}^{\text{ss}} = \frac{AUC}{\tau} = \frac{FD}{V_1 k_{10} \tau} = \frac{FD}{V_{\text{Area}} \lambda \tau} \tag{5.26}$$

See also Equation 4.42.

5.1.6 Concept of compartments

The concept of pharmacokinetic compartments may be difficult for anyone new to the topic. Understanding is not helped by the fact that on some occasions the volume of a 'pharmacokinetic' compartment may be identical to a known anatomical volume such as plasma or total body water, but more often than not the calculated volumes of distribution bear no comparison. Then there is the question of which tissues constitute a particular compartment. Generally, well-perfused tissues are components of the central compartment for lipophilic drugs because lipid membranes provide little in the way of a barrier to the movement of such drugs, and tissue and plasma concentrations rapidly equilibrate. Such tissues include liver and kidney, and often brain (Figure 5.9). Thus, although the concentrations in individual tissues may be very different, the kinetics describing the changes in concentrations are the same, and they decline in parallel on a semi-logarithmic plot, for example plasma and liver in Figure 2.12(a). For a lipophilic drug, the rate of delivery (blood flow) to the tissue is important and so equilibration with the less well-perfused tissue is slower and apparent from the plasma concentration–time plot. In this situation, such tissues constitute the peripheral compartment and might include fat and muscle (Figure 5.9).

It is worth remembering that (i) concentration–time plots, particularly with human subjects, are of plasma or blood, (ii) concentrations in the tissues of the central compartment are usually higher than those in plasma and (iii) the concentrations in the peripheral compartment(s) will rise as drug is distributed from the central compartment.

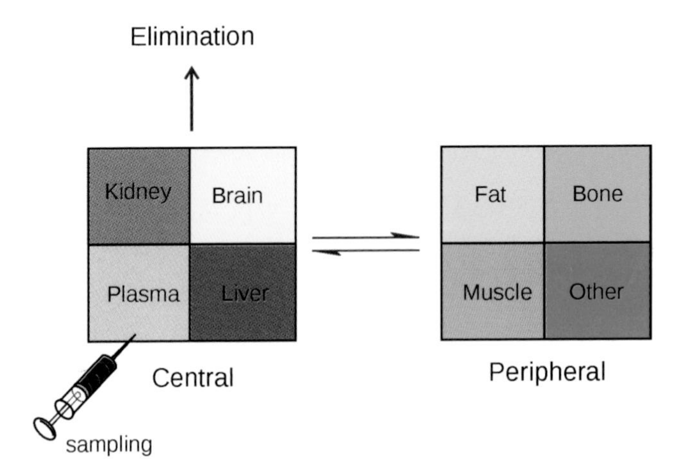

Figure 5.9 *Representation of tissues that might constitute the central and peripheral compartments for a lipophilic drug.*

5.1.7 Relationship between dose and duration of effect

The relatively simple relationship between dose and duration of effect for a single-compartment model was discussed in Section 4.8. The situation is more complex for multiple-compartment models but these are necessary to explain the duration of action of many drugs. This is best exemplified by thiopental, which clearly exhibits multiple-compartment kinetics (Figure 2.12(a)). Liver concentrations rapidly equilibrate with those in plasma whilst the concentrations in skeletal muscle rise initially and then equilibrate. The concentrations in adipose tissue rise for at least the first 3.5 h of the experiment, therefore the model for thiopental needs at least three compartments: a central compartment and two peripheral compartments (Figure 5.10).

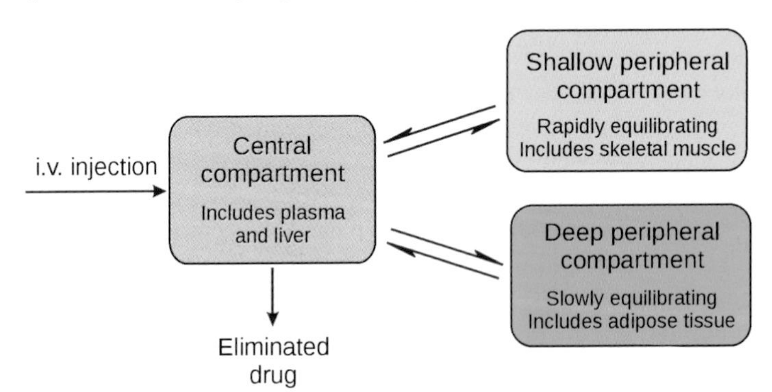

Figure 5.10 *The model for thiopental requires three compartments.*

As discussed in Section 2.4.3.1, following an i.v. bolus injection of thiopental the duration of action is short due to uptake of the drug into skeletal muscle and fat. When larger doses were injected or, more particularly, when doses were repeated, the duration of action was disproportionately long. When dosing was carried out cautiously, noting the patient's response and using a flexible dosing policy, long-term anaesthesia could be safely maintained. However, incautious administration, generally with a fixed-dose regimen unrelated to patient response, led to each successive dose exerting a duration of action longer than the previous one until an excessively long duration of action occurred, even when administration ceased. It has been incorrectly assumed that this phenomenon was due to saturation of tissue stores leading to increasing proportions of subsequent doses being unable to redistribute from plasma and brain. In fact there is no evidence of tissue saturation as shown by Table 5.2 in which the tissue:blood ratio (T/B) is more or less constant. However, the observation can be explained in terms of a multiple-compartment model.

Table 5.2 Duration of action and tissue localization of thiopental at four doses in humans (adapted from Curry, 1980)

Dose (g)	Administration time (min)	Duration (h)	Amount in blood at 1 h (g)	Amount in tissues at 1 h (g)	T/B
0.4	2	0.25	0.027	0.333	12.3
1	5	0.5–1	0.066	0.834	12.6
2	5	1.5–2.5	0.156	1.744	11.1
3.8	50	4–6	0.288	3.512	12.2

Using composite data for plasma concentrations in human subjects from Brodie's work and, for simplicity, *assuming* a two-compartment model (the original work could be fitted to a three-compartment model in agreement with the results for dogs) plasma concentration decay curves for 0.5 and 0.75 g doses were calculated (Figure 5.11). The subject from whom the data of Figure 5.11 were derived was anaesthetized with a dose of 0.5 g (i.v.) and regained consciousness after 20 min when the plasma concentration was 6.99 mg L^{-1}. The kinetic model predicts that the subject would be unconscious for 2.8 h had the dose been increased by only 1.5-fold to 0.75 g. This is clearly a different situation from that predicted for intravenous injection into a single-compartment model (Section 4.8).

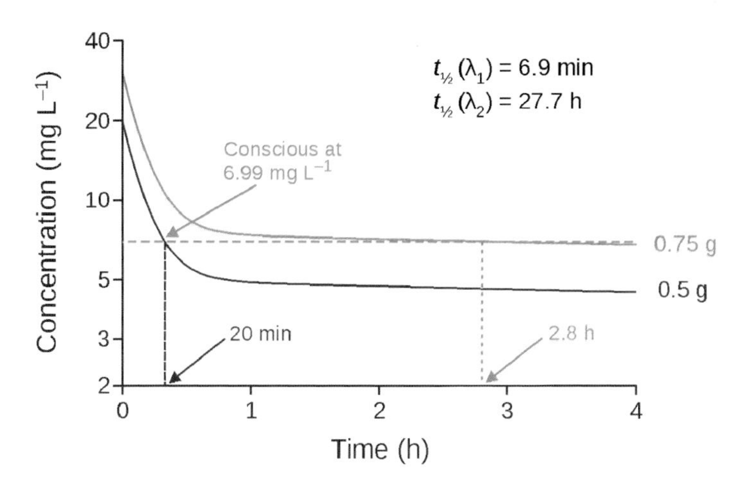

Figure 5.11 Red line: time course of thiopental in plasma calculated from the composite data of Brodie after 0.5 g dose; the subject regained consciousness after 20 min. Blue line: calculated time course for 0.75 g dose, with predicted recovery at 2.8 h.

Additionally, the mean terminal half-life of thiopental has been shown to be different in lean and obese subjects, being 6.33 and 27.9 h, respectively, the variations being due to differences in apparent volumes of distribution rather than clearance (Jung *et al.*, 1982). Thus, the curve of Figure 5.11 may represent a more extreme situation than the average, but it does illustrate a general phenomenon applicable to all multiple-compartment models. It is apparent from the figure that the relative duration of effects will vary depending on the concentration at which recovery occurs. If recovery occurred post-equilibrium, during the terminal phase, then the relationship would be the same as that for the single-compartment model. Similarly, if recovery were largely during the distribution phase the increment in duration would not be so great, until of course the dose was increased to the point that recovery occurred in the terminal phase.

The effect of repeating the dose at the time of recovery, and at the same interval subsequently, can be modelled using:

$$C_n = C_1\left(\frac{1-\exp(-n\lambda_1\tau)}{1-\exp(-\lambda_1\tau)}\right)\exp(-\lambda_1 t) + C_2\left(\frac{1-\exp(-n\lambda_2\tau)}{1-\exp(-\lambda_2\tau)}\right)\exp(-\lambda_2 t) \quad (5.27)$$

where C_n is the concentration at time t following the nth dose, C_1 and C_2 are the intercepts on the y-axis, following the intravenous bolus injection, and τ is the dosage interval (20 min in this example). This is illustrated in Figure 5.12.

While it is the case that with subsequent doses the rate of transfer from the central to the peripheral compartment will decline as the concentration in the peripheral compartment increases, this should not be confused with *saturation* of tissues. Saturation would lead to a change in the rate constants. Furthermore, the clinical situation will be even more complex because of the potential influence of tolerance, both receptor and pharmacokinetic, and the presence of other drugs.

The thiopental research described above was of seminal significance in the pursuit of pharmacokinetic understanding during the development of intravenous anaesthesia and analgesia, and thiopental continues to be an anaesthetic of significance. Similar pharmacokinetic

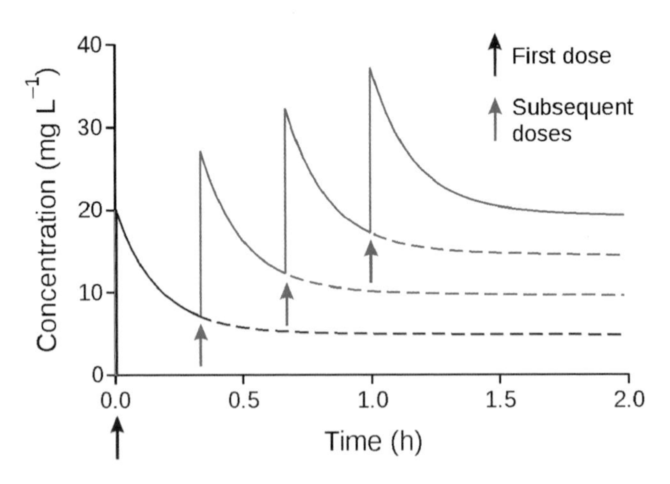

Figure 5.12 Modelled curves showing the effect of repeat injections (arrows) of thiopental at 20 min intervals using the parameters of Figure 5.11.

events are involved in the use of diazepam in status epilepticus, and in delirium tremens of alcohol withdrawal, with the added need in this case to avoid administration that is so rapid that the diazepam comes out of solution in the bloodstream before the injection solution is adequately diluted in the body. However, the most obvious modern example of a drug that follows the properties of the thiopental model is propofol, which is used as an intravenous sedative and anaesthetic, with or without other anaesthetics concurrently, particularly in outpatient procedures such as colonoscopies. Propofol concentrations rise rapidly during the early stages of infusion, then more slowly, even with constant rate infusion (as would be expected for a drug imparting the characteristics of a multiple-compartment model). The drug equilibrates very rapidly with the brain, so the effect is induced quickly, after which it is usually desirable to reduce the rate of infusion to maintain the effect whilst avoiding the excessive anaesthesia that could develop, relatively slowly, with prolonged administration. A distribution model analogous to that in Figure 5.10 has been devised for this drug. If drug infusion is discontinued after approximately 1–24 h, the concentrations in the plasma decline rapidly and recovery is fast. If administration is for longer, for example for as long as 1 week, in the intensive care unit, the effect declines slowly on cessation of dosing as the terminal half-life is approximately 2 days (Knibbe *et al.*, 2000).

5.2 Non-compartmental models

5.2.1 Calculation of V_{area} and clearance

Some of the pharmacokinetic parameters described above can be obtained without the need to define the number of compartments. Clearly, this applies to systemic availability, F, which is obtained by comparison of areas under the curve. The AUC, calculated by the trapezoidal method, can used to derive systemic clearance and V_{area} using Equations 4.16 and 5.14 or:

$$CL = F\frac{D}{AUC} \tag{4.30}$$

and

$$V_{area} = F\frac{D}{AUC\,\lambda_z} \tag{5.28}$$

for extravascular doses. For the intravenous case $F = 1$. The data have to be such that the rate constant of the terminal phase, λ_z, can be derived. The approach can be illustrated using the data of Figure 5.6. Although it is not possible to define the correct number of compartments in one of the cases, when $k_a < k_{21}$ (slow absorption), the agreement between the parameters is good (Table 5.3). Usually, it is not possible to calculate CL without knowing the systemic availability, so the value quoted is CL/F (apparent oral clearance). In this particular instance we know that $F = 1$ and the calculated value of CL is very close to the nominal flow rate 130 mL min^{-1}.

Table 5.3 Non-compartment approach to determine CL and V_{area} from the data of Figure 5.6

Parameter	Rapid absorption	Slow absorption
Terminal rate constant, λ_z (min^{-1})*	0.0679	0.0719
$AUC_{(0-30)}$ (mg min L^{-1})	68.5	67.9
C_{30}/λ_z (mg min L^{-1})	7.2	8.0
$AUC_{(0-\infty)}$ (mg min L^{-1})	75.7	75.9
CL/F (mL min^{-1})	132.1	131.8
V_{area}/F (mL)	1946	1833

* From last five data points

The average steady-state concentration that would be achieved on multiple dosing can be obtained from Equation 5.26.

5.2.2 Statistical moment theory

Although statistical moment theory (SMT) is sometimes described as being non-compartmental, it does assume that the drug is measurable in plasma and also that the rate of elimination (flux) of drug is proportional to the plasma concentration, that is, linear kinetics apply. The parameters that can be assessed are limited, but they are derived from measurements of areas under the curve without having to define compartments and assign what can be ambiguous rate constants to them. The method is useful when designing dosage regimens and comparing differences between species.

If one considers how long a single drug molecule stays in the plasma after administration, then it could be very short, very long or some intermediate interval, which is not particularly useful in itself. However, if large numbers of molecules are considered then there will be a mean residence time, *MRT*, which according to statistical moment theory is:

$$MRT = \frac{\int_0^{\infty} C\,t\,\mathrm{d}t}{\int_0^{\infty} C\,\mathrm{d}t} = \frac{AUMC}{AUC} \tag{5.29}$$

where *AUMC* is the area under the (first) moment curve. The relationship between these areas is depicted in Figure 5.13.

For a bolus i.v. injection into a single-compartment model:

$$MRT_{\mathrm{i.v}} = \frac{1}{\lambda} = \frac{V}{CL} \tag{5.30}$$

where λ is the rate constant of elimination. If comparing the results for a two-compartment model, then $MRT_{\mathrm{i.v.}} = V_{ss}/CL$, and the apparent first-order rate constant has a value between λ_1 and λ_2. It follows from Equation 5.30 that *MRT* is the time when 63.2% of the intravenous

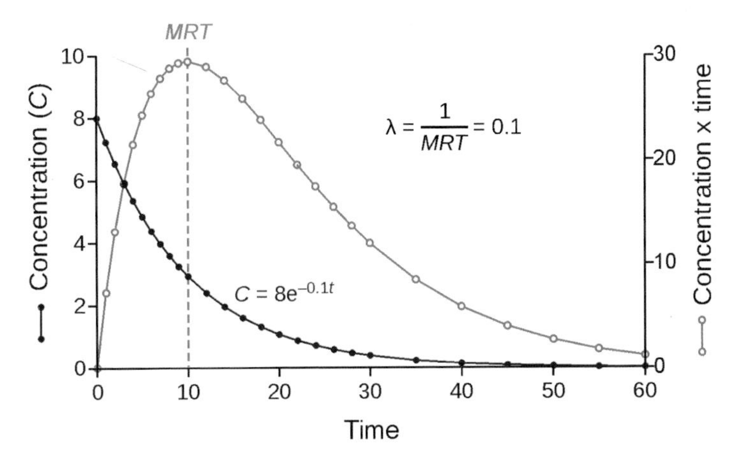

Figure 5.13 Model data for i.v. injection into a single compartment, showing C versus t and Ct versus t.

dose will have been eliminated. This can be shown by substituting Equation 5.30 into Equation 1.8 when $t = MRT = 1/\lambda$:

$$C = C_0 \exp\left[-\lambda(1/\lambda)\right] = C_0 \exp(-1) = 0.368\ C_0 \tag{5.31}$$

Note that exp(−1) = 0.368. Because 36.8% remains, 63.2% must have been eliminated. The apparent volume of distribution at steady state, V_{ss}, can be computed from:

$$V_{ss} = D\frac{AUMC}{AUC^2} = CL \times MRT \tag{5.32}$$

For first-order absorption, for example after an oral dose, the mean arrival time, *MAT*, can be used:

$$MAT = MRT_{\text{p.o.}} - MRT_{\text{i.v.}} \tag{5.33}$$

and

$$k_a = \frac{1}{MAT} \tag{5.34}$$

5.2.2.1 Estimating AUMC

The most obvious way of obtaining *AUMC* is to use the trapezoidal method, analogous to that used for *AUC* (Appendix 1). The area of the *n*th trapezium, $AUMC_n$, is:

$$AUMC_n = \frac{C_n + C_{n+1}}{2}\left(t_{n+1} - t_n\right)t_{n+1} \tag{5.35}$$

However, unlike *AUC*, the results are weighted by time, and the weight increases as the time increases. Also, Equation 5.35 overestimates the area, while using t_n underestimates *AUMC*. This can be reduced by using the midpoint for time:

$$AUMC_n = \frac{C_n + C_{n+1}}{2}\left(t_{n+1} - t_n\right)\frac{\left(t_n + t_{n+1}\right)}{2} \tag{5.36}$$

Even so, Equation 5.36 overestimates the area for the decay part of the curve. A more accurate equation is:

$$AUMC_n = \frac{C_n\left(t_{n+1}^2 - t_n^2\right)}{2} + \frac{C_{n+1} - C_n}{6}\left(2t_{n+1}^2 - t_n t_{n+1} - t_n^2\right) \tag{5.37}$$

and, despite its complexity, is easy to use once it has been entered into a spreadsheet. The area beyond the last time point is obtained by extrapolation:

$$AUMC_{(z-\infty)} = \frac{t_z C_z}{\lambda_z} + \frac{C_z}{\lambda_z^2} \tag{5.38}$$

where z denotes the last measured variable. Thus, despite being considered as a non-compartmental approach, it is necessary to calculate the terminal rate constant, which of course means identifying the last exponential phase.

5.2.2.2 *Example of application of SMT*

The data of Figure 5.2 was subjected to SMT (Figure 5.14). The first thing to note is that data have to be collected for a considerable time to define *AUMC* compared to that required to define two compartments from the *C* versus *t* data, as otherwise a large proportion of *AUMC* will have to be extrapolated. Obviously this has implications regarding the quality of the data because, in all probability, the errors will be highest at low concentrations and *AUMC* is weighted by the high values of *t*. Furthermore, the extrapolation of the remaining

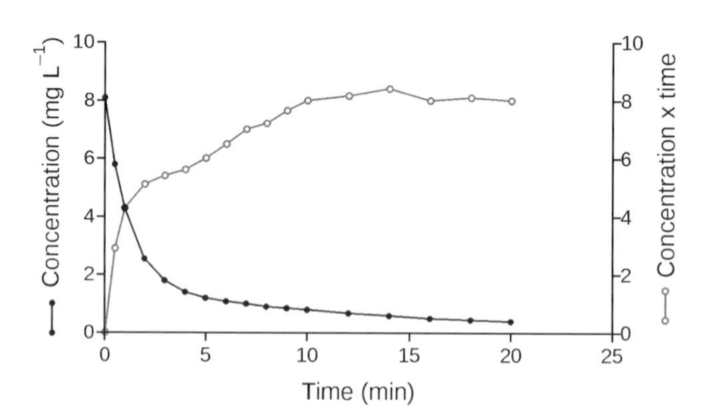

Figure 5.14 Statistical moment approach applied to the dye model data of Figure 5.2.

Table 5.4 SMT approach to calculating pharmacokinetic parameters

Time	Areas		Derived parameters	
0–20	141.4*	25.0	*MRT* (min)	11.08
20–∞	199.5†	5.8	*CL* (mL min^{-1})‡	130
0–∞	340.9	30.8	V_{ss} (mL)	1440

*Using Equation 5.37.
†From Equation 5.38.
‡From Equation 5.32.

area requires accurate assessment of λ_z. Despite this, the results for the data of Figure 5.14 are in good agreement with those derived previously, reflecting the high quality of data (Table 5.4) in this case.

AUMC can be calculated for a compartmental model:

$$AUMC = \Sigma C_i / \lambda_i^2 \tag{5.39}$$

where *i* is the number of compartments. In the example of Table 5.4, Equation 5.39 gives $AUMC = 338.0\,\text{mg}\,\text{min}^2\,\text{L}^{-1}$. Note the good agreement between figures derived using a two-compartment model (Table 5.1) and SMT (Table 5.4).

Thus the *MRT* is very easy to calculate, using only the *AUC* and the *AUMC* for a wide variety of pharmacokinetic systems. For one-compartment systems it provides a method for calculating the time for 63.2% of the dose to leave those systems, while for more complex systems it provides a straightforward means of evaluating changes caused by disease, age, interacting drugs and many more factors, on the properties of the drug in question.

5.3 Population pharmacokinetics

This is an analytical process designed to focus on variability and central tendencies in data, and to optimize the use of 'sparse data', which is the type of data most often obtainable during Phase III studies or after drugs have been marketed. Typically, one or two blood samples are occasionally available from any one patient at this stage in the life cycle of a new drug. However, sparse data may be available from large numbers of patients. Thus, in Phase I trials the investigations yield full pharmacokinetic profiles of investigational drugs in small numbers of healthy volunteers, often only male, and fitting a relatively narrow anthropomorphic profile. Phase II studies may extend this intensity of investigation to the target patient population, thus broadening the scope of the studies in regard to disease factors, interacting drugs and such factors as age, but still with relatively small numbers of subjects. Later, the scope broadens, the numbers of patients becomes large, but the limitations of sparse data become especially important.

Population pharmacokinetics employs statistical methods based, in part, on a Bayesian feedback algorithm. These methods utilize the gradually accumulating pool of data to draw conclusions about physiological, disease, drug interaction and other influences on the pharmacokinetic properties of the drug in the human population (Racine-Poon & Smith, 1990). Probability theory is used to facilitate step-by-step prediction of pharmacokinetic properties with ever greater precision, making possible better decisions on the choice of dose for various subpopulations, and eventually for individual patients, in the clinic. Because of this focus on the individual, some authors consider the label ‘population statistics’ to be unfortunate.

The methods of population statistics were pioneered by Sheiner & Beal (1982) and Whiting *et al.* (1986). Various statistical packages, in particular, and originally NONMEM (Beal & Sheiner, 1982), have become synonymous with this work, although many others have found application (Aarons, 1991).

5.4 Curve fitting and the choice of most appropriate model

Previously, reference has been made to resolving or fitting concentration–time data to obtain estimates of the pharmacokinetic parameters. However, discussion has been left until now because it is easier to understand the underlying principles by considering resolution of the decay curve of a two-compartment model into its component exponential terms.

5.4.1 Graphical solution: method of residuals

Before the ready availability of personal computers and relatively inexpensive curve-fitting software, pharmacokinetic parameters were often obtained graphically. Although rarely used these days, an understanding of the approach is important when assessing the quality of the data to be analysed. The method of residuals, as it is known, can be applied to the majority of compartment models, and also first-order input into single- or multiple-compartment models, and the post-infusion phase of zero-order infusions. The method is most easily understood from consideration of an intravenous bolus injection into a two-compartment model. It is necessary to collect data for long enough to define the terminal phase with negligible contribution from the distributional phase. This allows estimation of C_2 and λ_2. Values of $C_2\exp(-\lambda_2 t)$ are calculated for earlier time points (i.e. when $C_1\exp(-\lambda_1 t)$ is making a significant contribution to the plasma concentration) and subtracted from the experimental values at those times, giving estimates of $C_1\exp(-\lambda_1 t)$, which are referred to as *residuals*. Plotting the residuals allows estimation of C_1 and λ_1 (Appendix 3).

5.4.2 Iterative curve fitting

There are several commercially available curve-fitting programs that are sold specifically for pharmacokinetic analyses. However, other packages may be adapted to derive pharmacokinetic parameters. Many relationships do not have a mathematical solution and have to be solved iteratively, usually by computing the equation which gives the lowest residual sum of squares. On this occasion ‘residual’ refers to the difference between the observed value and the value calculated from the derived parameters. The advantages of iterative fitting are that statistical estimates of the quality of the fit can be computed and different models can be compared. Furthermore, the data can be ‘weighted’ to compensate for differences in errors associated with that data (Appendix 3).

5.4.2.1 *Choice of model*

The number of compartments required to fit the data is given by the number of exponential terms that describe the *declining* portion of the curve. The choice of how many compartments to fit should be dictated by the data. Statistical fitting allows the various equations to be compared. Simply choosing the equation which gives the lowest residual sum of squares (*SS*) is unhelpful because this will be the equation with the largest number of parameters. Consequently, most statistical packages compute 'goodness of fit' parameters which take into account the number of parameters in the equation.

The appropriateness of the model, including the weighting, should be tested by plotting the residuals as a function of concentration. These should be randomly distributed about zero. The correlation coefficient, *r*, is often a poor indicator of the goodness of fit, unless $r^2 = 1$!

5.4.3 Quality of the data

Whether the parameters are derived using the method of residuals or iterative computer fitting the results will be poor if the original data are poor. With modern analytical methods one would expect the concentration data to be reasonably accurate. The chief reasons for poor results are:

- insufficient number of data points
- incorrect timing of collection
- data not collected for sufficient time.

The three bullet points above are related. If the model requires a large number of parameters, for example absorption into a three compartment model will generate seven parameters, then, clearly, the number of points must be sufficient to ensure statistical significance – an absolute minimum of eight (7 + 1). However, these points must be spaced at appropriate intervals so that each phase is defined, which is extremely unlikely to be the case with only eight samples. Another issue with timing is that the study design will usually specify the collection times, and these may be printed on the sample tubes prior to collection. It is not always possible to adhere to these times, which need not be a problem provided the correct time is recorded and used in the calculation. Use of uncorrected times will be more significant when the half-life is short. If the duration of the study is too short then the estimate of terminal rate constant, λ_z, is likely to be in error and this will be reflected in the estimates of the other parameters. The rule of thumb is to collect data for at least $4 \times \lambda_z$. If a semi-logarithmic plot of the data does not show a more or less linear terminal phase then the results can be expected to be poor and using iterative fitting rather than the method of residuals will not improve them: it is a case of 'rubbish in = rubbish out'. For this reason data should always be plotted and inspected. In a study of temoporfin kinetics, the terminal half-life in blood was 13.9 h when concentrations up to 48 h were analysed, beyond which time the concentrations were too low to be quantified (Whelpton *et al.*, 1995). However, the decay in brain and lung was much slower, suggesting that 13.9 h was far too low an estimate. A second study using ^{14}C-labelled temoporfin showed that the terminal half-life in blood was ~10 days (Whelpton *et al.*, 1996), a value that was confirmed from faecal excretion data collected for up to 5 weeks (Section 6.4).

Summary

This chapter has illustrated the need for more complex models to explain the kinetics and effects of drugs that cannot be explained using the simpler single-compartment model, and approaches that can be adopted when a model cannot be defined. These include problems encountered with the use of loading doses and redistribution as a determinant of drug action. Further consideration of the fundamental role of clearance, physiological models and the influence of plasma protein binding are discussed in Chapter 7. More complex pharmacokinetic–pharmacodynamic (PK–PD) relationships are considered in Chapter 8. The use of other samples (urine and faeces) and metabolites to study pharmacokinetics is considered in the next chapter.

5.5 Further reading

Charles B. Population pharmacokinetics: an overview. *Australian Prescriber* 2014; 37: 210–3.

Curry SH. *Drug Disposition and Pharmacokinetics*. 3rd edn. Oxford: Blackwell Scientific, 1980.

Riviere JE. *Comparative Pharmacokinetics: Principles, Techniques, and Applications*. Chichester: Wiley-Blackwell, 2003.

5.6 References

Aarons L. Population pharmacokinetics: theory and practice. *Br J Clin Pharmacol* 1991; 32: 669–70.

Beal SL, Sheiner LB. Estimating population kinetics. *Crit Rev Biomed Eng* 1982; 8: 195–222.

Curry SH. *Drug Disposition and Pharmacokinetics*. 3rd edn. Oxford: Blackwell Scientific, 1980.

Jung D, Mayersohn M, Perrier D, Calkins J, Saunders R. Thiopental disposition in lean and obese patients undergoing surgery. *Anesthesiology* 1982; 56: 269–74.

Knibbe CA, Aarts LP, Kuks PF, Voortman HJ, Lie AHL, Bras LJ, Danhof M. Pharmacokinetics and pharmacodynamics of propofol 6% SAZN versus propofol 1% SAZN and Diprivan-10 for short-term sedation following coronary artery bypass surgery. *Eur J Clin Pharmacol* 2000; 56: 89–95.

Peter C, Hongwan D, Kupfer A, Lauterburg BH. Pharmacokinetics and organ distribution of intravenous and oral methylene blue. *Eur J Clin Pharmacol* 2000; 56: 247–50.

Racine-Poon AM, Smith AFM. *Population models*. In: Berry DA, editor. Statistical methodology in the pharmaceutical sciences. New York: Marcel Dekker, 1990: 139–62.

Sheiner LB, Beal SL. Bayesian individualization of pharmacokinetics: simple implementation and comparison with non-Bayesian methods. *J Pharm Sci* 1982; 71: 1344–8.

Vaughan DP, Tucker GT. General derivation of the ideal intravenous drug input required to achieve and maintain a constant plasma drug concentration. Theoretical application to lignocaine therapy. *Eur J Clin Pharmacol* 1976; 10: 433–40.

Whelpton R, Michael-Titus AT, Basra SS, Grahn M. Distribution of temoporfin, a new photosensitizer for the photodynamic therapy of cancer, in a murine tumor model. *Photochem Photobiol* 1995; 61: 397–401.

Whelpton R, Michael-Titus AT, Jamdar RP, Abdillahi K, Grahn MF. Distribution and excretion of radiolabeled temoporfin in a murine tumor model. *Photochem Photobiol* 1996; 63: 885–91.

Whiting B, Kelman AW, Grevel J. Population pharmacokinetics. Theory and clinical application. *Clin Pharmacokinet* 1986; 11: 387–401.

6

Kinetics of Metabolism and Excretion

Learning objectives

By the end of the chapter the reader should be able to:

- explain the time course of metabolites the concentrations and kinetics of which are formation rate limited
- sketch the concentration–time curves for a drug and metabolites that undergo interconversion
- discuss the criteria for dosing a drug that has active metabolite(s)
- discuss the effect of urine flow rate on renal clearance
- explain how excretion data can be used to investigate the kinetics of a drug.

6.1 Introduction

In previous chapters, discussion of metabolism and excretion has been largely qualitative, but it is also important to understand the kinetics of these processes. Because metabolism frequently results in metabolites that are more polar and more readily excreted, it might be expected that these metabolites would rapidly disappear from the body. In fact the situation is more complex and it is necessary to understand the factors that govern the time course of metabolites because this will have a bearing on drug monitoring, forensic analysis, toxicology and appropriate clinical use of many drugs.

Introduction to Drug Disposition and Pharmacokinetics, First Edition. Stephen H. Curry and Robin Whelpton.

Companion website: www.wiley.com/go/curryandwhelpton/IDDP

Studying the kinetics of a metabolite is obviously important in the case of prodrugs, for which the pharmacological properties reside in the metabolite. In some situations the drug and one or more of its metabolites may be active, although not necessarily in the same way. Indeed the metabolite may be responsible for toxicity and the US FDA has taken an interest in potentially toxic metabolites, particularly those unique to human beings, and issued its guidelines, *Safety Testing of Drug Metabolites*, in 2008. Metabolite concentrations may accumulate on repeated drug administration and sometimes it may be more appropriate to define the kinetics of a drug via its metabolite.

A similar situation applies to quantification of a drug in urine or faeces. The investigation may be to determine the proportion of drug that is metabolized as part of a mass-balance study or it may be important that the drug is excreted via the urine, for example in the case of diuretics or antimicrobical drugs intended for treating urinary tract infections. On the other hand, the excretion rate may be used to define the kinetics of the drug and/or its metabolite. As far back as 1929, Gold and DeGraff studied the intensity and duration of the effect of digitalis, which to a large extent reflects urinary excretion of digoxin, and demonstrated that elimination results from a fixed proportion, not a fixed amount, of the body content leaving the body in each 24-h period (i.e. first-order elimination).

6.2 Metabolite kinetics

It may not be possible to study the kinetics of all the metabolites of a drug. For example, it is unlikely that *all* of the metabolites of any drug will have been identified. When a metabolite is the product of sequential metabolism in an organ such as the liver, and the intermediate metabolites are not released into the bloodstream, then they may not be identified, let alone quantitated. Similarly, a metabolite may be formed in the liver and excreted in the bile without measurable amounts entering the general circulation. Thus, we will be considering metabolites that reach the systemic circulation and will assume, in the main, first-order input and output from a single compartment.

6.2.1 Basic concepts

The time course of a metabolite or metabolites, and hence its or their kinetics, can be influenced by a number of factors and consequently complex relationships are possible. The simplest case would be when *all* of the administered drug, D, is converted to one metabolite, M, which is then eliminated, either by further metabolism or excretion:

$$\mathrm{D} \xrightarrow{k_{\mathrm{m}}} \mathrm{M} \xrightarrow{k_{\mathrm{m\cdot z}}} \text{elimination}$$

where k_m is the rate constant of formation and $k_{m\cdot z}$ is the rate constant of elimination of the metabolite. If these reactions are first order, then the situation is analogous to first-order absorption into and elimination from a single-compartment model and the equation, which gives the concentration of metabolite, C_m, in plasma at any time, t, takes the same form as Equation 4.18:

$$C_{\mathrm{m}} = D\frac{k_{\mathrm{m}}}{V_{\mathrm{m}}\left(k_{\mathrm{m}} - k_{\mathrm{m\cdot z}}\right)}\left[\exp\left(-k_{\mathrm{m\cdot z}}\right) - \exp\left(-k_{\mathrm{m}}t\right)\right] \qquad (6.1)$$

where V_m is the volume of distribution of the metabolite. For Equation 6.1 to be valid the dose, D, should be in moles, or corrected for the differences in relative molecular masses of the drug and metabolite. If a proportion of the drug is excreted unchanged or metabolized via other pathways, then D needs to be corrected for the fraction of metabolite formed, f_m (analogous to systemic availability, F).

The elimination half-life of a metabolite may be longer or shorter than that of the drug. Because metabolites tend be more polar than the parent drug, the volume of distribution and the degree of reabsorption by the kidney are usually less than those of the parent drug. Thus, if purified metabolite was available and injected intravenously its half-life would be shorter than that of the parent drug (Figure 6.1(a)). When the drug is administered the metabolite cannot be eliminated faster than it is formed and so the drug and metabolite concentrations decline in parallel (Figure 6.1(b)). In this situation, $k_{m \cdot z} > k_m$ and the term $\exp(-k_{m \cdot z}t)$ approaches zero faster than $\exp(-k_m t)$ so that at later times the slope of the terminal phase of the metabolite curve is $-k_m$, that is, the same as that of the parent drug.

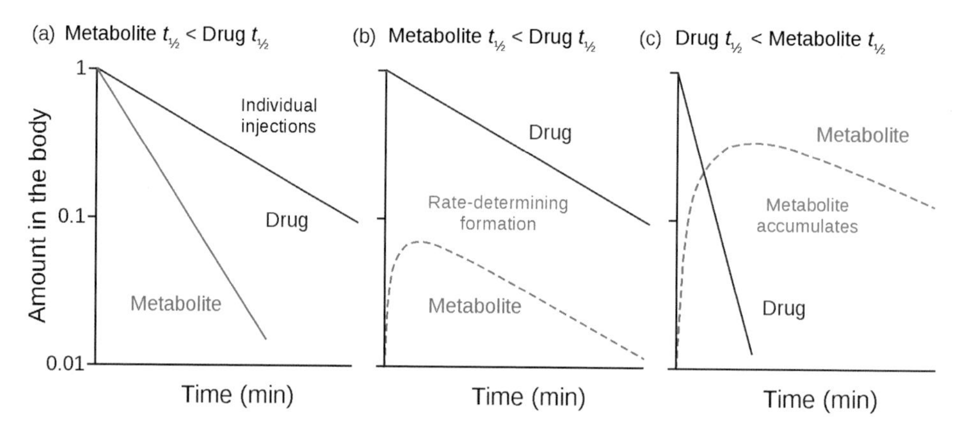

Figure 6.1 (a) Amount of drug and metabolite in the body (as a fraction of dose given) when the half-life of the metabolite is shorter than that of the drug following an i.v. dose of each. (b) Same drug and metabolite as in (a), but when only the parent drug is injected. (c) Situation when the metabolite has a longer half-life than the parent drug.

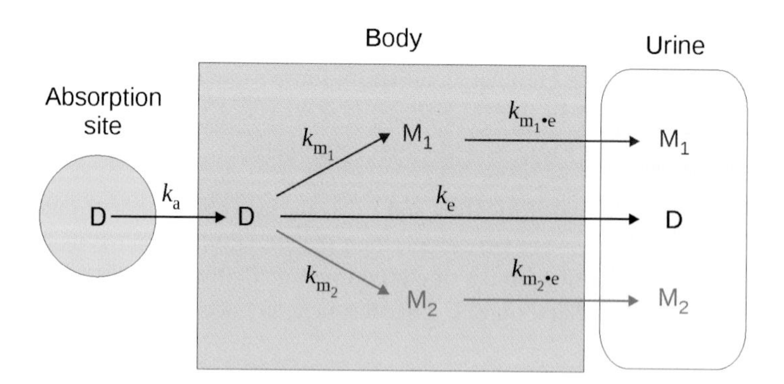

Figure 6.2 A model of metabolism where two metabolites, M_1 and M_2, are produced in parallel. Unchanged drug, D, and the metabolites are excreted into the urine.

The situation with regard to the metabolite concentrations is analogous to 'flip-flop' (Section 4.4.4) when, without further information, it is not possible to assign the rate constants. The kinetics of such metabolites are referred to as formation rate limited. When k_m is the larger rate constant, $\exp(-k_m t)$ approaches zero more quickly and so the slope of the terminal phase for the metabolite is $-k_{m \cdot z}$ (Figure 6.1(c)). These cases may be referred to as elimination dependent.

The examples of Figure 6.1 consider the *amount* of the substances in the body, and when formation is rate determining metabolite is removed almost as soon as it has been formed and so the amount of metabolite (A_m) at any time is less than that of the drug (Figure 6.1(b)). However, it is usual to measure the *concentrations* of metabolite, and these will be influenced by their apparent volumes of distribution, $C_m = A_m / V_m$. Increased polarity and, possibly, increased plasma protein binding, particularly of acidic metabolites to albumin, reduces the apparent volume of distribution relative to that of the parent drug. Consequently, the *plasma concentrations* of a metabolite can be higher than those of the drug, even if the *amounts* in the body are less.

6.2.2 Fraction of metabolite formed

The rate of change of amount of metabolite in the body, dA_m/dt, at any time will be the difference in the rates of formation and elimination. The rates are given by the plasma concentration multiplied by clearance (Equation 4.3). Thus:

$$\frac{dA_m}{dt} = C \times CL_f - C_m \times CL_m \tag{6.2}$$

where CL_f is the clearance associated with the formation of the metabolite. Integration of Equation 6.2 gives:

$$\frac{AUC_m}{AUC} = \frac{CL_f}{CL_m} \tag{6.3}$$

The amount of metabolite formed, A_m, can therefore be calculated from the area under the plasma concentration–time curve, AUC_m, provided the clearance, CL_m, is known. This will require intravenous administration of the metabolite. By analogy with Equation 4.15:

$$AUC_m = \frac{A_m}{CL_m} \tag{6.4}$$

so

$$A_m = AUC_m \times CL_m \tag{6.5}$$

and the fraction produced is:

$$f_m = \frac{A_m}{D} \tag{6.6}$$

An alternative method when it is not possible to administer metabolite is to use excretion rate data. This approach is only applicable when (i) the rate of formation is rate determining, (ii) all the metabolite is excreted via the urine and (iii) there is no metabolism by the kidney. Under these conditions the rate of metabolite excretion approximately equals the rate of metabolism:

$$\text{rate of renal excretion of metabolite} \approx C \times CL_{\text{f}} \tag{6.7}$$

Measuring the rate of excretion and the plasma drug concentration allows CL_{f} to be calculated and substituted into Equation 6.3 to solve for CL_{m}.

6.2.3 More complex situations

In the majority of cases, drug will have been given extravascularly, probably orally, and more than one metabolite will be formed. This formation may occur in parallel (Figure 6.2), in sequence (Figure 6.3) or a combination of the two (Figure 6.4). Additionally, a proportion of the dose may be excreted unchanged.

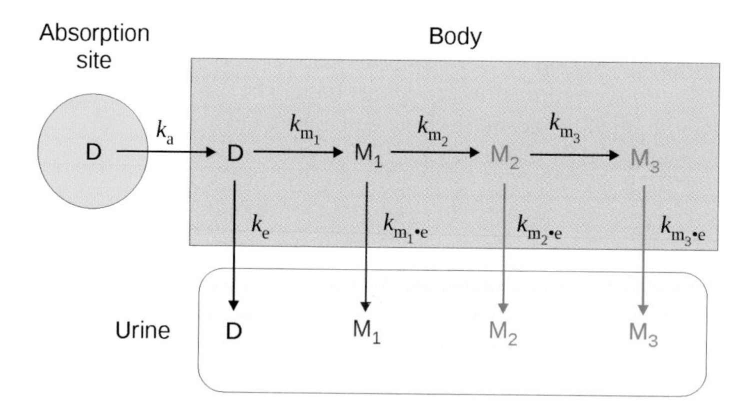

Figure 6.3 A model of excretion where the metabolites M_1, M_2 and M_3 are produced sequentially. Unchanged drug, D, and metabolites are excreted into urine.

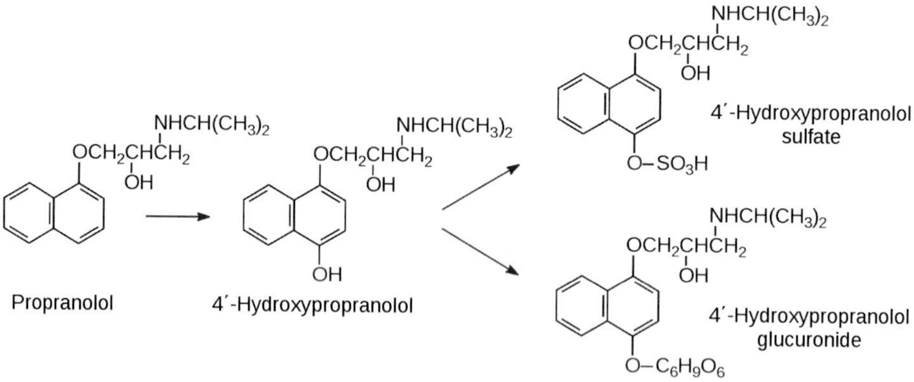

Propranolol → 4′-Hydroxypropranolol → 4′-Hydroxypropranolol sulfate; 4′-Hydroxypropranolol glucuronide

Figure 6.4 Propranolol is oxidized to 4′-hydroxypropranolol, which is conjugated to either the sulfate or glucuronide metabolite. Each of the three pathways is formation rate limited.

In the case of a single metabolite that is excreted into the urine, the equation relating the amount of metabolite in the body after an i.v. bolus injection of D is:

$$A_m = D\frac{k_m}{(k_{m\cdot e} - \lambda)}\left[\exp(-\lambda t) - \exp(-k_{m\cdot e}t)\right] \tag{6.8}$$

where $k_{m\cdot e}$ is the rate constant for urinary excretion of the metabolite and the elimination rate constant of the drug is $\lambda = k_m + k_e$. The equation for the amount of metabolite excreted into the urine is:

$$A_{m,urine} = D\frac{k_m}{\lambda}\left\{1 - \frac{1}{(k_{m\cdot e} - \lambda)}\left[k_{m\cdot e}\exp(-\lambda t) - \lambda\exp(-k_{m\cdot e}t)\right]\right\} \tag{6.9}$$

In the case depicted in Figure 6.3 any of the steps may be the rate-determining one, and so the plasma concentrations of any metabolites formed after the rate-determining step will decline in parallel. In either case, if the rate of absorption is the slowest step then the drug and all the metabolites formed from it will decline with a half-life equivalent to that of absorption.

Complex situations arise when a metabolite is further metabolized by more than one route. This is the case with 4′-hydroxypropranolol. The formation of the 4′-hydroxy metabolite of propranolol is rate limited, and so the propranolol and metabolite concentrations fall in parallel (as shown in Figure 6.6). The major routes of metabolism of 4′-hydroxypropranolol are sulfation or glucuronidation (Figure 6.4).

Figure 6.5(a) shows the time course of the sulfate and glucuronide metabolites after administration of 4′-hydroxypropranolol. The half-lives of the conjugates appear to be

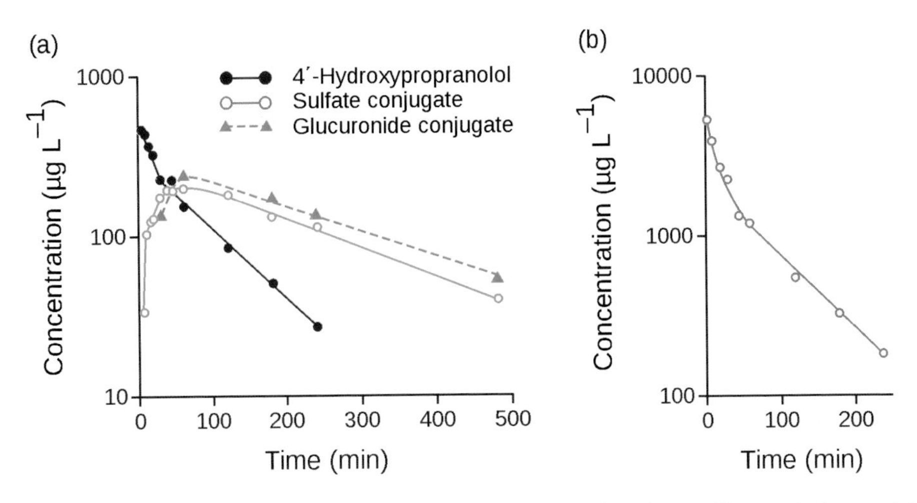

Figure 6.5 Plasma concentrations of 4′-hydroxypropranolol and its sulfate and glucuronide conjugates (a) after i.v. injection of 4′-hydroxypropranolol, 2 mg kg^{-1} in dogs. The kinetics of the conjugates are formation rate limited, but because 4′-hydroxypropranolol is metabolized by two routes in parallel its half-life is less. (b) Plasma concentrations of 4′-hydroxypropranolol sulfate after i.v. injection of 2 mg kg^{-1} in dogs. Adapted with permission from Christ et al. *(1990).*

greater than that of 4′-hydroxypropranolol even though these metabolites would be expected to be more readily excreted. In fact the longer elimination half-lives arise because the kinetics are formation rate limited and the decline in plasma concentrations is controlled by the relatively slow rates of formation. That the apparent elimination half-life of sulfate was formation rate limited was confirmed by administering the conjugate on a separate occasion, when the terminal half-life was 82 ± 12 min compared to 156 ± 21 min when it was a metabolite of 4′-hydroxypropranolol. Note that with this example the conjugates do not decline in parallel with the parent compound because 4′-hydroxypropranolol is metabolized by the two routes and so has a shorter half-life (Figure 6.5(a)).

6.2.4 Effect of pre-systemic metabolism

For a metabolite that displays formation rate-limited disposition it would be expected that when the drug is taken orally, drug and metabolite peak plasma concentrations would occur at the same time and then decline in parallel. However, when the drug undergoes extensive pre-systemic metabolism, both drug and metabolite appear in the systemic circulation together. This is like taking a mixture of drug and metabolite, and the initial concentrations of metabolite in the plasma are not dependent on rate-limited formation from the parent drug. The metabolite introduced into the systemic circulation as a result of first-pass metabolism declines according to the disposition kinetics of the metabolite, that is, as if the metabolite had been administered rather than the drug. Additional metabolite that is formed from the parent drug once it has been absorbed is eliminated according to formation rate-limited kinetics. This results in a bi-exponential decline in metabolite concentrations after oral administration of the parent drug. It is important not to confuse this bi-exponential decay with that seen for two-compartment models. This phenomenon is illustrated by Figure 6.6, which shows the time course of propranolol and its 4′-hydroxy metabolite after oral ingestion. The concentration of the metabolite declines rapidly at

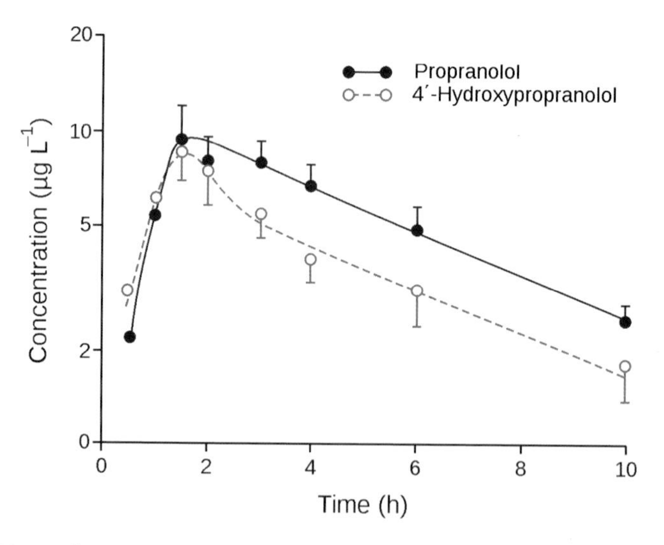

Figure 6.6 Mean plasma concentrations of propranolol and its 4′-hydroxy metabolite after oral administration of propranolol (20 mg) to six normal volunteers. Error bars represent mean ± SEM. Redrawn with permission from Walle et al. *(1980).*

first and then declines in parallel with the parent drug, as would be expected for a formation rate-limited metabolite, as described earlier.

When first-pass metabolism occurs primarily in the liver then the AUC_m values after oral and intravenous doses should be similar, provided, of course, that all the dose of drug is absorbed. This is because the drug can enter the liver whether it is given orally or by injection. Consequently, for a drug with low oral bioavailability, the areas under the metabolite plasma concentration–time curves can be used to differentiate poor absorption from extensive hepatic first-pass metabolism. If the AUC_m is higher after an oral dose than when the drug is injected, this is indicative that pre-systemic metabolism is occurring in the GI tract.

6.2.5 Interconversion of drug and metabolite

Some metabolites may be converted back to the parent drug:

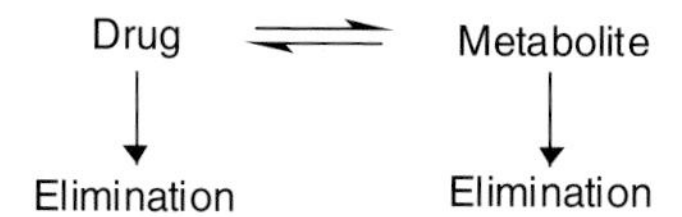

The drug and metabolite may also be eliminated via other routes, either further metabolism or excretion. This model is analogous to a two-compartment model that has loss from both the central and peripheral compartments. Initially drug concentrations fall rapidly but as drug and metabolite concentrations equilibrate interconversion has a major influence on the half-life of the drug and metabolite. Several drugs exhibit this phenomenon, including cortisol–cortisone, haloperidol–reduced haloperidol, prednisone–prednisolone and vitamin K–vitamin K epoxide. The effect of interconversion has been nicely demonstrated for prednisone and prednisolone (Figure 6.7).

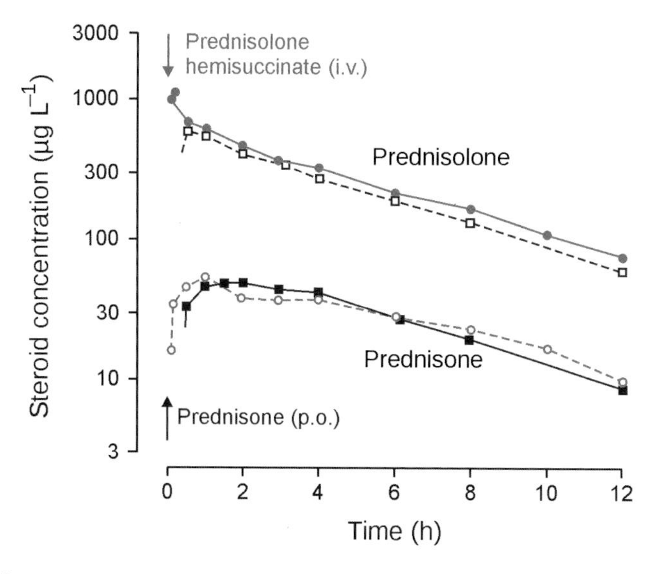

Figure 6.7 Plasma concentrations of prednisone and prednisolone after a single oral dose of prednisone (red) or an i.v. dose of prednisolone hemisuccinate ester (blue) to a healthy male volunteer. Adapted from Rose et al. *(1980), with permission.*

After administration of prednisone, the concentrations of the active metabolite, prednisolone, rose to over 10 times those of prednisone. When prednisolone hemisuccinate (a prodrug ester of prednisolone, which is very rapidly hydrolysed to prednisolone) was given intravenously the prednisolone concentrations fell rapidly until the prednisolone:prednisone ratio equilibrated at the same value as that after administration of prednisone, demonstrating that interconversion produces the same drug to metabolite ratio irrespective of which drug is administered.

In the case of sulindac, the sulphide metabolite, which is oxidized back to the parent drug, is active whereas the sulfone metabolite is not interconverted and is inactive (Figure 6.8). Thus, sulindac is a prodrug, and it has been suggested that this is why it is less prone to cause GI upsets compared to some other non-steroidal anti-inflammatory drugs as the active form is not produced until after absorption.

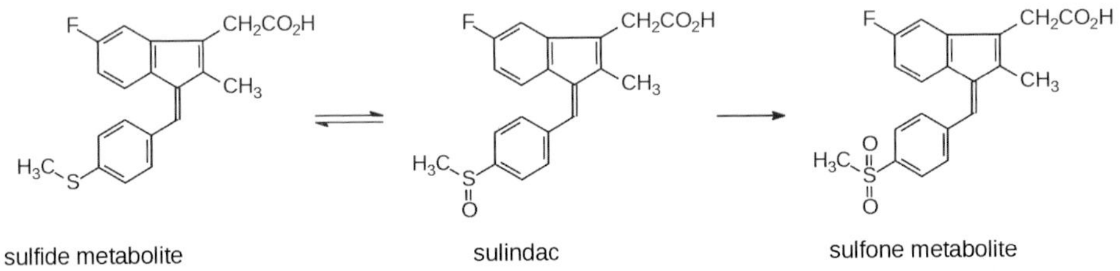

sulfide metabolite ⇌ sulindac → sulfone metabolite

Figure 6.8 Sulindac (a sulfoxide) is reversibly reduced to pharmacologically active sulfide or oxidized to inactive sulfone metabolite.

6.2.6 Active metabolites

The significance of an active metabolite will depend on whether the elimination kinetics of the metabolite are formation dependant or elimination dependant. In the former case the half-life of the metabolite will be the same as that of the parent drug and so it will accumulate at the same rate as the parent compound. Dosing can therefore be based on the disposition parameters for the drug and it is not necessary to be concerned with the kinetic parameters of the metabolite. This even applies to prodrugs, where the metabolite is the active species, although in practice a prodrug is likely to have a very short elimination half-life and the rate of formation of the metabolite will not be rate limiting. Under these circumstances the metabolite will be monitored and the dosing based on the kinetics of the metabolite.

When the kinetics are elimination dependant, the half-life of the metabolite is greater than the drug and so the metabolite continues to accumulate after the drug has reached steady-state conditions, for example during a continuous infusion or on multiple dosing. Because the metabolite takes longer to reach steady state, dosing should be determined by the disposition characteristics of the metabolite. The average concentration of metabolite at steady state can be calculated from the AUC_m after a single i.v. dose:

$$C_{av}^{ss} = \frac{AUC_m}{\tau} \tag{6.10}$$

where τ is the dosage interval (cf. Equation 4.42). Furthermore, the frequency of dosing should be based on the half-life of the metabolite, that is, it is not necessary to give the drug as frequently as the half-life of the parent drug might suggest.

The *N*-desmethyl metabolites of several centrally acting drugs, including imipramine and amitriptyline, show pharmacological activity similar to that of the parent drug and tend to accumulate to higher concentrations than the parent drug on multiple dosing. Nordazepam is a metabolite of several 7-chlorobenzodiazepine drugs and, because of its longer elimination half-life, takes longer to reach steady state than diazepam. Furthermore, and in keeping with what has been discussed in previous chapters, there is less fluctuation in peak to trough concentration during multiple dosing (Figure 6.9). In humans steady-state concentrations of nordazepam are approximately twice those of diazepam after repeated administration of the latter.

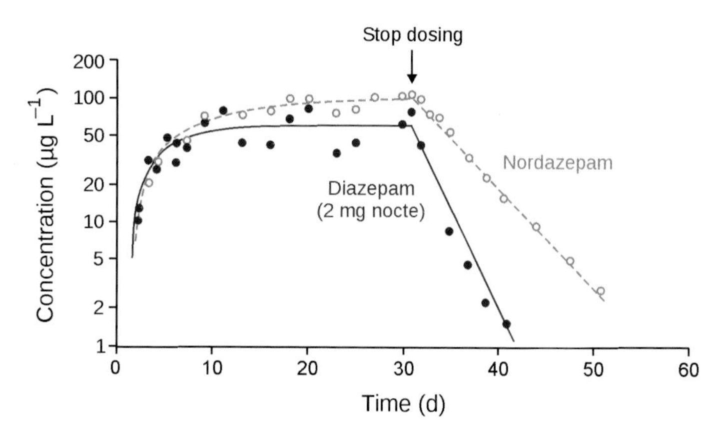

Figure 6.9 Plasma diazepam (red) and nordazepam (blue) concentrations in a healthy female volunteer after 2 mg diazepam at night for 30 days. Redrawn with permission from Abernethy et al. *(1983).*

Drug metabolites are not usually licenced for human investigations of their kinetics. However, there are examples of active metabolites that have been identified, developed and approved for medical use. These drugs allow comparison of the kinetics of the metabolite when it is produced from the parent drug and when it is administered directly. Table 6.1 is a compilation of pharmacokinetic data for compounds related in this way. For the pairs, carbamazepine and its epoxide metabolite and codeine and morphine, the half-life of the metabolite administered in its own right is shorter than that of the parent drug. These examples show data of the type in Figure. 6.1(a) after administration of the two drugs separately, but of the type in Figure 6.1(b) after administration of the parent drugs. For caffeine and theophylline, amitriptyline and nortriptyline, and imipramine and desipramine, in each pair the half-life of the metabolite is longer than that of the parent drug. These examples show data of the type in Figure 6.1(c) after administration of the parent drugs.

The other data in Table 6.1 are for four benzodiazepines that are related to each other in various ways as precursors and metabolites (Figure 3.7). In keeping with the other examples in the table, *N*-desmethylation appears to cause a lengthening of the half-life. This is largely because of a reduced clearance of nordazepam compared with that of

Table 6.1 *A compilation of pharmacokinetic data for examples of drug pairs used in patients where there is a relation within each pair of precursor and metabolite*

Drug example	Protein binding (%)	Half-life (h)	*V* (L kg^{-1})	*CL* (ml min^{-1} kg^{-1})	Urinary excretion (%)	Metabolic reaction
Carbamazepine	74	15 (36)*	1.4	1.3 (0.36)*	<1	
Carbamazepine-10, 11-epoxide	50	7.4	1.1	1.7†	<1	Epoxide formation
Codeine	7	2.9	2.6	11†	0	
Morphine	35	1.9	3.3	24	4–14	*O*-desmethylation
Caffeine	36	4.9	0.61	1.4	1.1	
Theophylline	56	9.0	0.5	0.65	18.0	*N*-desmethylation
Amitriptyline	94.8	21 (19.5)‡	15	11.5	<2	
Nortriptyline	92	31 (41.5)‡	18.4	7.2	2	*N*-desmethylation
Imipramine	90.1	12	18	15	<2	*N*-desmethylation
Desipramine	82	22	20	10	2	
Benzodiazepines						
1. Diazepam	98.7	43	1.1	0.38	<1	See Figure 3.7
2. Nordazepam	97.5	73	0.78	0.14	<1	
3. Temazepam	97.6	11	0.95	1.0	<1	
4. Oxazepam	98.8	8	0.6	1.05	<1	

All data from Goodman and Gilman (see Hardman *et al.*, 1996), and so represent interpretations of multiple publications except data in parentheses for amitriptyline.
* Data in parentheses from single doses – other data from long-term treatment (carbamazepine is a self-inducer).
† Data from *CL/F*; clearance data for carbamazepine metabolite from renal clearance.
‡ Data in parentheses from amitriptyline and nortriptyline measured after amitriptyline doses (Curry *et al.*, 1988).

the parent compound. The apparent volume is less, as would be expected for a more polar metabolite. Diazepam may be cleared more rapidly because it is metabolized in parallel to nordazepam and temazepam. The 3-hydroxy metabolites, temazepam and oxazepam, have much shorter half-lives; they are glucuronidated and have the highest clearance values, suggesting that conjugation is relatively rapid compared to phase 1 oxidation reactions. Oxazepam has the lowest apparent volume of distribution of the four benzodiazepines.

In the 1960s and 1970s when the benzodiazepines were being developed, nordazepam was not licenced for human use. About this time, clorazepate, a prodrug which spontaneously decarboxylates to nordazepam in gastric acid, was introduced. If the half-life of nordazepam produced from diazepam is elimination dependent then the nordazepam rapidly produced from clorazepate should have the same half-life. This principle is illustrated in Figure 6.10. As predicted, the half-life of the nordazepam was the same irrespective of which precursor drug was given. The amount of nordazepam produced following i.p. injection of clorazepate was markedly reduced compared with oral administration, illustrating the role of gastric acid in the formation of the active moiety.

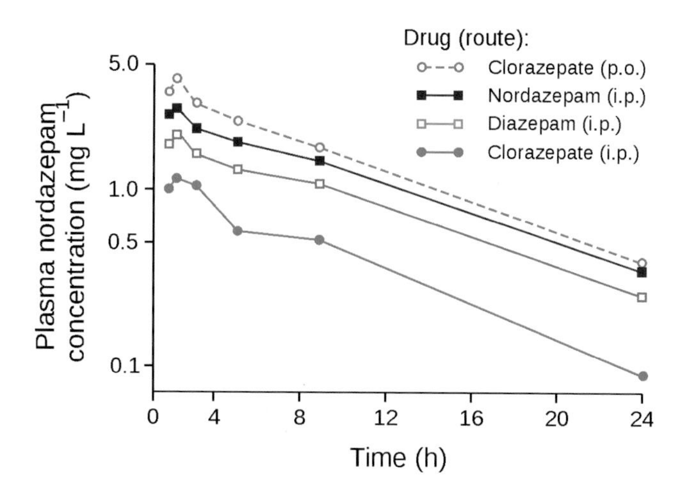

Figure 6.10 Plasma nordazepam concentrations after dosing with nordazepam ($3.0\,mg\,kg^{-1}$), diazepam ($3.0\,mg\,kg^{-1}$) and clorazepate ($4.5\,mg\,kg^{-1}$).

6.3 Renal excretion

The factors contributing to urinary excretion of a compound are:

- glomerular filtration
- passive reabsorption from tubular fluid
- secretion into renal tubular.

These were discussed in Section 3.3.1.

6.3.1 Kinetics of urinary excretion

Glomerular filtration and diffusion across the renal tubular epithelium are generally first-order processes so that the rate of transfer of drug is related to the amount of drug in plasma. The excretion of a drug in urine is complicated by further factors:

- The concentration of drug in renal tubular fluid is influenced by changes in urine volume so there is a tendency for the concentration to increase as water is reabsorbed by the kidney.
- Changes in urinary pH can influence the rates at which weak electrolytes are excreted.

It is therefore usual to relate the *rate* of renal elimination to the concentration of drug in plasma. The differential equation for the rate of appearance of drug in urine is:

$$\frac{dAe}{dt} = k_e A \tag{6.11}$$

where Ae is the amount of drug excreted in urine at time t and k_e is the first-order rate constant for urinary elimination. For an i.v. dose, D, A is given by:

$$A = D\exp(-\lambda t) \tag{6.12}$$

so

$$\frac{\mathrm{dAe}}{\mathrm{d}t} = k_{\mathrm{e}} D \exp(-\lambda t) \tag{6.13}$$

Integrating Equation 6.13 gives:

$$\mathrm{Ae} = \frac{k_{\mathrm{e}}}{\lambda} D\left[1 - \exp(-\lambda t)\right] \tag{6.14}$$

The total amount of drug excreted in the urine, $Ae(\infty)$ is:

$$\mathrm{Ae}(\infty) = \frac{k_{\mathrm{e}}}{\lambda} D \tag{6.15}$$

that is, the fraction of an intravenous dose that is eventually excreted into the urine is given by the ratio of the rate constants for urinary excretion and elimination.

A plot of ln(*excretion rate*) versus time should give a straight line of slope $-\lambda$ (Equation 6.13). In practice the excretion rate is calculated from urine samples collected at discrete intervals and the data are either plotted as a histogram or the mid-points of the collection period are used. This approach assumes that there are no changes in the rate of renal excretion as a result of fluctuations in urinary pH, urine volume or unknown factors. To reduce the effects of fluctuations seen in excretion rate plots, an alternative approach is the 'sigma-minus' method. Urine is collected for sufficiently long to allow estimation of $Ae(\infty)$ – this should be for up to ~7 elimination half-lives. Substituting Equation 6.15 into Equation 6.14 and rearranging gives:

$$\mathrm{Ae}(\infty) - \mathrm{Ae} = Ae(\infty)\exp(-\lambda t) \tag{6.16}$$

A semilogarithmic plot of percentage of drug remaining to be excreted against time is a straight line of slope $-\lambda$.

Similar approaches can be used to derive equations for urinary excretion following first-order absorption into a single-compartment model, elimination from a two-compartment model and non-linear kinetics.

6.3.1.1 Renal clearance

The rate of elimination of a drug is the systemic clearance multiplied by the plasma concentration (Equation 4.3). The urinary excretion rate is the renal clearance, CL_{R}, multiplied by the plasma concentration, so rearranging and substituting Equation 6.11:

$$CL_{\mathrm{R}} = \frac{\mathrm{dAe}/\mathrm{d}t}{C} = \frac{k_{\mathrm{e}} A}{C} \tag{6.17}$$

Note that the average urinary excretion rate over a collection period ($t_1 - t_2$) is the amount in the urine sample (concentration × volume) divided by the collection time:

$$\frac{dAe}{dt} \approx \frac{C_{urine} V_{urine}}{(t_2 - t_1)} \approx C_{urine} \times \text{urine flow rate} \tag{6.18}$$

Substitution of Equation 6.18 into Equation 6.17 yields the familiar the physiologists' definition of renal clearance (Equation 3.1):

$$CL_R = \frac{U}{P} \times \text{urine flow rate} \tag{3.1}$$

demonstrating that extrarenal clearance may be calculated from the difference between systemic clearance, CL and CL_R, as discussed previously.

Because A/C is the apparent volume of distribution, Equation 6.17 can be rewritten:

$$CL_R = k_e V \tag{6.19}$$

Equation 6.19 gives a route to determining k_e, provided that the apparent volume of distribution of the drug is known (from an intravenous injection) and the renal clearance is measured using Equation 3.1.

6.3.1.2 *Effect of urine flow rate*

The effect of urine flow rate will depend on whether the plasma and urine concentrations have equilibrated. At equilibrium, U/P (Equation 3.1) will be constant and renal clearance will be directly proportional to urine flow rate. Furthermore, for a neutral molecule that binds to plasma protein the urine concentration and the unbound concentration of drug in plasma will be equal at equilibrium so renal clearance will be:

$$CL_R = f_u \times \text{urine flow rate} \tag{6.20}$$

where f_u is the fraction of unbound drug in plasma. An alternative way of visualizing the effect of flow rate is that if the urine and plasma concentrations have more or less equilibrated then by whatever proportion the urine flow rate is increased then the *amount* of drug in urine must increase by the same proportion to maintain the required urine concentration. Ethanol, which is not bound to plasma proteins, does not ionize and usually equilibrates rapidly, shows a nearly linear relationship between renal clearance and urine flow rate with a slope close to 1 (Figure 6.11(a)). Correlations between urine flow and clearance have been shown for other drugs, including phenobarbital, sulfafurazole and glutethimide (Figure 6.11(b)). For glutethimide, the clearance is less than the urine flow rate, which in part can be explained by the fact that drug is approximately 50% bound to plasma proteins. Urine flow will not have much influence on the clearance of drugs for which there is little renal tubular reabsorption.

Despite the increased clearance of some drugs with increased urine flow, forced acidic or alkaline diuresis is no longer employed for treating drug overdose because the changed fluid balance is considered potentially dangerous. Sodium bicarbonate may be used to increase the renal clearance of salicylate and chlorophenoxyacetic acid herbicides, such as 2,4-dichlorophenoxyacetic acid (2,4-D) and 2,3,5-trichlorophenoxyacetic acid (2,3,5-T), but an additional diuretic is not used. Several drugs, including amfetamine, may cause rhabdomyolysis, a life-threatening condition in which breakdown of striated skeletal muscle leads to myoglobinuria with subsequent kidney failure. Acidification of urine exacerbates the

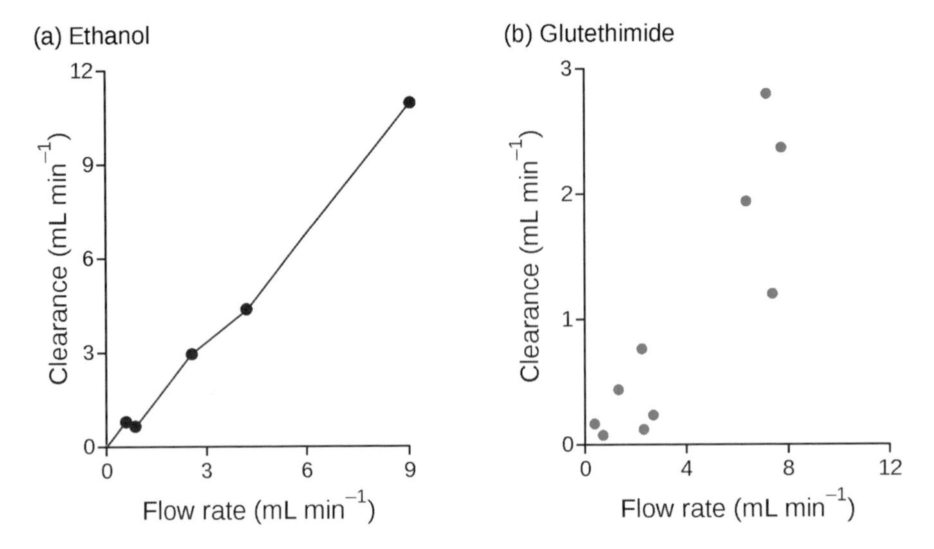

Figure 6.11 Renal clearance as a function of urine flow rate: (a) ethanol, data are five independent estimates in one subject, and (b) glutethimide, data are 10 independent estimates.

condition, indeed bicarbonate infusions may be used in the treatment of rhabdomyolysis. Consequently, despite the fact that studies have shown that increasing urine flow and reducing urine pH increases renal excretion of amfetamine, forced acid diuresis is not used to treat overdose. In fact most of a dose of amfetamine is eliminated via metabolism.

6.3.2 Specific drug examples

6.3.2.1 Ethanol

Ethanol distributes freely with total body water (Section 2.4.1.1). In spite of wide variations in the volume of urine produced, the concentration of ethanol in urine in the elimination phase is closely related to its concentration in blood (Figure 6.12). In the case of ethanol, diffusion of the drug between urine and blood apparently occurs up to a time not long before the urine is removed from the body. The near constant ratio of blood to urine concentrations allowed urine to be used as an alternative to blood for the purposes of drink-driving laws, 1.07 $g L^{-1}$ in urine being equivalent to 0.8 $g L^{-1}$ (0.08%) in blood. However, blood to breath ratios are considered more consistent and breath alcohol concentrations (BrAC) are preferred for medico-legal cases.

The data of Figure 6.12 show much more than the near-constant ratio (1.3–1.5) for urine to blood concentrations; the diuretic effect of ethanol is clearly seen, as is the change in the order of elimination as a function of blood concentration. From the peak concentration (~1.2 $g L^{-1}$) to about 0.4 $g L^{-1}$ the decline is approximately linear, after which the slope decreases as the kinetics increasingly approach first order. Of course the whole of the curve can be described using Michaelis–Menten kinetics.

6.3.2.2 Fluphenazine

Fluphenazine is usually administered as an i.m. depot injection of its decanoate or enanthate ester (Section 4.4.4), prodrugs which are hydrolysed to fluphenazine and released into the

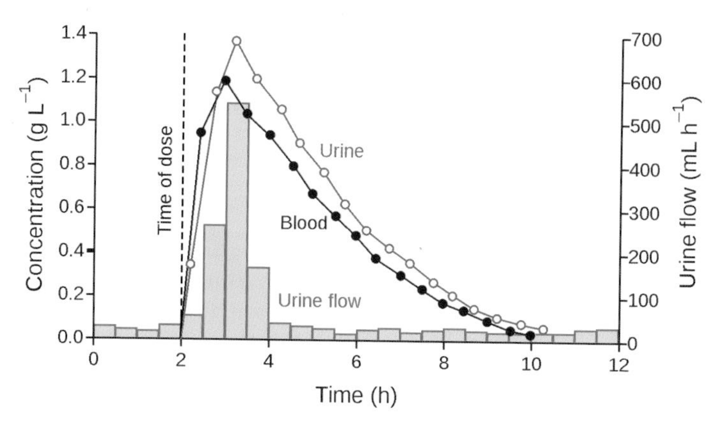

Figure 6.12 Concentrations of ethanol in whole blood (•) and urine (○) after an oral dose of 64 g of ethanol with 116 mL of water. Note that the urine concentrations are plotted at the midpoint of the collection interval. Redrawn with permission from Haggard et al. (1941).

plasma. Before the advent of sufficiently sensitive assays, fluphenazine kinetics in plasma could only be studied by administration of radiolabelled drug. However, urinary excretion rates could be used to define the kinetics. In humans, fluphenazine is present in urine as fluphenazine plus conjugates that can be released by hydrolysis with β-glucuronidase. Using ^{14}C-labelled drug the relationship between the plasma concentration and urinary excretion rate of fluphenazine plus conjugates was demonstrated (Figure 6.13(a)). Having established this relationship, the kinetics of fluphenazine after intramuscular injection of the decanoate ester (50 mg weekly) to a psychiatric inpatient was investigated by following the renal excretion of either the drug or drug plus conjugated metabolites (Figure 6.13(b)). The excretion rates of conjugated metabolites parallel those of fluphenazine, as would be expected for a

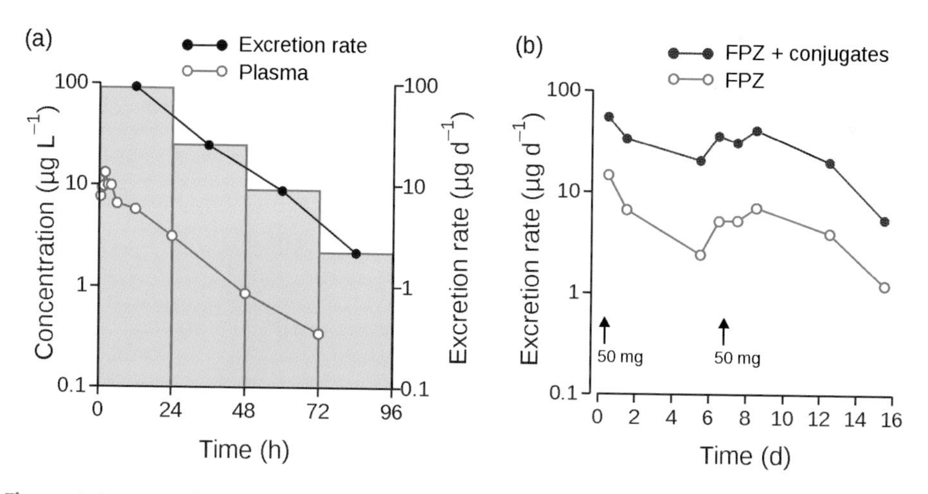

Figure 6.13 (a) Plasma concentrations and urinary excretion rate of fluphenazine after an intramuscular injection of 25 mg. (b) Urinary excretion rate of fluphenazine and conjugated fluphenazine in a patient receiving 50 mg fluphenazine decanoate (i.m.) weekly. Redrawn with permission from Whelpton & Curry (1976).

metabolite whose disposition kinetics are formation rate limited. This is an example where the parent drug was a prodrug and the kinetics were studied by monitoring the urinary excretion of a metabolite of the pharmacologically active compound, fluphenazine.

6.4 Excretion in faeces

Drugs and their metabolites that are excreted via the bile are either reabsorbed or remain in the GI tract to be removed in the faeces, as explained in Section 3.3.2. Although it is rarely done, it is possible to determine drug kinetics from faecal excretion data, as exemplified below for the photodynamic agent temoporfin. The same arguments as those made for renal excretion apply. Temoporfin was particularly suitable because (i) it is administered by intravenous injection so any drug or metabolite in the faeces is there because it has been excreted, (ii) over 99% is eliminated in faeces and (iii) it has a long elimination half-life. A study with non-radiolabelled drug indicated that the terminal half-life in BALB/c mice was at least 13.9 h, but because it was possible to measure the drug in blood for only 48 h this was considered to be an underestimate (Whelpton *et al.*, 1995). The slow decline in temoporfin concentrations in some tissues, notably lung and kidney, which were monitored for 96 h, confirmed this view. When ^{14}C-temoporfin became available the study was repeated and the drug monitored for up to 35 days. Faecal excretion rates are shown in Figure 6.14. The decay was bi-exponential.

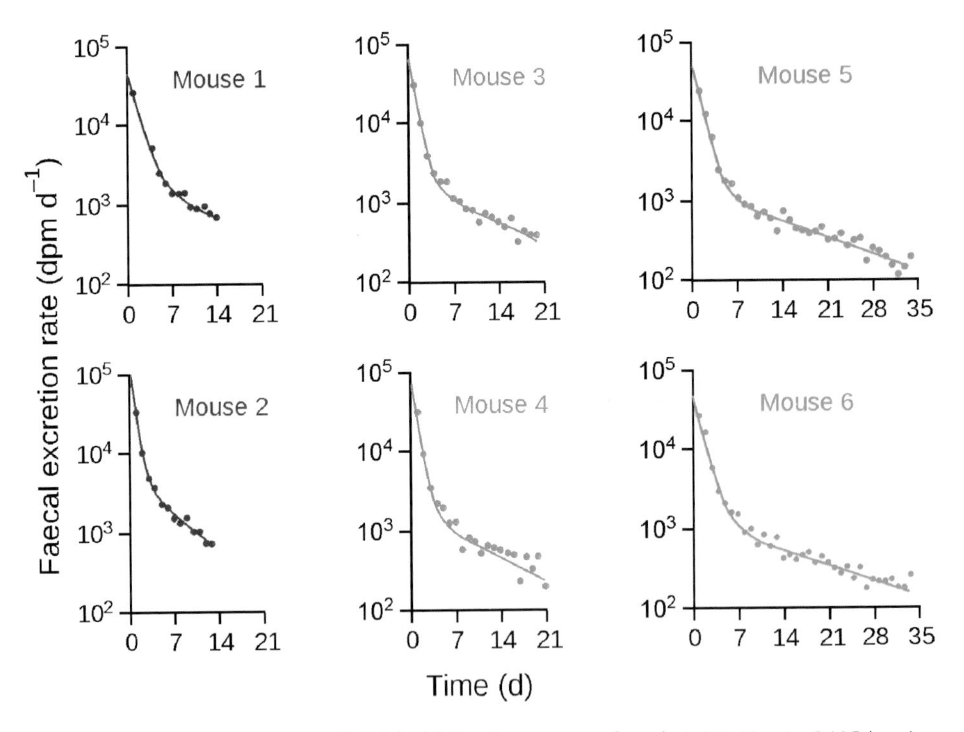

Figure 6.14 Faecal excretion of ^{14}C-label following temoporfin administration to BALB/c mice; solid lines are the least squares fit of the data to a two-compartment model. Tissues samples were assayed after 2, 3 and 5 weeks. Redrawn with permission from Whelpton et al. *(1996).*

Table 6.2 Elimination half-lives derived from the faecal excretion data of Figure 6.14

Mouse	Duration 0–14 days*		Duration 0–21 days		Duration 0–28 days	
	$t_{½}$ (λ_1) (h)	$t_{½}$ (λ_2) (d)	$t_{½}$ (λ_1) (h)	$t_{½}$ (λ_2) (d)	$t_{½}$ (λ_1) (h)	$t_{½}$ (λ_2) (d)
1	22.7	9.0				
2	13.0	5.0				
3	15.4	8.7	14.6	7.3		
4	14.9	6.2	15.4	7.5		
5	18.0	6.5	19.9	10.5	20.2	11.0
6	18.2	5.8	21.2	10.2	21.1	10.0
Mean	17.0	6.9	17.8	8.9	20.7	10.5

*Time over which kinetic parameters were calculated.

A putative metabolite was detected in liver and faeces but not in blood or any other tissue. This is in keeping with a metabolite that is produced in, and excreted by, the liver, probably via bile. The metabolite concentrations declined in parallel with those of temoporfin, indicating that its disposition was formation rate limited. Thus, it was possible to define the disposition of temoporfin by monitoring the rates of faecal excretion for up to 7 weeks.

The data of Figure 6.14 illustrate another important point – the issue of the quality of the data (Section 5.4.3). The study was designed for tissue samples to be assayed after 2, 3 and 5 weeks, which determined the durations of the excretion studies. However, it is necessary to ensure that the data are collected for sufficient time to derive a reliable estimate of the half-life of the terminal phase. Failure to do so will result in an underestimate (Table 6.2). With an elimination half-life in excess of 10 days, measuring for only 14 days was clearly inadequate.

Summary

This chapter has explored the factors that determine the time course of a metabolite relative to that of the parent drug, including the effect of pre-systemic metabolism. The accumulation of metabolites that have longer elimination half-lives than the parent drug and appropriate dosing when such metabolites are pharmacologically active have been considered. The use of excretion data to define the kinetics of drug has been exemplified.

6.5 Further reading

Houston JB, Taylor G. Drug metabolite concentration-time profiles: influence of route of drug administration. *Br J Clin Pharmacol* 1984; 17: 385–94.

http://www.fda.gov/downloads/Drugs/GuidanceComplianceRegulatoryInformation/Guidances/ucm079266.pdf, accessed 14 March 2016.

6.6 References

Abernethy DR, Greenblatt DJ, Divoll M, Shader RI. Prolonged accumulation of diazepam in obesity. *J Clin Pharmacol* 1983; 23: 369–76.

Christ DD, Walle UK, Oatis JE, Jr., Walle T. Pharmacokinetics and metabolism of the pharmacologically active 4'-hydroxylated metabolite of propranolol in the dog. *Drug Metab Dispos* 1990; 18: 1–4.

Curry SH, DeVane CL, Wolfe MM. Hypotension and bradycardia induced by amitriptyline in healthy volunteers. *Human Psychopharmacology* 1988; 3: 47–52.

Gold H, Degraff AC. Studies on digitalis in abulatory cardic patients: II. The Elimination of Digitalis in Man. *J Clin Invest* 1929; 6: 613–26.

Haggard HW, Greenberg LA, Carroll RP. Studies in the absorption, distribution and elimination of alcohol: VIII. The diruesis from alcohol and its influence on the elimination of alcohol in the urine. *J Pharmacol Exp Ther* 1941; 71: 349–57.

Hardman JG, Limbird LE, Molinoff PB, Ruddon RW, Gilman AG (editors). *Goodman and Gilman's The Pharmacological Basis of Therapeutics*. 9th edition. New York: McGraw-Hill, 1996.

Rose JQ, Yurchak AM, Nsko WJ, Powell D. Bioavailability and disposition of prednisone and prednisolone from prednisone tablets. *Biopharmaceutics & Drug Disposition* 1980; 1: 247–58.

Walle T, Conradi EC, Walle UK, Fagan TC, Gaffney TE. 4-Hydroxypropranolol and its glucuronide after single and long-term doses of propranolol. *Clin Pharmacol Ther* 1980; 27: 22–31.

Whelpton R, Curry SH. Methods for study of fluphenazine kinetics in man. *J Pharm Pharmacol* 1976; 28: 869–73.

Whelpton R, Michael-Titus AT, Basra SS, Grahn M. Distribution of temoporfin, a new photosensitizer for the photodynamic therapy of cancer, in a murine tumor model. *Photochem Photobiol* 1995; 61: 397–401.

Whelpton R, Michael-Titus AT, Jamdar RP, Abdillahi K, Grahn MF. Distribution and excretion of radiolabeled temoporfin in a murine tumor model. *Photochem Photobiol* 1996; 63: 885–91.

7

Clearance, Protein Binding and Physiological Modelling

Learning objectives

By the end of this chapter the reader should be able to:

- describe how microsomes and hepatocytes are used to determine *in vitro* values of clearance
- discuss the prediction of *in vivo* pharmacokinetic parameters from data collected from *in vitro* studies
- explain the differences between capacity-limited (restrictive) and flow-limited (non-restrictive) clearance
- discuss the circumstances under which hepatic clearance and elimination half-life are affected by protein binding
- compare compartmental and physiological approaches to modelling.

7.1 Introduction

Clearance has already been defined and discussed in previous chapters, particularly its importance, along with apparent volume of distribution, as a determinant of the elimination half-life of a drug. The concept of extraction, which was introduced with regard to elimination, can also be applied to the uptake of drugs by non-eliminating organs.

Introduction to Drug Disposition and Pharmacokinetics, First Edition. Stephen H. Curry and Robin Whelpton.

Companion website: www.wiley.com/go/curryandwhelpton/IDDP

The term 'clearance' is sometimes used to describe the phenomenon of removal of a drug from the body as a whole, when the term 'elimination' would be better. It should be remembered that clearance always refers to a volume of fluid from which a substance is removed in unit time, and thus will always have units of flow, for example mL min^{-1}, L h^{-1}. Clearance can be used to describe the metabolism of a drug *in vitro,* as well as in *in vivo* systems. Clearance *in vitro*, sometimes referred to as metabolic stability, is important in drug discovery and development. Before one can appreciate how protein binding affects the elimination of drugs, it is necessary to understand how extraction is related to intrinsic clearance and liver blood flow, as well as the influence of binding on drug distribution. Clearance and extraction are fundamental to physiologically based pharmacokinetic (PBPK) modelling. These topics are discussed in this chapter.

7.2 Clearance

The study of renal clearance dates from the 1930s, when pioneering renal physiologists discovered that kidney function could be assessed in terms of the removal of substances from the blood in the renal artery (Section 3.3.1.4). Pharmacokineticists have extended this to embrace all processes of drug elimination, including *in vitro* investigations with microsomes and hepatocytes.

7.2.1 Clearance studies *in vitro*

Studying the elimination of drugs *in vitro* provides an assessment of the ability of the enzymes involved to remove the drug from the incubation medium in the absence of any delivery (by blood) or availability (e.g. restrictions imposed by protein binding) influences that are usually present *in vivo.* During the screening of potential new drugs (new chemical entities, NCEs) metabolic stability studies may show that a candidate drug is too rapidly metabolized to warrant further development or the *in vitro* results may be used to predict the *in vivo* properties.

7.2.1.1 Microsomal intrinsic clearance

The experimental measurement of microsomal intrinsic clearance, CL_{mic}, *in vitro* involves incubation of drug in a fixed volume of fluid in which is suspended a known quantity of liver microsomes. The decay of drug concentration, C, is monitored using a suitable analytical method. First-order decay is ensured by using an appropriately low drug concentration and an appropriately high microsome concentration. The first-order rate constant, k, is obtained from the slope of a ln (concentration)–time plot and CL_{mic} from:

$$CL_{mic} = V_{inc} k \tag{7.1}$$

where V_{inc} is the volume of the incubation solution. Microsomes are considered to be 100% viable, and so the activity is usually expressed in terms of the microsomal protein concentration: mL min^{-1} mg $protein^{-1}$. The rate of metabolism is $CL_{mic} \times C$ (see Equation 4.3).

At relatively high drug concentrations, when the kinetics are non-linear, Equation 7.1 can be written in terms of the Michaelis–Menten equation:

$$CL_{mic} = \frac{V_{max}}{Km + C} \tag{7.2}$$

which is analogous to Equation 4.45. At very high concentrations, when the enzymes are saturated with drug, the rate of reaction is V_{max}, so:

$$CL_{mic} = V_{max}/C \tag{7.3}$$

which is analogous to the zero-order case, Equation 4.49. Obviously, the microsomes contain liver enzyme systems in which only microsomally catalysed chemical change occurs. However, microsomal reactions include oxidations, reductions, hydrolyses and some phase 2 reactions, so multiple chemical changes can occur. Only by measuring the concentrations of the different products can pure, single-reaction kinetics be studied. This is not commonly done, as pharmacokineticists have, historically, measured disappearance of substrate, rather than appearance of products because interest was primarily in the disappearance of pharmacologically active molecules. Also, until metabolites have been identified, it is not possible to develop assays for them.

7.2.1.2 Hepatocytes

Analogous experiments can be performed using hepatocytes instead of microsomes. The clearance is expressed in terms of the numbers of cells: mL min^{-1} million $cells^{-1}$. Because hepatocytes are not necessarily 100% viable, a viability correction determined in a separate experiment with a compound whose properties are known may be needed. Also, hepatocytes reproduce a somewhat larger collection of metabolic reactions, microsomal and otherwise, so that the result with hepatocytes assesses a somewhat larger collection of product-formation reactions. Again, separate assays of products are needed if the kinetics of individual reactions are to be studied.

When working *in vitro* there is a maximum concentration of hepatocytes that can be incorporated into a suspension without the stirring damaging the cells. Consequently, the suspension of hepatocytes must be relatively dilute. In contrast, microsomal suspensions can contain higher protein concentrations than is the case with the hepatocyte suspension. Experimentally, therefore, drug half-life values are often shorter in the conditions of the microsomal suspensions than in those of the hepatocyte suspensions, in spite of the fact that more reactions can take place in the hepatocyte incubations.

7.2.2 Clearance *in vivo*

The concept of extraction was introduced in Chapter 4 with regard to drug elimination, but a non-eliminating organ can remove drug molecules from the blood passing through it until equilibrium between the tissue and plasma concentrations is reached, after which elimination in the liver and kidney reduces the concentrations in both blood and tissue. In this situation:

$$\begin{aligned} \text{rate of removal from plasma} &= QC_a - QC_v \\ &= Q(C_a - C_v) \end{aligned} \tag{7.4}$$

where Q is blood flow, C_a is the afferent arterial concentration and C_v is the efferent venous concentration (see Figure 4.4). In this situation, the extraction ratio, E, can be viewed as assessing organ uptake:

$$E = \frac{C_a - C_v}{C_a} \tag{7.5}$$

This approach has been used in the search for an understanding of brain uptake in particular, where, in an appropriately designed experiment, carotid artery and jugular vein concentrations can be measured. For an eliminating organ, the organ clearance is the elimination rate divided by C_a, and this provides the basis for assessment of renal clearance in particular (Section 3.3.1.4).

7.2.2.1 Effects of systemic clearance on kinetic parameters

Systemic clearance can be derived from Equation 4.16 as shown previously:

$$CL = \frac{D}{AUC} \tag{4.16}$$

An advantage of this equation is that it is not necessary to define the number of compartments, and it can be applied to intravenous infusions as well as bolus injections. In the case of an infusion *AUC* is the total area under the curve, that is, the *AUC* during the infusion plus the post-infusion area extrapolated to $t=\infty$, and D is the total dose that was administered. A practical issue when applying Equation 4.16 is that sufficient concentration–time points have to be collected, and for sufficient time, to allow an accurate estimate of $AUC_{(0-\infty)}$. The extrapolation should not exceed 10% of the total *AUC*. This is not always possible. However, if it is possible to infuse the drug to steady-state conditions then Equation 4.37 can be rearranged to:

$$CL = \frac{R_0}{C^{ss}} \tag{7.6}$$

so that a single plasma sample taken during the infusion after $5–7 \times t_{½}$ can be used to calculate *CL*. Note that after administration of multiple doses (regardless of route of administration), rearrangement of Equation 4.42 gives:

$$CL = F\frac{D}{C_{av}^{ss}\tau} = F\frac{\text{dosing rate}}{C_{av}^{ss}} \tag{7.7}$$

where dosing rate = dose/dosage interval (D/τ). Even after intravenous doses when $F=1$, Equation 7.7 is not a useful route to estimating *CL* because the average steady-state concentration has to be determined from *AUC* after a single dose (Equation 4.42). For oral doses, *F* has to be measured by comparing the *AUC* values after oral and i.v. doses, if an accurate value of *CL* is to be obtained.

Of course, *CL* is the independent variable, and the other variables change as a consequence of changes in *CL*, therefore clearance is a useful way to calculate maintenance doses to achieve a particular steady-state concentration via rearrangement of Equations 7.6 and 7.7. If a loading dose is required then this is calculated from the other independent variable, *V* (Equation 4.36).

It has already been shown that the elimination rate constant is directly proportional to *CL* (Equation 4.5) and that *AUC* is inversely proportional to *CL* (Equation 4.15). Thus, if clearance increases for any reason (e.g. enzyme induction) the elimination half-life will decrease, as will the area under the curve. In the case of oral administration, there will be

Table 7.1 Effect of systemic clearance on kinetic parameters

CL (L h^{-1})	λ (h^{-1})	$t_{½}$ (h)	C_{max} (mg L^{-1})	t_{max} (h)	AUC (mg h^{-1} L^{-1})
1	0.00714	97.0	0.583	28.4	100.0
2	0.0142	48.5	0.516	22.7	50.0
4	0.0286	24.3	0.433	17.5	25.0
8	0.0571	12.1	0.339	13.1	12.5

$F=1$, $V=140$ L, $k_a=0.1$ h^{-1}.

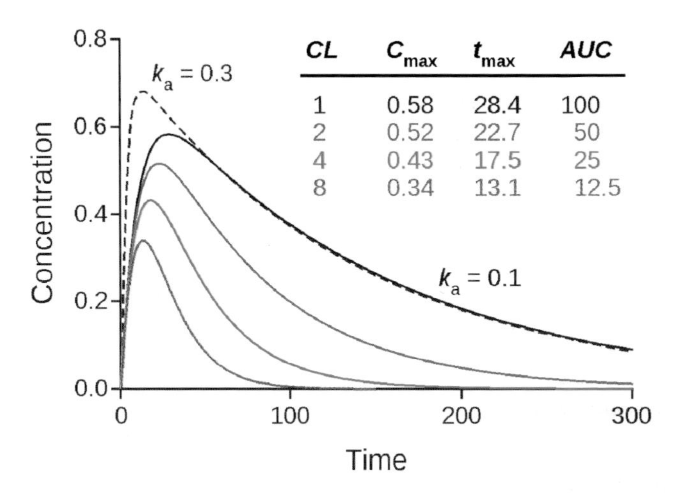

Figure 7.1 Effect of clearance on AUC, C_{max} and t_{max}. The curves (solid lines) were modelled using dose = 100, $V=140$, $F=1$, $k_a=0.1$ and $CL=1$–8. Broken line: $CL=1$ and $k_a=0.3$, other parameters unchanged.

a corresponding decrease in C_{max}, but because λ will be greater, the time at which the rate of input equals the rate of output will be sooner, that is, t_{max} will be decreased. This can be demonstrated via Equations 4.5, 4.15, 4.21 and 4.22. Table 7.1 shows a simple example where clearance increases from 1 to 8 L min^{-1} with corresponding decreases in AUC. The data were modelled using a dose of 100 mg, $V=140$ L, $F=1$ and $k_a=0.1$ h^{-1}. Note that in this example all the changes listed in Table 7.1 are as a result of the changes in clearance. Under different circumstances, reductions in C_{max} and AUC could be because of a reduction in systemic availability, F, but t_{max} would be unchanged if that were the case (Equation 4.21). An increased rate of absorption would reduce t_{max}, but without changing AUC if there was no change in the extent of absorption. The effect is shown by the broken line of Figure 7.1, in which k_a has been increased from 0.1 to 0.3 h^{-1}, $CL=1$ and the other parameters are unchanged. C_{max} is increased but AUC remains the same.

7.2.2.2 *Apparent oral clearance*

Following an oral dose CL is given by Equation 4.30:

$$CL = F\frac{D}{AUC} \tag{4.30}$$

and so CL cannot be calculated without knowing the value of F. However, there are situations in which it is not possible to administer an intravenous dose of the drug being studied, and Equation 4.30 has been rearranged to:

$$\frac{CL}{F} = \frac{D}{AUC} = CL_{\text{oral}} \tag{7.8}$$

and the value referred to as 'oral clearance' or '*apparent* oral clearance' or sometimes even just 'clearance', a potential cause of confusion. To avoid ambiguity, any value obtained for D/AUC from extravascular doses should be referred to as CL/F or CL_{oral}. Obviously, a value for systematic clearance, CL, cannot be derived without knowing the proportion of the dose which reaches the systemic circulation, F. When the bioavailability is thought to be very high it is sometimes assumed that $F=1$. CL_{oral} may be used in comparative studies when it is usually assumed that F does not change in the groups of subjects being compared. Apparent oral clearance serves no purpose in characterizing a drug, when i.v. and oral data would normally be available, but it is commonly used in studies of special populations in which i.v. data are only available from literature or F is believed to be close to unity (Chapter 12), but any conclusions must be made with caution because in these populations it is likely that any changes in AUC are a result of changes in CL and/or F, and possibly V.

7.2.2.3 Hepatic intrinsic clearance

As discussed earlier, organ clearance is the blood flow through the organ multiplied by the extraction ratio, so hepatic clearance, CL_{H}, is:

$$CL_{\text{H}} = Q_{\text{H}}E \tag{7.9}$$

where Q_{H} is the blood flow through the liver. Several models have been devised to describe metabolism by the liver, including the parallel-tube model and the dispersion model, and it is also possible to invoke multiple plate ideas, as in chromatographic columns. However, the homogeneous well-stirred pool concept has a simplicity that facilitates our understanding of a broad range of pharmacokinetic observations, and it is the one most generally used. It assumes that the drug-metabolizing enzymes are distributed evenly throughout the liver, and that the hepatic portal vein and the hepatic artery are equivalent in providing blood flow and therefore drug delivery to the liver. Clearly, during drug absorption, the drug concentrations are much higher in the hepatic portal vein than they can ever be in the hepatic artery, and this would be expected to affect the drug concentrations at the enzyme surfaces, which could in turn affect the likelihood of first-order metabolism occurring, that is, initially the enzymes could be saturated.

For an orally administered drug hepatic extraction:

$$E_{\text{H}} = 1 - F_{\text{H}} \tag{7.10}$$

where F_{H} is the fraction of drug not metabolized by the liver. If the drug is not metabolized by the GI mucosa as it is being absorbed, then $F_{\text{H}} = F$, allowing an estimate of E. Equation 7.10 can be useful in scaling up from *in vitro* to *in vivo*, and in understanding clearance

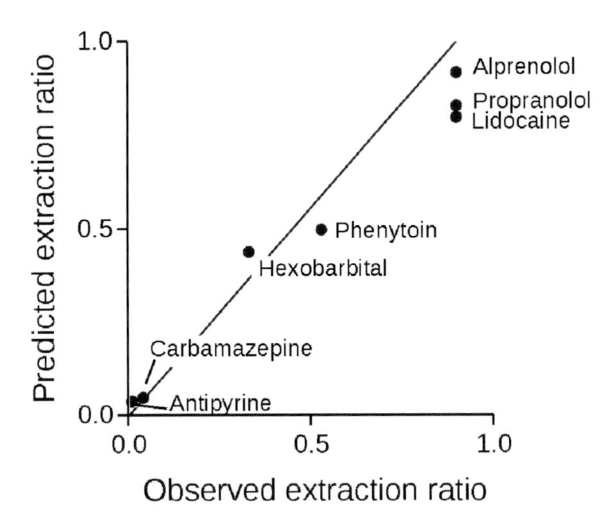

Figure 7.2 *Relationship between observed extraction ratios in perfused rat liver and the values predicted using* V_{max} *and Km values from metabolism in rat liver homogenates. The solid line is the 'line of identity', that is, slope = 1. Redrawn with permission from Rane* et al. *(1977).*

calculations with data from oral doses, which are commonly exposed to the liver before they reach the remainder of the body. Rane *et al.* (1977) predicted the hepatic extraction ratio from V_{max} and *Km* estimates *in vitro* using rat liver homogenates (Figure 7.2).

Hepatic intrinsic clearance, CL_{int}, is the maximal ability of the liver to remove drug irreversibly without any restrictions due to flow limitations or binding to proteins, and so takes the form of Equation 7.2. However, when the substrate concentration is very low compared with *Km*, the equation can be written:

$$CL_{int} = \frac{V_{max}}{Km} \tag{7.11}$$

Note that the variables of Equation 7.11 are measured with different units in different situations, most obviously in the case of V_{max}, which can have either mass/time or concentration/time units. When the expression V_{max}/Km is shown as identifying a first-order rate constant λ the units are reciprocal time. Values will be 'real' or 'apparent' depending on whether purified enzymes are used. This becomes especially important when V_{max} and *Km* concepts are applied to plasma concentrations of phenytoin (Section 4.7.1). Equation 7.11 is usually used for intrinsic clearance of a drug exhibiting first-order elimination kinetics. As hepatic blood flow increases, hepatic clearance increases to a maximum, the value of which depends on CL_{int}:

$$CL_H = Q_H \frac{CL_{int}}{(Q_H + CL_{int})} \tag{7.12}$$

Comparing Equations 7.9 and 7.12, it follows that:

$$E = \frac{CL_{int}}{Q_H + CL_{int}} \tag{7.13}$$

Thus, this predicts that the extraction ratio is a function of flow rate: the larger the blood flow, the smaller the extraction ratio. However, if the intrinsic clearance is small relative to the flow rate, then the denominator in Equation 7.12 approximates to Q_H, so:

$$CL_H \approx CL_{int} \tag{7.14}$$

Such drugs are referred to as 'capacity-limited' or 'restricted'. However, for a drug with $CL_{int} >> Q_H$, Equation 7.9 reduces to:

$$CL_H \approx Q \tag{7.15}$$

These drugs are referred to as 'flow rate-limited' or 'non-restricted' drugs. If a drug is entirely removed from the body by hepatic clearance, then $CL_H = CL$.

Drugs with $E > 0.7$ are therefore considered to be flow rate limited whilst those with $E < 0.3$ are considered to be capacity limited and simplified equations such as Equations 7.14 or 7.15 can be applied. This distinction is important when predicting the effects of changes in hepatic blood flow, intrinsic clearance and protein binding, as described below.

7.2.3 Effect of plasma protein binding on elimination kinetics

There is little doubt that binding to plasma proteins can affect the rate of elimination of a drug, but in what way and to what extent are other issues. The equations in Section 2.5 apply to equilibrium between bound and unbound drug, and because plasma protein binding influences the apparent volume of distribution, it would be expected to affect rate of elimination as:

$$t_{½} = \frac{0.693V}{CL} \tag{4.7}$$

Protein binding may also influence CL, but there have been misunderstandings as to the effects of this influence. At one time, a widely held belief was that plasma protein binding inevitably delayed elimination because less drug was available to the drug-metabolizing enzymes. This erroneous generalization was 'supported' by some studies which demonstrated a correlation between percentage bound and the degree of metabolism, or more usually the elimination half-life, which may be additionally affected by renal clearance. However, it was pointed out that metabolism is dynamic and when unbound drug is metabolized bound drug dissociates to maintain the equilibrium:

$$\text{drug-protein complex} \underset{k_1}{\overset{k_{-1}}{\rightleftharpoons}} \text{protein} + \text{drug} \xrightarrow{k_m} \text{metabolite}$$

so for binding to delay metabolism, k_{-1} would have to be smaller than k_m (Curry, 1977). In a series of theoretical calculations, Gillette (1973) reasoned that 'it seems probable that the rate of dissociation of the drug–protein complex seldom becomes rate limiting in the metabolism of drugs', indeed Gillette demonstrated that it is possible for plasma protein binding to hasten metabolism by the efficient transport of drug to the liver. It transpired that those who believed that protein binding always has a major influence on the rate of elimination were working, for the most part, with capacity-limited drugs.

7.2.3.1 *Influence of protein binding on hepatic clearance*

Wilkinson and Shand (1975) showed that there is only a delay in metabolism if protein binding is high and intrinsic clearance is low. They did this by a modification of their original equations, so that Equation 7.13 becomes:

$$E = \frac{f_u CL'_{int}}{Q_H + f_u CL'_{int}} \tag{7.16}$$

where f_u is the fraction unbound and

$$CL'_{int} = \frac{EQ_H}{f_u(1-E)} \tag{7.17}$$

CL'_{int} is the intrinsic clearance of the unbound drug. Modifying Equation 7.12 to take account of protein binding gives:

$$CL_H = Q_H\left(\frac{f_u CL'_{int}}{Q_H + f_u CL'_{int}}\right) \tag{7.18}$$

If the intrinsic unbound clearance is very small compared with the flow, Q_H, then Equation 7.18 approximates to

$$CL_H = f_u CL'_{int} \tag{7.19}$$

Equation 7.19 is usually used for drugs with hepatic extraction ratios <0.3, whereas for highly extracted drugs (E_H>0.7) Equation 7.18 reduces to Equation 7.15, and protein binding does not influence CL_H. For drugs with E_H less than 0.7 but greater than 0.3 Equation 7.18 is applicable. An equation for oral clearance (CL/F) has been derived:

$$CL_{Oral} = f_u CL'_{int} \tag{7.20}$$

Drugs with a low intrinsic clearance (capacity limited) include warfarin and diazepam, and, as predicted by Equation 7.19, the elimination of these drugs is affected by the degree of plasma protein binding (Figure 7.3). If the liver is the major route of elimination for these drugs, changes in CL'_{int}, resulting from enzyme induction or inhibition (Chapter 13), may markedly affect their elimination half-lives.

Some drugs, such as propranolol and lidocaine, have intrinsic clearances greater than liver blood flow and when the intrinsic unbound clearance is very large compared with the hepatic flow then Equation 7.18 reduces to Equation 7.15. The clearance of these drugs will be unaffected by changes in plasma protein binding, but will be affected by changes in hepatic blood flow, as might occur with heart or liver disease or drugs that affect cardiac output (Chapter 12). Enzyme induction or inhibition, which will change intrinsic clearance,

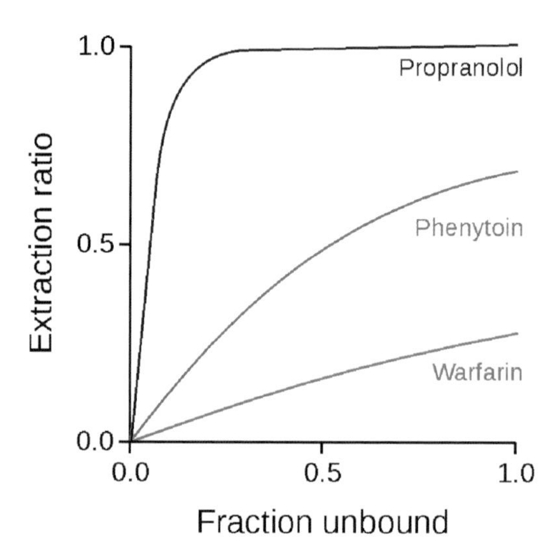

Figure 7.3 *Effect of plasma protein binding on the extraction of a highly extracted drug (propranolol), a poorly extracted drug (warfarin) and one with intermediate extraction (phenytoin). Reproduced with permission from Shand* et al. *(1976).*

should have less impact on the kinetics of these drugs. However, for a constant rate infusion the steady-state concentrations will be:

$$C^{ss} = \frac{R_0}{Q_H} \tag{7.21}$$

by rearrangement of Equation 4.36 and substitution of Equation 7.15. A clinically important point is that Equation 7.21 predicts that the steady-state total concentrations will be unaffected by alterations in protein binding. There may be some change due to a change in volume of distribution because of changes in redistribution between plasma and tissue concentrations, but if the change in V is small, it is possible that total concentrations remain reasonably constant when the unbound concentration increases. Thus dosing should be based on the unbound concentrations. Most drugs fall between the extremes of capacity limited and flow limited (Figure 7.3).

7.2.3.2 Influence of protein binding and volume of distribution on half-life

Wilkinson and Shand also examined the significance of tissue distribution and protein binding on half-life by using the following definition of volume of distribution:

$$V = V_b + V_t \frac{f_u}{f_t} \tag{7.22}$$

where V_b is the blood volume, V_t is the apparent volume of distribution made up of other tissues of the body, and f_u and f_t are the fractions of unbound drug in the blood and tissues, respectively. It was shown that the half-life is a function of volume of distribution, hepatic blood flow, fraction unbound and unbound intrinsic clearance:

$$t_{½} = 0.693\left(\frac{V}{Q_H} + \frac{V}{f_u CL'_{int}}\right) \tag{7.23}$$

Increases in the left-hand term within the brackets (increased volume of distribution or reduced liver blood flow) will tend to increase $t_{½}$, as might be expected. An increase in intrinsic clearance will decrease $t_{½}$. The effect of binding is more complex, depending on whether the drug has a high or low intrinsic clearance. Basically, however, with increased binding (i.e. decreased f_u) the right-hand term of the part of Equation 7.23 in brackets will tend to increase $t_{½}$. For a drug with low intrinsic clearance, as f_u decreases from unity, half-life increases to become very long as f_u approaches zero, whereas for a drug with high intrinsic clearance the increase in $t_{½}$ is very much less marked (Figure 7.4). Using the data of Figure 7.4, the predicted increase in half-life for the drug with the highest intrinsic clearance is less than two-fold (~1.98 times) as binding increases from 0 to 99%, whereas for the drug with the lowest intrinsic clearance the half-life increases by nearly 100-fold. Perhaps a more clinically relevant comparison would be the change in $t_{½}$ from 95 to 99% binding, when the changes are 1.66 and 5 times, respectively.

Wilkinson and Shand also considered the effect of tissue binding:

$$t_{½} = 0.693\left(\frac{V_b}{f_u CL'_{int}} + \frac{V_t}{f_t CL'_{int}}\right) \tag{7.24}$$

The precise effect on half-life will be determined in each case by the interplay of binding, intrinsic clearance, hepatic blood flow and tissue binding. At one extreme, the half-life of propranolol was shown to be relatively short in the presence of high protein binding.

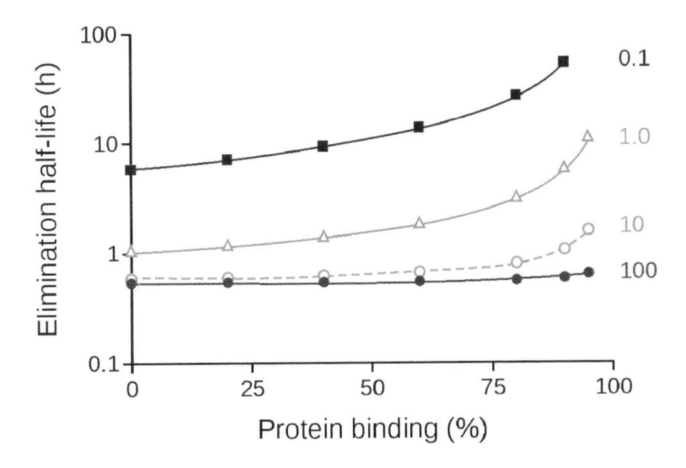

Figure 7.4 Effect of plasma protein binding and elimination half-life for four values of intrinsic unbound clearance. Calculated from Equation 7.23, using $Q_H = 1.5\,L\,min^{-1}$, $V = 70\,L$ and intrinsic clearance = 100, 10, 1 and 0.1 times Q_H. Note that the proportionate changes in $t_{½}$ are much greater for capacity-limited drugs.

In contrast, for drugs at the other extreme (e.g. warfarin and tolbutamide), the consequence of high plasma protein binding will be a long half-life. The combination of low V, low f_u and low CL'_{int} occurs with tolbutamide, warfarin and non-steroidal anti-inflammatory drugs, and it is only with these drugs and a small number of others with similar properties that protein binding effects on elimination are recorded (Chapter 13).

7.2.4 *In vivo* to *in vivo* extrapolation

Intrinsic clearance has been used to help determine the expected systemic clearance, and hence the half-life, in human investigations. The strategy for this *in vitro*/*in vivo* scaling is relatively straightforward:

1. Measure the half-life of the metabolism of the drug *in vitro*.
2. Calculate microsomal *in vitro* intrinsic clearance.
3. Calculate microsomal *in vivo* intrinsic clearance from *in vivo* CL_{mic} = *in vitro* CL_{mic} × microsomal protein (mg g of protein^{-1}) × g liver kg of body weight^{-1}.
4. Calculate hepatic clearance using hepatic blood flow (use the Wilkinson–Shand equation, Equation 7.12, and literature values for hepatic blood flow; this assumes that microsomal intrinsic clearance is the only contributor to hepatic clearance).
5. Use the volume of distribution to calculate the hepatic contribution to CL and hence the contribution to the half-life.
6. If the percentage of the dose that is excreted unmetabolized is known use the additivity of clearance to calculate the anticipated CL.
7. Correct the calculation for protein binding using $CL_{int} = f_u CL'_{int}$.

In regard to the fifth point above, if the apparent volume of distribution is known then an *in vivo* experiment and assessment of half-life has already been done. In fact, it is likely that the *in vivo* kinetics in a suitable animal species, including renal clearance, and the approach described above to determine the microsomal intrinsic clearance contribution to total clearance, will have been carried out. This collection of data can be used to make predictions for human beings, in combination with allometric scaling. Allometric scaling is a term to describe the mathematical modelling of parameters derived for one species to predict the parameters in another; this is usually extrapolation from animals to human beings (see Curry & Whelpton, 2011).

Ideally, the intrinsic clearance obtained *in vitro* would equal that observed *in vivo*. Various investigators have studied correlations between *in vitro* and *in vivo* values for both rats and humans. Typically, in quite complex studies, *in vitro* microsomal intrinsic clearance accounted for, on average, only about one-fifth of *in vivo* intrinsic clearance in humans (Naritomi *et al.*, 2001). Similarly, *in vitro* hepatocyte intrinsic clearance accounted for, on average, one-fifth to one-quarter of *in vivo* intrinsic clearance in rats (Lavé *et al.*, 1997). Further, *in vitro* hepatocyte intrinsic clearance accounted for, on average, approximately one-fifth of *in vivo* intrinsic clearance in humans (Lavé *et al.*, 1997). Among the possible reasons for these results were the following:

- Intrinsic clearance *in vivo* includes all processes of elimination, including renal excretion and non-hepatocyte, non-microsomal metabolism.
- While the calculations can allow for blood flow and protein binding effects, they do not allow for variations in the fine detail of liver perfusion local to enzyme surfaces.

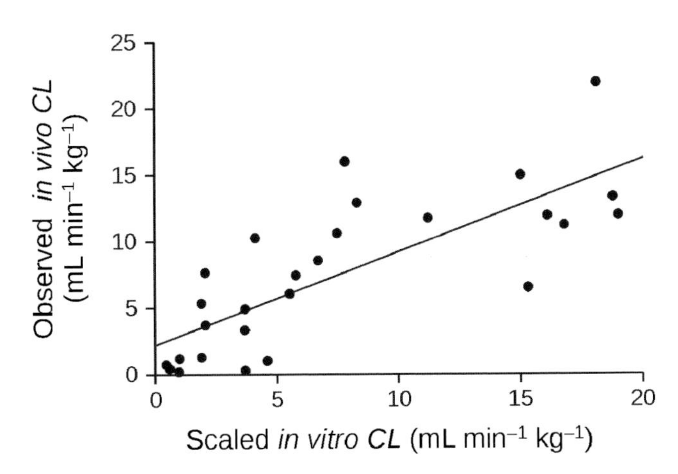

Figure 7.5 *Correlation between scaled* in vitro *clearance values from human hepatocytes and observed* in vitro *hepatic clearance for 26 drugs,* $r^2 = 0.61$. *Redrawn with permission from Chao* et al. *(2010).*

- Non-specific binding effects could reduce the actual drug concentrations at enzyme surfaces.
- There could be a lack of homogeneity of distribution of the enzymes through the liver.

As a result, *in vivo* intrinsic clearance is found in fact to correlate quite well with extraction ratio. The best *in vitro* predictor of human *in vivo* data appears to be human hepatocytes (Figure 7.5).

7.2.5 Limiting values of clearance

Conceptually, it seems to be obvious that clearance numbers will relate to blood flow properties of organs and will have upper limits, such as:

- CL cannot exceed cardiac output: $5.3\,\mathrm{L\,min^{-1}}$ (or $75\,\mathrm{mL\,min^{-1}\,kg^{-1}}$)
- CL_H cannot exceed hepatic blood flow: $1.5\,\mathrm{L\,min^{-1}}$
- $\mathrm{CL_R}$ cannot exceed renal blood flow: $1.2\,\mathrm{L\,min^{-1}}$
- CL_R for drugs for which there is no renal tubular membrane transfer cannot exceed glomerular filtration rate ($125\,\mathrm{mL\,min^{-1}}$ plasma $\cong 230\,\mathrm{mL\,min^{-1}}$ blood).

Many measurements of clearance conform to these concepts and indeed the clearance of some compounds may be used to estimate plasma/blood flows, for example *p*-amino-hippuric acid to measure plasma flow (Section 3.3.1.4). However, Table 7.2 shows a selection of clearance values, together with data for apparent volume of distribution and half-life, which show that in certain cases systemic clearance can exceed cardiac output.

Examples of very high clearance drugs also include physostigmine, esmolol, loratidine, misoprostol, spironolactone and, according to some reports, selegiline. Some of the explanations as to why CL can appear to, or actually, exceed cardiac output could include:

- experimental errors in measurement of CL or calculation errors resulting from use of inappropriate models
- widespread non-enzymatic chemical degradation of the drugs throughout the body

Table 7.2 Values of systemic clearance, apparent volume of distribution and elimination half-life for selected drugs in a 70 kg subject

Drug	CL (L min^{-1})	V (L kg^{-1})	Half-life (h)
Glyceryl trinitrate	16.1	3.3	2.3 (min)
Prazepam	9.8	14.4	1.3
Triamterene	4.4	13.4	4.2
Azathioprine	4.0	0.81	0.16
Hydralazine	3.9	1.5	0.96
Isosorbide	3.2	1.5	0.8
Cocaine	2.5	2.1	0.71
Desipramine	2.1	34	18.0
Nicotine	1.3	2.6	2.0
Propranolol	0.84	3.9	3.9
Diltiazem	0.81	5.3	3.2
Chlorpromazine	0.60	21.0	30.7

- a major contribution from non-hepatic and non-renal elimination. (This seems to occur with glyceryl trinitrate, which is extensively metabolized in blood vessel walls and something similar could occur with drugs metabolized by plasma enzymes, e.g. physostigmine and cocaine are metabolized by plasma cholinesterase. In these cases metabolism or chemical degradation occurs continuously, independent of blood flow to any particular organ of the body.)

7.3 Physiological modelling

The concepts of organ clearances can be applied to what are known as 'physiological models'. In the previous chapters most of the pharmacokinetic models have been based on the concept of the body as one or more 'compartments' or 'pools' which are treated as if they contain homogenous solutions of the drug. These models are useful for such things as deriving dosing schedules, but these compartments may have little or no relationship to anatomical spaces or organs. Furthermore, tissues in the same compartment can have markedly different concentrations, for example the plasma and liver concentrations depicted in Figure 2.12(a). In physiologically based pharmacokinetic (PBPK) modelling, the tissues and organs that play a role in the disposition of the drug being investigated are included in the model. Thus, there is no general physiological model; individual models will be dictated by the nature of the drug and to some extent by the route of administration. The various organs are connected by arterial and venous blood flows. Because of their importance in drug disposition the liver and kidney are usually included. If the lungs are included in the model, for example when the drug is administered and/or exhaled via the lungs, then they are placed in series with the right and left heart (Figure 7.6).

Each organ or tissue type has an associated blood or plasma flow, Q_t, and volume, V_t. A further complexity is that each organ is modelled as consisting of plasma, interstitial and intracellular components. Thus, many physiological factors can be incorporated in the model, including the effects of plasma and tissue binding, and the effects of the drug on blood flows to the organs, should that be appropriate, for example the effect of propranolol

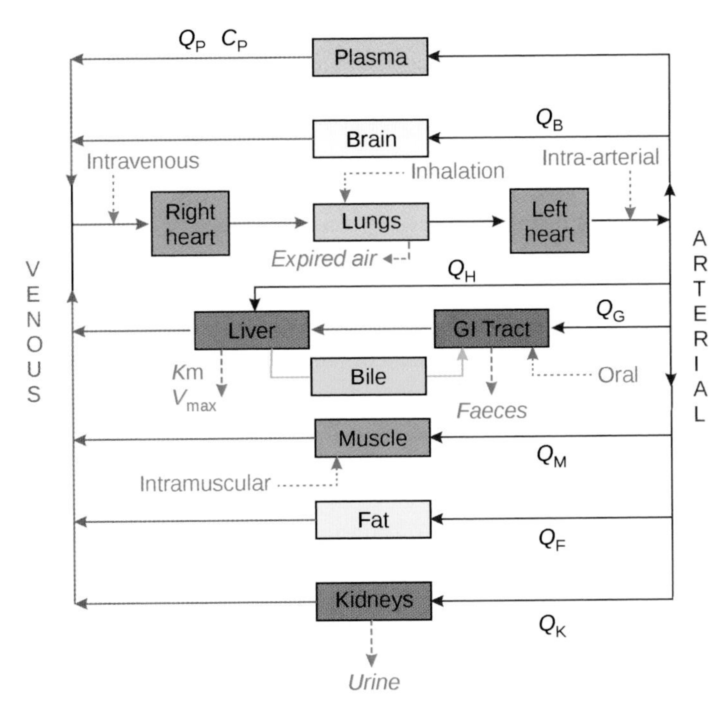

Figure 7.6 *Hypothetical physiological model demonstrating how relevant organs are connected by arterial and venous blood flows, and how various routes of administration can be depicted as required by the model.*

on cardiac output. *In vitro* data such as *K*m and V_{max} values from metabolism studies and partitioning between the components of a tissue can be incorporated.

In many situations the distribution of drug between tissue and blood is flow limited. As blood flows through a tissue, t, drug is extracted so that at equilibrium the tissue to blood concentration (C_B) is given by a partition coefficient, R_t:

$$R_t = C_t / C_B \tag{7.25}$$

The differential equation describing the rate of change in the tissue is:

$$\begin{aligned}\frac{dC_t}{dt} &= \frac{Q_t C_B - Q_t C_t / R_t}{V_t} \\ &= \frac{Q_t \left(C_B - C_t / R_t\right)}{V_t}\end{aligned} \tag{7.26}$$

where V_t is the apparent volume of distribution of the tissue. However, if the tissue is an eliminating organ, such as the liver, then drug is also removed by elimination (rate = concentration × clearance) and Equation 7.26 becomes:

$$\frac{dC_t}{dt} = \frac{Q_t \left(C_B - C_t / R_t\right)}{V_t} - C_t CL_t \tag{7.27}$$

If the kinetics of elimination are non-linear then the tissue clearance, CL_t, can be related to the apparent Km and V_{max} values using an equation analogous to Equation 4.48.

Having decided which tissues and organs should be included in the model, the rate of change of drug concentration in the blood (or plasma) can be modelled by summing all the component terms. If the modelling is done in terms of plasma concentrations, then the overall apparent volume of distribution of the drug at steady state is given by all the apparent volumes of distribution of the tissues plus the volume of plasma, V_P:

$$V_{ss} = V_P + \Sigma V_t R_t \tag{7.28}$$

The situation is more complex when the tissue distribution is membrane limited, rather than flow limited. Under these circumstances, equations describing the net flux of drug through the membrane have to be derived. The movement may be by simple diffusion or saturable carrier-mediated transport.

7.3.1 Practical considerations

To utilize PBPK models it is necessary to know the volumes of the appropriate tissues, the partition coefficients of the drug between blood and those tissues, as well as the blood flow through them for the species under investigation. It may be possible to use published data for these values (Table 7.3) or it may be necessary to measure them. Sometimes allometric scaling is used. Blood flows may be determined using such techniques as microsphere, laser Doppler velocimetry or tracer dilution. It should be remembered that the total blood flow through the tissues cannot exceed cardiac output. Values of R_t can be obtained by infusing drug to steady-state conditions, after which the animals are killed for analysis of tissue concentrations so that Equation 7.25 can be used. Non-linearity of R_t values with increasing drug doses indicates binding or complex diffusion in the tissue being studied.

Once all the parameters have been obtained these can be used in the model. The plasma concentration data are not fitted statistically as in other models but the physiological parameters are adjusted to obtain the most appropriate model. Tissues with large blood flows and volumes will have the greatest influence on the model while smaller tissues may have little influence on the overall quality of fit. Thus models, unsurprisingly, are likely to be heavily dependent on the liver and kidney.

Table 7.3 Physiological parameters for several species (Bischoff et al., 1971)

Parameter	Mouse	Rat	Monkey	Dog	Man
Body weight (g)	22	200	5,000	17,000	70,000
Volume (mL)					
Plasma	1.0	9.0	220	650	3,000
Muscle	10	100	2,500	7,500	35,000
Kidney	0.34	1.9	30	76	280
Liver	1.3	8.3	135	360	1,350
Gut	1.5	11	230	640	2,100
Plasma flow rate (mL min^{-1})					
Muscle	0.5	3.0	50	140	420
Kidney	0.8	5.0	74	190	700
Liver	1.1	6.5	92	220	800
Gut	0.9	5.3	75	190	700

Summary

This chapter has considered further aspects of clearance and how *in vitro* determination of intrinsic clearance can be of use in predicting pharmacokinetic parameters, something that is particularly important during drug development. The relationship between *CL*, which is affected by blood flow to the elimination organs and restrictions due to protein binding, and intrinsic clearance has been explained. The concept of flow-limited and capacity-limited extraction was used to consider when plasma protein binding is likely to influence pharmacokinetic parameters such as elimination half-life. The importance of extraction and clearance in PBPK modelling was exemplified at the end of the chapter.

7.4 Further reading

Curry SH, Whelpton R. *Drug Disposition and Pharmacokinetics: From Principles to Applications* Chichester: Wiley-Blackwell, 2011.

Riviere JE. *Comparative Pharmacokinetics: Principles, Techniques, and Applications* Iowa: Iowa State Press, 1999.

7.5 References

Bischoff KB, Dedrick RL, Zaharko DS, Longstreth JA. Methotrexate pharmacokinetics. *J Pharm Sci* 1971; 60: 1128–33.

Chao P, Uss AS, Cheng KC. Use of intrinsic clearance for prediction of human hepatic clearance. *Expert Opin Drug Metab Toxicol.* 2010; 6: 189–98.

Curry SH. *Drug Disposition and Pharmacokinetics*. 2nd edn. Oxford: Blackwell, 1977.

Curry SH, Whelpton R. *Drug Disposition and Pharmacokinetics: From Principles to Applications* Chichester: Wiley-Blackwell, 2011.

Gillette JR. Overview of drug-protein binding. *Ann N Y Acad Sci* 1973; 226: 6–17.

Lavé T, Dupin S, Schmitt C, Chou RC, Jaeck D, Coassolo P. Integration of in vitro data into allometric scaling to predict hepatic metabolic clearance in man: application to 10 extensively metabolized drugs. *J Pharm Sci* 1997; 86: 584–90.

Naritomi Y, Terashita S, Kimura S, Suzuki A, Kagayama A, Sugiyama Y. Prediction of human hepatic clearance from in vivo animal experiments and in vitro metabolic studies with liver microsomes from animals and humans. *Drug Metab Dispos* 2001; 29: 1316–24.

Rane A, Wilkinson GR, Shand DG. Prediction of hepatic extraction ratio from in vitro measurement of intrinsic clearance. *J Pharmacol Exp Ther* 1977; 200: 420–4.

Shand DG, Cotham RH, Wilkinson GR. Perfusion-limited effects of plasma drug binding on hepatic drug extraction. *Life Sci* 1976; 19: 125–30.

Wilkinson GR, Shand DG. Commentary: a physiological approach to hepatic drug clearance. *Clin Pharmacol Ther* 1975; 18: 377–90.

8

Quantitative Pharmacological Relationships

> **Learning objectives**
>
> By the end of the chapter the reader should be able to:
>
> - draw annotated diagrams to demonstrate the relationship between drug concentration and effect
> - explain sigmoidicity and why it is important
> - explain how dose (concentration) and time combine to control the intensity and duration of drug effect
> - discuss direct response and indirect response models, including hysteresis
> - outline the principle of using effect compartments to integrate PK-PD data.

8.1 Pharmacokinetics and pharmacodynamics

As stated in Chapter 1, it is quite common to read that there are two branches of pharmacology: pharmacodynamics and pharmacokinetics. At a more specific level, many authors prefer to distinguish between the description of overt effects of drugs and mechanistic studies (as sub-divisions of pharmacology), limiting the use of the word 'pharmacodynamics' to studies of the relationship between drug concentrations and mechanisms at

Introduction to Drug Disposition and Pharmacokinetics, First Edition. Stephen H. Curry and Robin Whelpton.

Companion website: www.wiley.com/go/curryandwhelpton/IDDP

sites of action. Also in previous chapters, particularly Chapter 4, we showed that, after single doses of drugs, while concentrations in plasma increase and decrease, so the drug effects are presumed to increase and decrease. This is a fundamental expectation. No particular models, linear or non-linear, need to be invoked in stating this principle, which can be considered descriptively. However, we can now examine in greater detail the ways in which this principle does and does not apply, and study the application of mathematical models to what has come to be known as the pharmacokinetic/pharmacodynamic (PK-PD) relationship.

It is generally accepted that the effects of drugs are usually greater when exposure is greater, and that the effects of drugs that are rapidly removed from the body disappear quickly. Often 'exposure' equates to 'dosage', but there are situations, particularly with drugs that have large inter-subject variability, for example in half-life and bioavailability, that dosage may not be the best measure. In these situations exposure may be accessed from the determination of *AUC* values. Reducing exposure is thought to reduce effect, either by reduction in dose or by use of a drug that is rapidly removed from the body. However, when these relationships are studied in detail relatively complex relationships are observed. During the last 25 years there has been explosive growth in our understanding of the PK-PD relationship. PK-PD modelling uses measurements of effects and of drug concentrations in plasma and at other sites to explore mechanisms. This is conveniently studied through:

- fundamentals of concentration–effect relationships (dose–response curves,) recorded principally *in vitro*
- study of similarities and differences between *in vitro* and *in vivo* observations
- specific examples of dose–effect and concentration–effect relationships *in vivo*
- PK-PD modelling, which integrates pharmacokinetic concepts with concentration–effect relationships.

8.2 Concentration–effect relationships (dose–response curves)

Concentration–effect relationships follow from the nature of the reversible interaction between a drug and its receptor. A drug may be an agonist and exert a stimulatory effect on the receptor, or it may be an antagonist and act by binding to the receptor to prevent the effects of an agonist. The effects of a drug do not increase indefinitely with increasing doses because there is a finite number of receptors. The reversible binding of a drug, D, to its receptor, R, can be assessed by the law of mass action:

$$D + R \rightleftharpoons DR$$

where DR is the drug–receptor complex. The equilibrium constant (Section 1.3.1), $K_{D,}$, is related to the molar concentrations of the species at equilibrium:

$$K_D = \frac{[D][R]}{[DR]} \tag{8.1}$$

The total concentration of receptors is the sum of the bound and non-bound receptors, [DR] + [R], and the proportion, or fraction, of receptors bound, f_A, can be obtained by rearrangement of Equation 8.1:

$$f_A = \frac{[DR]}{[DR]+[R]} = \frac{[D]}{[D]+K_D} \tag{8.2}$$

This is the equation of a rectangular hyperbola and a plot of f_A versus [D] is the same shape as those defining plasma protein binding, enzyme kinetics and binding of substrates to transport proteins, which have been discussed earlier (Figure 8.1(a)).

K_D has units of concentration and is numerically equal to the concentration at which 50% of the receptors are occupied. The lower the value of K_D the greater the *affinity* (i.e. the tendency to bind) of the agonist for its receptor.

The measured response to a drug is related, but not necessarily directly proportional, to receptor occupancy. The equation relating the effect, E, to the concentration is:

$$E = \frac{E_{max}C}{EC_{50} + C} \tag{8.3}$$

where EC_{50} is the concentration that produces 50% of the maximum effect possible. Pharmacodynamic models that use Equation 8.3 are referred to as Emax models.

Agonists that interact with their receptors to produce the maximum response possible are referred to as *full* agonists whereas agonists that are incapable of producing the maximum response, even when all the receptors are occupied, are referred to as *partial agonists* (Figure 8.2). A partial agonist in the presence of a full agonist may antagonize the effects of the latter and so partial agonists are sometimes called *agonist-antagonists*.

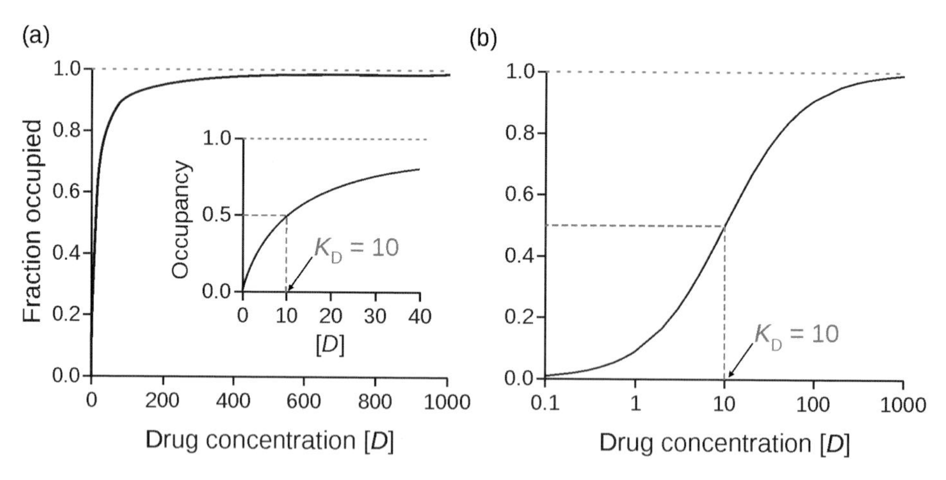

Figure 8.1 (a) Fraction of receptors occupied (occupancy) as a function of drug concentration. K_D is numerically equal to the concentration at which half the receptors are occupied (inset). (b) Logarithmic transformation converts the hyperbola into a symmetrical sigmoid shape typical of log(dose)–response curves.

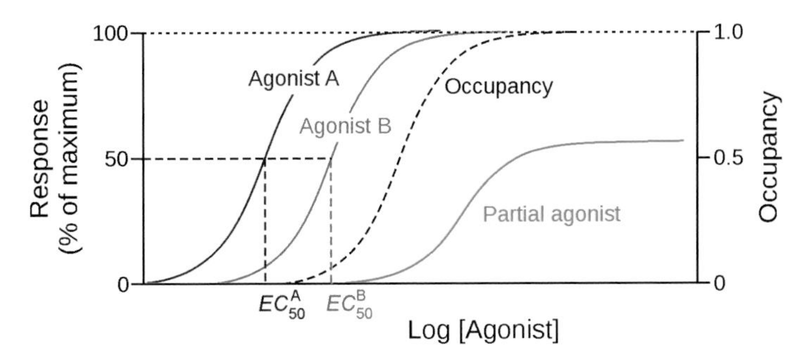

Figure 8.2 *Log(agonist concentration)–response curves. Agonist A is more potent than B because it has a lower EC_{50}. The partial agonist cannot elicit the maximum response even when all the receptors are occupied.*

Figure 8.2 illustrates some important features of agonists and log(concentration)–response curves. First, it is usual to plot the response as the percentage of the maximum that a full agonist can produce. If the responses are from an *in vitro* experiment using isolated tissues, say guinea pig ileum, then different pieces of tissue will give different responses, but normalizing the results to the maximum obtainable with each tissue allows comparison between experiments and average responses can be plotted. Such dose–response curves, where the response is continuously variable between the minimum and maximum response, are 'graded' response curves. For *in vivo* data 'quantal' response curves may be used. With these, some measurable endpoint is defined, for example pain relief, and then the numbers of subjects from a test population that attain the endpoint are recorded for each concentration or dose of drug. The *y*-axis for a quantal dose–response curve would be 'percentage of subjects responding'; the maximum (100%) being when all the subjects responded to the treatment (see Figure 8.4).

In Figure 8.2, the curves for agonists A and B have been positioned to the left of the occupancy curve, illustrating that not all the receptors have to be occupied for the response to be maximal. Stephenson (1956) introduced the term *efficacy* to describe the way in which agonists vary in the response they produce, even when they occupy the same number of receptors. The partial agonist is less efficacious than the full agonists because it cannot elicit as great a response. The full agonists, A and B, are considered equi-efficacious as they both have maximal efficacy. The responses to agonist A occur at lower concentrations than those for agonist B. Agonist A is more *potent* than agonist B, that is, less drug is required to obtain a defined effect. Potency is usually assessed from the concentration producing 50% of the maximum response, EC_{50}. For a quantal response, EC_{50} or ED_{50} is the dose at which 50% of the *test* population respond. It may seem paradoxical, but partial agonists can be highly potent. This is true of the opiate partial agonist buprenorphine, which has high affinity for the μ-opioid receptor. As a consequence, it can displace other opiate analgesics such as morphine, reducing the pain relief and possibly causing withdrawal symptoms in addicts.

A further consideration is that the slopes of dose–response curves may differ, and this is accounted for with the sigmoidicity parameter, slope factor or Hill coefficient, γ, which accounts for differences in the slope:

$$E = \frac{E_{max} C^{\gamma}}{EC_{50}^{\gamma} + C^{\gamma}} \tag{8.4}$$

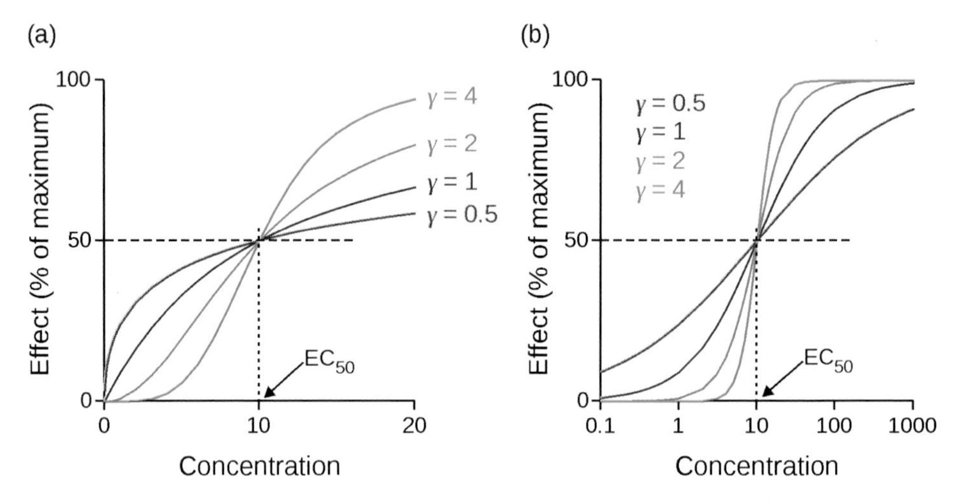

Figure 8.3 (a) The influence of the Hill coefficient, γ, on the shape of the effect–concentration curve. All the curves have the same $E_{max} = 100$ and $EC_{50} = 10$. (b) The same data with concentration plotted on a logarithmic scale.

Pharmacodynamic models based on Equation 8.4 may be referred to as sigmoidal Emax models. The reason is clear from Figure 8.3, which shows curves modelled using $E_{max} = 100$ and $EC_{50} = 10$. As the size of the exponent, γ, increases the shape of the effect–concentration curve becomes increasingly sigmoidal in shape. This arises because of the relationship between indices and logarithms (Appendix 1).

At the molecular level, $\gamma > 1$ indicates cooperativity of binding, indeed Hill introduced the coefficient to explain the binding of oxygen to haemoglobin ($\gamma \sim 2.8$). However, *in vivo* it is more difficult to explain why the value of the Hill coefficient is what it is. The slopes of dose–response curves vary between subjects and some of the variation may reflect differences in plasma protein binding, as it is generally the non-bound drug that can diffuse to the site of action. From a practical point of view, the larger the value of the exponent, the steeper the dose–response curve (Figure 8.3). A very high value of γ can be viewed as indicating an all-or-none effect (e.g. the development of an action potential in a nerve). Within a narrow concentration range, the observed effect goes from all to nothing or *vice versa*.

Another notable feature of Figure 8.3(a) is that unless a sufficient range of concentrations is studied, it may be difficult to accurately estimate E_{max}, which in turn will lead to an erroneous estimate of EC_{50}. Looking at the curve for $\gamma = 0.5$, it might be thought that E_{max} is approximately 60–70%, and yet the data were modelled using $E_{max} = 100$. Even concentrations ranging over four orders of magnitude were insufficient to encompass all of the effect–concentration range when γ equalled 0.5 (Figure 8.3(b)). This is yet another illustration that poor data, in this case an inadequate range of concentrations studied, will produce poor results.

Sometimes there is a baseline effect or there may be a response when no active drug has been administered – the placebo effect (Figure 8.4).

Baseline effects, E_0, can be added to Equation 8.4 to give:

$$E = E_0 + \frac{E_{max} C^{\gamma}}{EC_{50}^{\gamma} + C^{\gamma}} \tag{8.5}$$

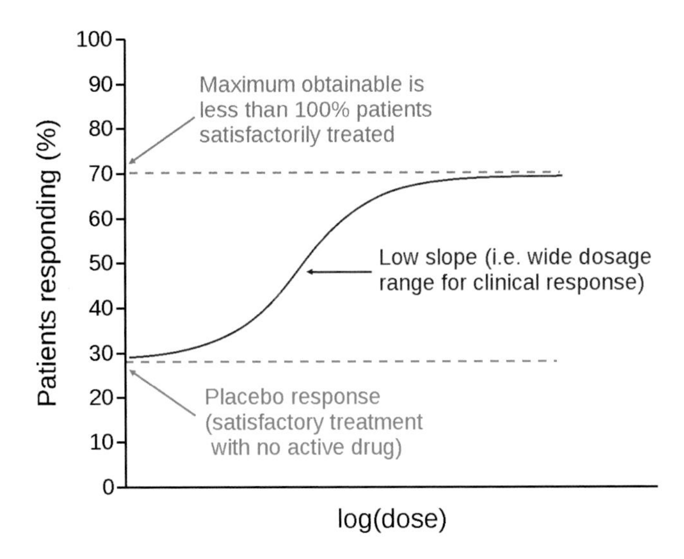

Figure 8.4 *Model quantal dose–response curve for psychopharmacology.*

The equivalent equation for the relationship between concentration and effect for an antagonist is:

$$E = E_0 + \frac{I_{max} C^{\gamma}}{IC_{50}^{\gamma} + C^{\gamma}} \quad (8.6)$$

where IC_{50} is the concentration causing 50% inhibition.

8.2.1 Dose–response curves *in vivo*

Although it is not difficult to demonstrate a 'near perfect' dose–response curve *in vitro*, demonstration of such perfect relationships in many other situations is much more difficult. One of the obvious constraints is that it may not be possible to define a maximum response if this requires using doses that cause intolerable effects. Figure 8.5 illustrates a dose–response curve that could be constructed over a limited range of concentrations. The figures shows 'sleeping times', actually the duration of anaesthesia, for phenobarbital over a relative narrow range of concentrations, 120–160 mg kg^{-1}. Below the lowest dose the mice were only sedated while at doses above the highest, the animals may have died. This also exemplifies the difficulty of working with drugs with a small therapeutic window. Further considerations of dose–response curves are listed in Box 8.1.

Under steady-state conditions, the concentration of drug at the site of action will be in equilibrium with the plasma concentration, so the latter can be used to model the concentration–effect relationship. The choice of model will be dictated by the data, and may be one of the following:

- linear
- log-linear
- Emax
- sigma Emax.

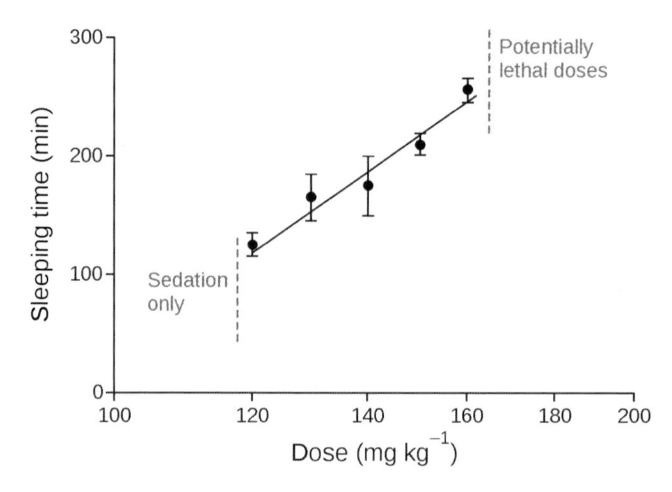

Figure 8.5 Dose–response relationship for phenobarbital concentrations (logarithmic scale) and sleeping time in mice. Each point is the mean of 6–8 mice ± SEM.

Box 8.1 Considerations for dose-response curves *in vivo*

- In human pharmacology it is often necessary to work at the lower end of the dose–response curve because of the prime importance of the safety and comfort of the subjects involved.
- In chemotherapy it is usual to work at the upper end of the curve, as antimicrobial drugs usually have a wide margin of safety and the maximum effect against the microbes is required.
- In psychopharmacology it is commonly impossible to define or achieve the maximum possible effect, so the upper limit is less than 100% in a quantal curve. Additionally, there is a large subjective contribution (placebo effect) and some response is obtainable from a placebo preparation, so the graph does not necessarily go through the origin (Figure 8.4).
- Compensatory reflexes play a considerable part in modifying dose–response curves when the measurable effect is a change in blood pressure; this makes the observations highly time dependent.
- Latent pharmacological effects can show unusual dose–response relationships. For example, the effect of warfarin on clotting time occurs after a considerable delay and after the peak plasma level of the drug has passed. In such a situation, the perfect dose–response relationship is obscured.
- Preventive treatment with drugs is common in therapeutics. A relationship between degree of prevention (which desirably is 100%) and dose is very difficult to demonstrate, especially as many diseases requiring preventative treatment, notably mania and epilepsy, are episodic in their occurrence.

For a drug–receptor interaction, one would expect the sigma Emax model be applicable, the Emax model being a special case of the sigma Emax model when $\gamma = 1$. The linear and log-linear models are used when the concentration range is limited to only part of the dose–response curve.

The simplest effect–concentration model is where (i) the drug distributes into a single compartment, so concentration is represented by the plasma concentration, and (ii) the effect is an instantaneous, direct function of that concentration. In this situation, the relationship between drug concentration (C) and a pharmacological effect (E) can be simply described by the linear function. If the measured effect has some baseline value (E_0) when drug is absent (e.g. a physiological property such as diastolic blood pressure) then the model may be expressed as:

$$E = E_0 + SC \tag{8.7}$$

which is a straight line with intercept $= E_0$ and slope $= S$. This model does not contain any information about efficacy and potency, cannot identify the maximum effect and thus cannot be used to find the concentration for 50% effect or for a defined effect in 50% of patients, that is, the EC_{50}. Equation 8.7 may be suitable when there is a limited range of concentrations and the observations are made towards the bottom end of a dose–response curve. At higher concentrations, plotting log(concentration) commonly linearizes the data within the approximate range of 20–80% of the maximal effect (Figure 8.1(b)). For larger ranges of concentration it may be necessary to use an Emax or sigmoidal Emax model, which will allow estimation of apparent E_{max} and EC_{50} values; apparent because they are derived from plasma concentrations rather than the concentrations at the receptor site.

8.3 Time-dependent models

In previous chapters it was assumed that *in vivo* the effect–time curve parallels the plasma concentration–time curve (e.g. Figures 4.1 and 4.2) because this adequately explains *concepts* such as onset and duration of effect. The effects of time on the drug response have been classed as:

- *direct response*: the effect is proportional to the concentration, but the peak may occur before or after the peak plasma concentration
- *indirect response*: the effect is the result of subsequent biochemical changes, for example the effect of warfarin on prothrombin clotting times when the effect can be related to the decreased rate of formation of clotting factors.

Modelling time-dependent relationships requires integration of pharmacokinetic and pharmacodynamic data (PK-PD modelling). The most common approach to *in vivo* PK-PD modelling involves sequential analysis of the concentration versus time and effect versus time data, such that the kinetic model provides an estimate of concentration that is then related to the effects. This has been referred to as the kinetics *driving* the dynamics because usually it is the concentration of drug at the site of action that is causing the effect. In dose–response–time models, the underlying assumption is that pharmacodynamic data give us information on the kinetics of the drug in the *biophase*, that is, the tissue or compartment precisely where the drug exhibits its effect.

8.3.1 Relationship of concentration, effect and time

There are three basic types of relationship, which may be further classified as direct or indirect responses, as exemplified in Figure 8.6. When the effect parallels the plasma concentration there is a direct relationship between concentration and effect, which may be linear over the range studied (Type A_1) or saturable, as with the Emax model (Type A_2). Because there is no time influence, the same intensity of effect is observed at the same concentration, whether it is on the raising or declining phase of the concentration–time curve, as shown in the right-hand panel of Figure 8.6(a). The filled symbols represent data on the rising phase and the open symbols show data collected from the declining phase of the concentration–time plot.

Relationships of Type B may be encountered. Generally, there is an early intense effect occurring at relatively low concentrations in plasma. As time goes by, the effect becomes maximal and then decreases before the plasma concentrations return to zero. This can occur when distribution to tissues plays a large part in the time course of the action of the drug, such as with alcohol, or when acute tolerance (Section 8.4.2.1) causes the effect of the drug to wear off while the drug is still present. In this situation the effect–concentration plot produces a curve and if a line is drawn linking the data points *in the order in which they were recorded*, it is clear that the order goes in a clockwise direction (Figure 8.6(b)). This phenomenon is described as hysteresis, a word derived from the Greek, meaning 'deficiency' or 'lagging behind'.

Probably the most encountered situation is that of Type C, when the effect lags behind the plasma concentration. This results in an anticlockwise hysteresis (Figure 8.6(c)). The effect develops relatively slowly so that early high plasma concentrations elicit disproportionately small effects. This occurs when the drug effect is caused by an active metabolite, or when the receptors are in a slowly perfused compartment, or when there is a specific effect compartment (Section 8.4.1), or when the effect that is measured is dependent on a sequence of biochemical changes (e.g. warfarin).

8.3.2 Examples of drugs showing direct and indirect effects

8.3.2.1 Ebastine and carebastine

Ebastine is a non-sedating antihistamine (H_1-antagonist). It is extensively metabolized to its carboxylic acid analogue carebastine. Investigations into the pharmacokinetics, the plasma concentrations and the effects of carebastine, which appears to account for most, if not all, of the pharmacological potency, were conducted in healthy volunteers after oral doses (Vincent *et al.*, 1988). The pharmacokinetics of carebastine exhibited exponential growth and decay, with an elimination half-life of 10–12 h. The pharmacodynamic effect studied was the change in wheal area induced by intradermal injection of histamine (2 μg). The time to maximum wheal inhibition, which occurred between 2 and 6 h after dosing with ebastine, corresponded with the time to peak plasma concentration of carebastine (Figure 8.7(a)). The plasma concentrations were linearly and significantly correlated with the absolute percentage of the histamine-induced wheal area in some but not all of the subjects. There was no clear indication of hysteresis, the most appropriate model being that of Type A_1 in Figure 8.6. The wheal area was not 100% at the earliest time point, when the concentration was presumably zero, suggesting that a simple linear equation, such as Equation 8.7, would be the most appropriate mathematical model.

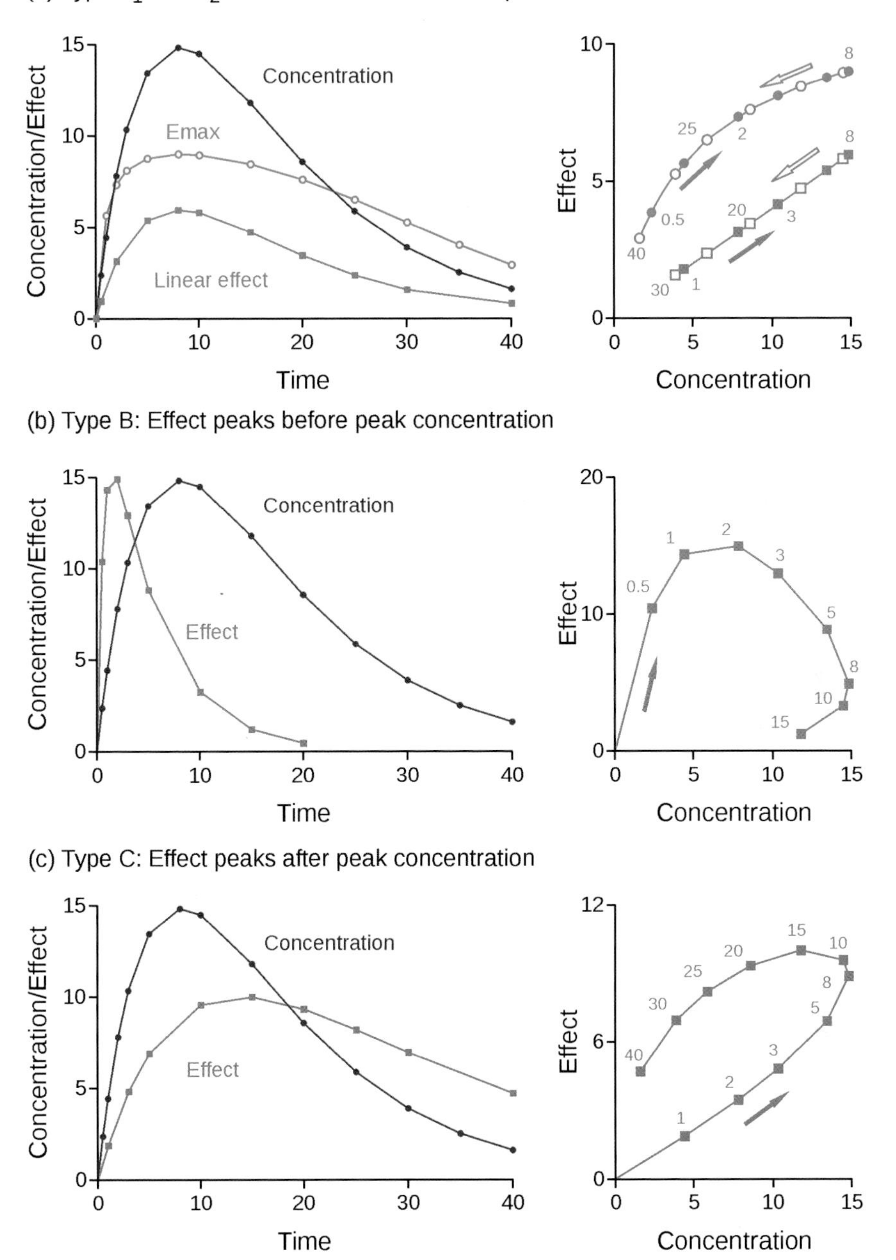

Figure 8.6 Types of concentration–effect–time relationship. (a) Type A: The effect parallels the concentration and the same effect occurs at the same concentration irrespective of the time at which it occurs. The concentration–effect curve may be linear (A_1) or saturated if it is an Emax model (A_2). The numbers represent the times at which the data were sampled. The closed symbols represent data on the rising phase and open symbols those on the declining phase. (b) Type B: The peak effect occurs before the peak plasma concentration. The concentration–effect curve shows a clockwise hysteresis (see text for details). (c) Type C: The effect lags behind the concentrations. The concentration–effect curve shows an anticlockwise hysteresis and the same effect may occur at different concentrations.

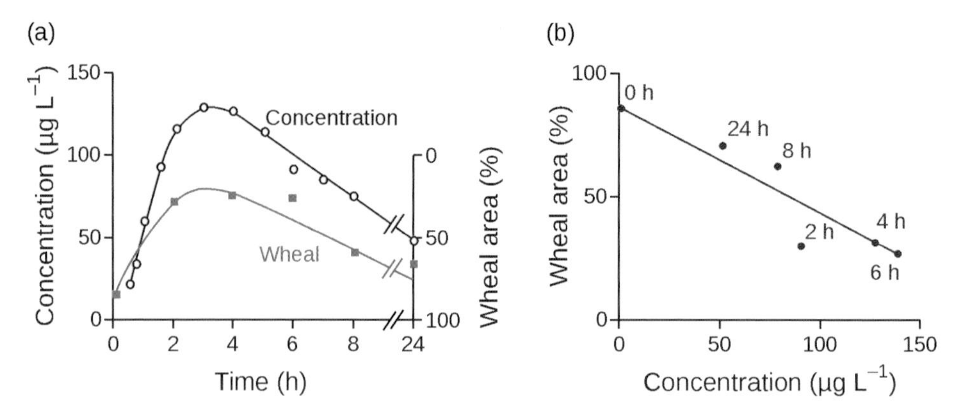

Figure 8.7 (a) Time course of the plasma concentrations of carebastine (red) and the histamine-induced wheal area (blue) in a representative subject after a single oral dose of 10 mg of ebastine. (b) Relationship between plasma concentration and effect – the times of collection are shown. Redrawn with permission from Vincent et al. *(1988).*

8.3.2.2 Gluthethimide

Observations made with glutethimide (Figure 8.8) suggest that it is an example of a Type B relationship. The ability of a subject to follow a moving light with the subject's head in a fixed position was studied. This is known as a smooth tracking test. The inhibition of this ability reached its peak when the concentration of glutethimide was still rising. Drugs absorbed from the GI tract pass via the liver and the hepatic vein to the heart and then in arterial blood to the organs. Initially the arterial concentrations will be high and lipophilic drugs can rapidly enter the brain, leading to early onset of effects. That the effect of alcohol is greater during the rising phase of the blood alcohol–time curve has long been known. It is believed that during absorption, the ethanol in arterial blood rapidly passes into the brain, and because the blood that is sampled for kinetic measurements is venous blood, there is a discrepancy between the effects of ethanol and measured plasma concentrations. This probably applies to several centrally acting drugs, including, in this example, glutethimide.

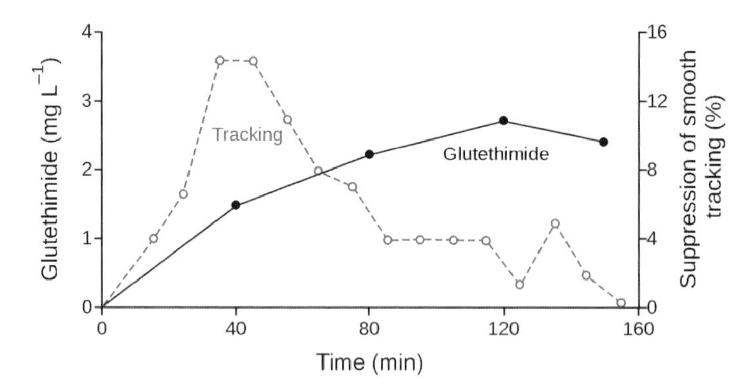

Figure 8.8 Plasma concentrations of glutethimide and percentage suppression of smooth tracking. From the data of Curry & Norris (1970).

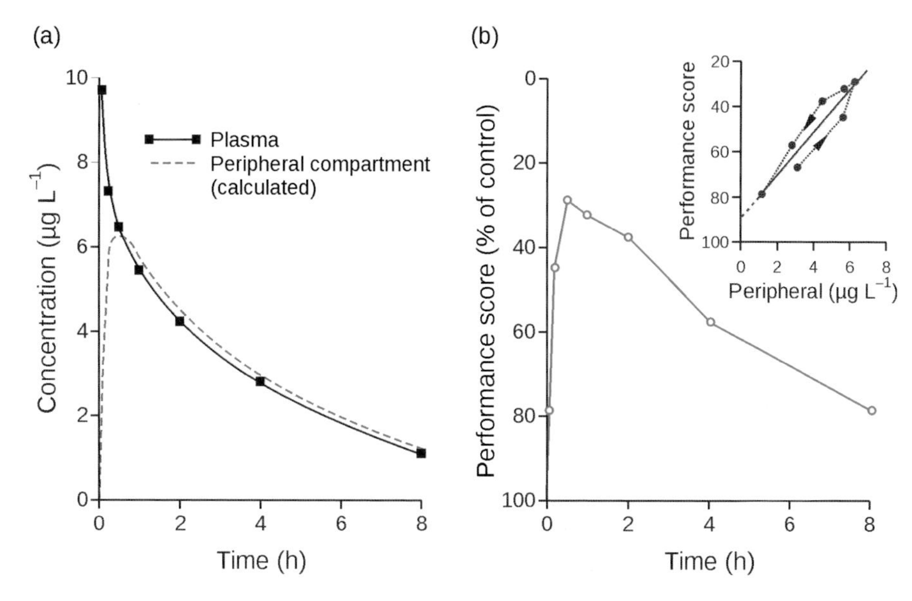

Figure 8.9 *(a) Mean plasma concentrations of LSD in five individuals following i.v. bolus doses of 2 μg kg⁻¹, together with calculated concentrations in the peripheral compartment. (b) Performance score based on arithmetic tests in the same experiment. The insert in (b) shows the performance score plotted against concentration in the peripheral compartment, indicating a straight line relationship with evidence of modest hysteresis. Redrawn from Wagner* et al. *(1968), with permission.*

8.3.2.3 Lysergic acid diethylamide

Of the explanations for Type C relationships, one was that it is possible that the receptors are in the peripheral compartment, in which case there should be better correlation between effect and the concentration in the peripheral compartment. Wagner *et al.* (1968) showed that after i.v. injections of lysergic acid diethylamide (LSD), the concentrations of LSD showed the typical bi-exponential decay associated with distribution through a two-compartment system (Figure 8.9(a)). During the initial steep phase of decay, during which the drug was transferring from the plasma and other parts of the central compartment to the peripheral compartment, the effect, assessed using an arithmetical performance test, was rising. At the same time, the calculated concentrations in the peripheral compartment were rising. There was a linear relationship between effect and the concentration in the peripheral compartment, with a baseline performance score (Figure 8.9(b)). Such models are known as compartmental or distributional models. The model of Equation 8.7 would be appropriate in this case with an anticlockwise hysteresis curve with no saturation.

8.4 PK-PD modelling

Integrating pharmacokinetics and pharmacodynamics is important during drug development. The challenge is to be able to predict the effect of a drug from the plasma concentration–time data, particularly under non-steady-state conditions. There are various approaches, some of which are more amenable to some drugs than others. One approach is the use of 'link models', so-called because they link the pharmacokinetic data to the effect model. In principle the aim is simple: it is to derive the concentrations at the site of action so that they can be used in the pharmacodynamic model. If there is a direct link, then the plasma concentrations may be used, but with an indirect link model there is the issue of hysteresis, as described above. A hypothetical 'effect compartment' was introduced to deal with this (Sheiner *et al.*, 1979).

8.4.1 Effect compartment models

The compartment model, as described above for LSD, cannot always be used to explain the time course of observed effects. When a small but pharmacologically significant subset of the dosed molecules reach the active site this is unlikely to be observable in the plasma level decay curve or available for calculation, as is the case with the pharmacokinetic distribution models. However, that site (the biophase) must have a volume, albeit small, and rate constants can be assigned to entry of drug and cessation of effect. It is in essence a compartment – an effect compartment.

The simplest of these models is depicted in Figure 8.10, where the compartment of a pharmacokinetic model is linked to an effect compartment, with first-order rate constants k_{1e} and k_{e0}. Note the use of the same convention of labelling the parameters, e represents the effect compartment.

For an intravenous bolus injection (so as to avoid the complication of input into the body) the equation describing the concentration, C_e, in the effect compartment takes the same form as a single compartment with first-order input and output:

$$C_e = \frac{k_{1e}D}{V_e(k_{e0} - \lambda)}\left[\exp(-\lambda t) - \exp(-k_{e0}t)\right] \tag{8.8}$$

It can also be shown that plasma concentration–time data can be used to obtain an estimate of k_{e0}, using the usual curve-fitting techniques:

$$C_1 = \frac{k_{e0}D}{V_1(k_{e0} - \lambda)}\left[\exp(-\lambda t) - \exp(-k_{e0}t)\right] \tag{8.9}$$

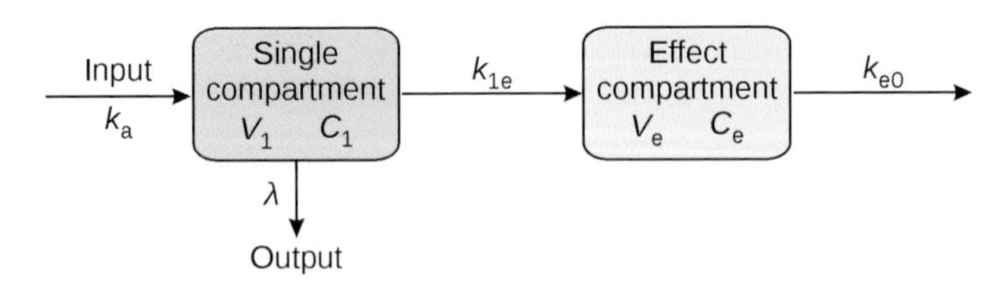

Figure 8.10 Addition of an effect compartment to a single-compartment model. The same approach is applied to two-compartment models and the effect compartment is usually linked to the central compartment.

Once values of C_e have been derived these can be plotted against the observed effects (see Figure 8.11). In the so–called 'soft-link' model E_{max} and EC_{50} are estimated from curve fitting of the E versus C_e data, giving:

$$E = E_0 + \frac{E_{max} C_e}{EC_{50} + C_e} \tag{8.10}$$

In the 'hard-link' model, estimates of E_{max} and EC_{50} are obtained via *in vitro* studies.

Note that the effect equilibration rate constant (k_{e0}) may be viewed as a first-order distribution rate constant. It can also be thought of in terms of the rate of presentation of a drug to a specific tissue, determined by, for example, tissue perfusion rate, apparent volume of the tissue and eventual diffusion into the tissue. It should be emphasized that the link model applies to situations where a small but pharmacologically significant subset of the dosed molecules penetrates a separate compartment with its own input–output characteristics, distinct from anything observable in the plasma level decay curve or available for calculation, as in the case with the pharmacokinetic distribution models (contrast LSD an example of a compartment distribution model application).

Lorazepam has provided us with a remarkable demonstration of the power of PK-PD modelling with an effect compartment. Gupta *et al.* (1990) used a battery of psychomotor and cognitive tests in a group of subjects given oral lorazepam and compared the data with plasma concentrations of the drug. The plasma concentrations were best described by a two-compartment model with first-order absorption. There was an anti-clockwise hysteresis loop for each effect. Fitting the time course of an integrated PK-PD model required an effect compartment with a finite equilibrium rate constant between it and the plasma compartment. The magnitude of the temporal lag was quantified by the half-time of equilibration, and the CNS effect was characterized by a mean estimate of maximum predicted effect. Sigmoid concentration effect curves were generated relating the concentration in the effect compartment and the measured pharmacological effects (Figure 8.11).

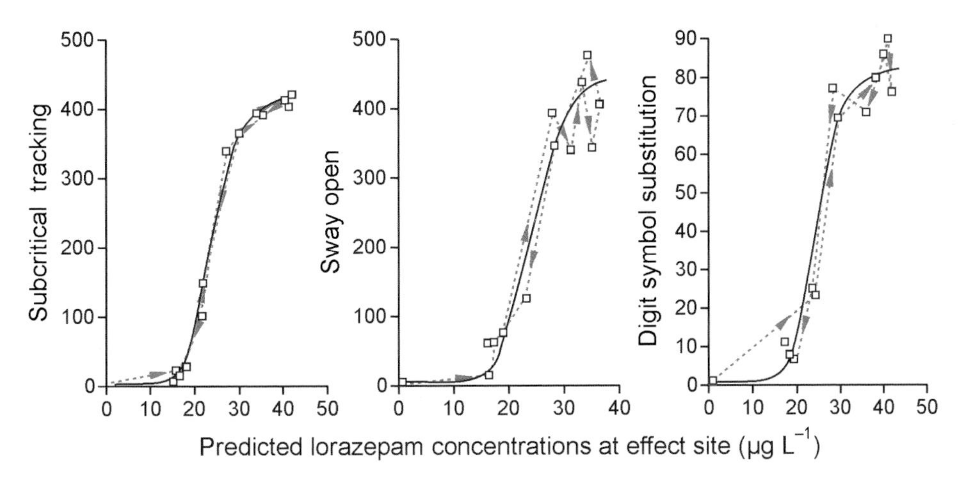

Figure 8.11 *Relationship between predicted effect site concentration (C_e) of lorazepam (□) and three psychomotor effects. Solid lines fitted to the PK-PD model as a function of C_e; arrows show hysteresis. Redrawn with permission from Gupta* et al. *(1990).*

8.4.2 Time-related changes in pharmacodynamic parameters

The discussion so far has assumed that only concentration and effects change with time, but there are situations which require the models to be adapted to take into consideration time-related changes in E_{max} and EC_{50}. These changes arise from the development of tolerance and changes in receptor function.

8.4.2.1 Tolerance and resistance

Tolerance describes when, over time, the effects of a drug diminish even though the concentration or dose remains the same. Tolerance may develop to some effects and not others. For example, treatment with some antidepressants (amitriptyline, desipramine) is started at a low dose, which can be increased as tolerance to the potentially dangerous cardiovascular effects develops. Usually when patients become tolerant to the desired effects of a drug these can be restored by increasing the dose.

Several alternative names may be used for tolerance, each conveying a subtle difference. The onset of rapid tolerance may be referred to as 'acute tolerance' or tachyphylaxis (from the Greek, meaning rapid protection) or 'desensitization'. The effect of β-agonists, such as isoprenaline (isoproterenol), very quickly decline after relatively short exposure – in fact during a single exposure. This is believed to be because of desensitization caused by two mechanisms, phosphorylation and internalization of the G-protein receptors. Tachyphylaxis is often used to describe the loss of effect caused by depletion of mediators. Indirectly acting sympathomimetic amines (amfetamine, ephedrine) displace neurotransmitters from aminergic nerve terminals, with subsequent loss of activity until transmitter stores are replenished. The term 'acute tolerance' has been applied to the observation that lipophilic CNS drugs appear to have greater effects at low plasma concentrations during the absorption phase than when the concentration has risen to higher levels. This is thought to be attributable to differences between arterial and venous blood concentrations shortly after administration, as discussed in Section 8.3.2.2. Arguably, this is a distributional phenomenon rather than tolerance in the true sense of the word.

The mechanisms whereby tolerance develops more slowly are not always known. Sometimes, drugs induce homeostatic changes. Antihypertensive treatment often results in renin being released from the kidney, the resultant increase in angiotensin elevating blood pressure. In the case of agonists tolerance may be because of down-regulation of receptors and reductions in receptor density. Long-term exposure to antagonists will increase the numbers of receptors. Down-regulation of receptors, that is, a reduced number of receptors, reduces E_{max} and desensitization reduces ED_{50}.

For those developing and using antimicrobial drugs there is the issue of resistance. There are several mechanisms by which microorganisms become resistant, but of major concern is the rate at which resistant strains spread. This is largely because of the ability of bacteria to share genetic material via plasmids, short lengths of DNA which in some instances carry the genes that confer resistance. Alarmingly, plasmids can be exchanged between different types of bacteria.

Thus tolerance- and time-related changes to the pharmacokinetic parameters, such as those caused by enzyme induction (sometimes referred to as metabolic tolerance), complicate the both PK and PK-PD modelling.

Summary

This chapter has examined the use of drug–response relationships to quantify the intensity of drug effects and how differences in the slope of the dose–response curve have practical implications for both deriving E_{max} and EC_{50}, and for clinical use. Drugs displaying different types of concentration–effect–time relationships have been discussed. The chapter has outlined the approaches to modelling non-steady-state concentration–effect relationships, including the use of link models to deal with hysteresis. Integrating dose-related and time-related effects presents intellectual challenges of a high order for those involved in applying PK in the world of new drug discovery, for those involved in characterizing dosage forms, and in the clinical management of drug response.

8.5 Further reading

Brodie BB. Physicochemical and biochemical aspects of pharmacology. *Jama* 1967; 202: 600–9.

Curry SH. *Drug Disposition and Pharmacokinetics*. 3rd edition. Oxford: Blackwell Scientific, 1980.

8.6 References

Curry SH, Norris H. Acute tolerance to a sedative in man. *Br J Pharmacol* 1970; 38: 450P–451P.

Gupta SK, Ellinwood EH, Nikaido AM, Heatherly DG. Simultaneous modeling of the pharmacokinetic and pharmacodynamic properties of benzodiazepines. I: Lorazepam. *J Pharmacokinet Biopharm* 1990; 18: 89–102.

Sheiner LB, Stanski DR, Vozeh S, Miller RD, Ham J. Simultaneous modeling of pharmacokinetics and pharmacodynamics: application to d-tubocurarine. *Clin Pharmacol Ther* 1979; 25: 358–71.

Stephenson RP. A modification of receptor theory. *Br J Pharmacol Chemother* 1956; 11: 379–93.

Vincent J, Liminana R, Meredith PA, Reid JL. The pharmacokinetics, antihistamine and concentration-effect relationship of ebastine in healthy subjects. *Br. J. Clin Pharmacol.* 1988; 26: 497–501.

Wagner JG, Aghajanian GK, Bing OH. Correlation of performance test scores with "tissue concentration" of lysergic acid diethylamide in human subjects. *Clin Pharmacol Ther* 1968; 9: 635–8.

9

Pharmacokinetics of Large Molecules

> **Learning objectives**
>
> By the end of the chapter the reader should be able to:
>
> - explain how the drugs discussed in this chapter are administered
> - compare the differences between the mesenteric capillary and lymphatic networks, with regard to the transport of drugs
> - briefly describe the disposition of large molecules
> - using mipomersen as an example, explain the development and use of antisense oligonucleotides
> - explain why stable forms of butytylcholinesterase have been developed.

9.1 Introduction

It is difficult to find a single term that adequately characterizes all of the drugs described in this chapter. Many of them are, or at least were, 'biological' in origin. Some may be described as new, others old. The monoclonal antibodies and oligonucleotides have been recently introduced, but others, insulin and heparin for example, have been established as drugs for several generations. Some are peptides and proteins, including hormones and enzymes. Others are polysaccharides, oligonucleotides and antibodies. The one tenuous

Introduction to Drug Disposition and Pharmacokinetics, First Edition. Stephen H. Curry and Robin Whelpton.

Companion website: www.wiley.com/go/curryandwhelpton/IDDP

connection between them is that they all have relative molecular masses >1000, and so it is that we have grouped this important collection of compounds under the one heading. The importance of this disparate group of drugs can be appreciated from inspection of Table 9.1, which lists some of the types of molecule in this category, with some examples of their use and some of their properties.

The invention and manufacture of recent products of the biotechnology industry have been made possible by dramatic advances in the scientific ability to chemically modify molecules in molecular weight ranges up to 30,000 g mol^{-1}, such as proteins and nucleotides. Synthesis of peptides and proteins stable to proteolytic enzymes has resulted in molecules with half-life values measured in days, rather than minutes – very important with injected drugs. Modification of amino acid sequences has facilitated ‘humanization’ of molecules, as with humanized insulin, without diminishing efficacy or potency. ‘Cutting’ DNA with endonucleases, use of virus vectors, and recombinant techniques has facilitated formation of re-ordered nucleotides optimized for their effects and pharmacokinetic properties. Combination with polyethylene glycol (PEGylation) has also been used with similar results, particularly changing conformation and electrostatic charges, thereby modifying potency and efficacy, and also reducing polarity and therefore renal clearance, resulting in prolonged persistence in the body.

Working with many of these compounds is often made difficult by their chemical properties. In some cases they are not only standardized, but also dosed, in terms of internationally accepted units, based on biological potency, rather than on the basis of weight. This is because different batches of the drug may vary in chemical constitution, while retaining the required potency. Pharmacokinetic studies often require biological assay methods, such as radioimmunoassay, rather than the more usual chromatographic techniques. A curiosity of this is that antibody drugs are commonly assayed by ELISA methods which thus utilize the interaction of antibodies with the antibodies.

9.2 Pharmacokinetics

9.2.1 Administration and dosage

Most of the compounds in this category are poorly absorbed from the GI tract and so are administered by injection. The problems of the oral route include instability in gastric acid, or metabolism by peptidases in the intestines and/or the liver, or from their relative polarity, molecular size and charges, resulting in little transcellular diffusion. Which parenteral route is used will depend on how quickly the pharmacological effect is wanted. Insulin is given intramuscularly or subcutaneously because that allows the release and onset to be controlled as required (Section 4.4.4) and the patient can self-administer it. In some instances the administration may need to be supervised by medical personnel, for example intravenous infusions of trastuzumab are supervised in case of drug-related allergic responses.

Not all peptides and proteins are destroyed in the GI tract. Ciclosporin is resistant to acid and peptidases, and is given orally. A lethal oral dose of botulinum toxin in human beings has been estimated as ~70 μg. It survives exposure to gastric acid and peptidases and, once absorbed by endocytosis, is transported via the lymphatic system as it is too large to enter the mesenteric capillary network. Absorption of nutrients is reduced in sufferers of cystic fibrosis, and when pancreatic enzymes are administered they are given

Table 9.1 *Some examples of peptides, proteins and other large molecules used as drugs*

Group	Example	Dosing route	Uses	Comments
Hormones	Vasopressin and desmopressin	Intranasal	Control of urine in diabetes insipidus	Nonapeptides given by any parenteral route; first intranasal drugs
	Insulin	s.c.	Diabetes mellitus	Polypeptide hormone of 51 amino acids
	Oxytocin	i.v.	Induction of labour	Nonapeptide, given by any parenteral route
	Human growth hormone	i.v.	Promoting growth in children	Somatrophin – single polypeptide chain of 191 amino acids
Enzymes	Pancreatic enzymes	Oral	Promoting digestion of food	Enteric coated granules – used in cystic fibrosis
	Thrombolytic agents (streptokinase)	i.v.	Dissolving blood clots	Degrades fibrin and breaks up thrombus
	Asparaginase	i.v.	Leukemia	Can be PEGylated (see text)
	Cocaine hydrolase	i.m.	Cocaine toxicity and dependence	Two modified butyrylcholinesterases in development
Synthetic peptides	Copaxone	s.c.	Suppression of autoimmune encephalomyelitis, MS	Mixture of synthetic tetrapeptides
Botanical peptides	Ciclosporin	Oral	Preventing transplant rejection	Natural fungal cyclic peptide – non-polar and P-450 substrate
Protein-bound drugs	Paclitaxel/protein complex	i.v./i.m.	Cancer treatment	Paclitaxel is highly insoluble and must be solubilized to facilitate injection
Polysaccharides	Heparin	i.v.	Anticoagulation	Complex drug used in management of risk of blood clots in several medical conditions
Proteins (natural or modified)	Interferon	i.v.	Multiple sclerosis and other conditions	Non-glycosylated proteins with several applications; some products PEGylated.
	Erythropoietin	i.v.	Anaemia of renal failure	Product of recombinant technology – standardized in international units
	Interleukin	i.v.	Prevention of abortion and/or miscarriage	Rh_o(D) immune globulin

Monoclonal antibodies	Trastuzumab	i.v.	Breast cancer (HER2 type)	‘Humanized’ monoclonal antibody drug example
	Rituximab	i.v.	Non-Hodgkin’s lymphoma	Targets phosphorprotein CD 52 on T- and B-lymphocytes
	Adalimumab	s.c.	Rheumatoid arthritis, Crohn’s disease and ulceratice colitis	IgG_1 monoclonal antibody
	Infliximab	i.v.	Crohn’s disease	Inhibits TNF-alpha
	Basiliximab	i.v.	Acute rejection of kidney transplants	Inhibits IL-2 on activated T-cells
Oligonucleotides	Mipomersen	Injection	Familial hypercholesterolaemia	One of two drugs of this chemical type available

as enteric-coated granules to protect them from degradation in the stomach. For peptides that cannot be administered orally other routes must be used. Vasopressin in particular can be given by the intranasal route for systemic action. Insulin has been studied for decades, and chemically modified, in the hope of finding an effective preparation that can be given intranasally, by inhalation, by means of suppositories, through the skin, or orally, with little, if any, impact on its use by patients.

The lymphatic system has a significant role in the absorption of proteins. This system is a network of vessels draining from the many tissues of the body into the left and right subclavian veins. Although it is a one-way system, together with the cardiovascular system it serves to deliver nutrients and oxygen to tissues, and to remove metabolic waste products from them. Simply put, the lymphatic system returns fluid (lymph) and its contents to the circulatory system from intercellular tissue spaces. Because drainage from the intestinal tissue is *towards* the venous system, materials from the intestine become dissolved in lymph and absorbed into the cardiovascular system.

The lymphatic system bypasses the portal system of the liver (Figure 9.1) and it is important to the absorption of dietary lipids. It also undoubtedly plays a part in the absorption of all orally administered drugs, although with so much potential for absorption into the cardiovascular system via the mesenteric blood capillaries and the hepatic portal system this contribution is, in many cases, quite small. However, part of the lymphatic system is made up of thin-walled lymphatic capillaries within the tissues of the GI tract. These vessels are more permeable than blood capillaries. Also, the lymphatic vessels are well adapted for the movement of large molecules, particularly proteins, and this includes any peptide and protein drugs that survive the acidity of the stomach and the proteases of the

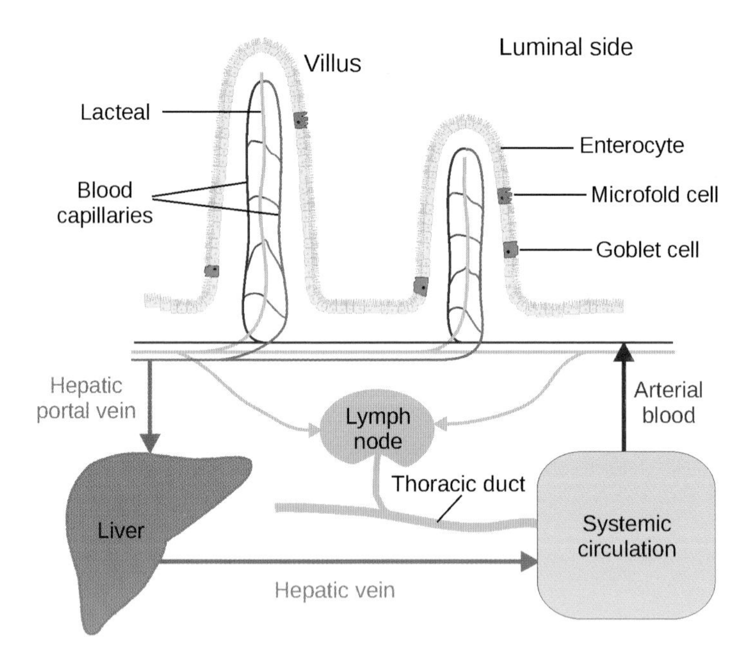

Figure 9.1 Lymphatic and capillary networks. Drugs absorbed into the lymphatic system avoid the hepatic portal vein and escape first-pass metabolism. Molecules too large to enter the capillaries are transported via the lymphatic system. Microfold cells can absorb antigens.

intestine. For example, polypeptides of size greater than $M_r = 20,000$ are able to reach the blood via the lymphatic system, even though they are unable to traverse blood capillary membranes. The lymphatic system also plays a significant part in the absorption of subcutaneously administered drugs of all kinds.

Microfold cells (M cells), found in discrete anatomical regions of the ileum known as Peyer's patches, absorb antigens as part of the body's immune response. They can absorb nanoparticles, which are being investigated as a way of administering drugs that either have poor or no oral bioavailability (Lopes *et al.*, 2014). Absorption via the lymphatic system is a way of avoiding first-pass metabolism.

9.2.2 Bioequivalence

Concepts of bioequivalence have to be applied carefully with these drugs. In some cases the chemical properties and methods of analysis allow the application of standards similar to those for small molecules. However, different batches within the processes of a single manufacturer can vary with protein drugs, let alone between manufacturers, and regulatory authorities permit a rational protocol of the best available practices in manufacturing, chemical analysis and bioassay to be used in ensuring batch-to-batch and manufacturer-to-manufacturer consistency. Hence the concept of 'biosimilars' (FDA, 2016).

9.2.3 Distribution

Because most of the compounds in this category are polar, their apparent volumes of distribution tend to be low, often similar to the volume of plasma (Table 9.2). For some it may be higher, a little over 1 L kg^{-1}, indicating the occurrence of some specific tissue uptake.

Distribution can be very specific. After intranasal administration, cholecystokinin is found only at sites in the stomach associated with the neuronal pathways to the brain that are connected with feelings of satiety. Trastuzumab shows specific localization in tumour tissue because it forms a complex with the extracellular domain of HER-2.

9.2.4 Metabolism

Peptides are mostly metabolized by proteases in the liver and kidney. Half-life values vary from a few minutes to as much as 28 days, depending on the ease with which the peptide bonds interact with the peptidase active sites. Just exactly why particular peptides and proteins vary so much in their sensitivity to proteases has been the subject of a considerable amount of research. Proteases are broadly classified into endopeptidases, which

Table 9.2 *Apparent volumes of distribution of selected large molecules*

Drug	Volume of distribution (L kg^{-1})
Trastuzumab	0.055
L-Asparaginase	0.055
Heparin	0.058
Streptokinase	0.08
Desmopressin	0.30
Ciclosporin	1.3

cleave the internal peptide bonds in substrates, and exopeptidases, which cleave the terminal peptide bonds. Exopeptidases can be further subdivided into aminopeptidases and carboxypeptidases. There is a further subclassification based on models applicable to describing the physical nature of the interactions themselves between the substrate and the active site of the enzyme. Proteases can also be classified as aspartic proteases, cysteine proteases, metalloproteases, serine proteases and threonine proteases, depending on the nature of the active site. Whether or not a particular substrate interacts with a particular active site will depend on the degree to which the peptide bond(s) in the substrate lack(s) hindrance to a close fit, and whether or not the substrate and enzyme are to be found in close proximity in the same body fluids. Thus, small, straight chain peptides have little chance of survival in the presence of intestinal proteases, while the peptide bonds of ciclosporin are so hindered that this drug barely behaves like a peptide and is primarily metabolized by cytochrome P-450 enzymes. The therapeutic proteins that are given by intravenous injection do not come into contact with intestinal proteases. Whenever possible, protein drugs will be designed to resist attack by proteases.

9.2.5 Excretion

The standard concepts apply (Section 3.3). In particular, insulin (M_r ~6,000) is filtered at the glomerulus and is then metabolized to its amino acids by enzymes in the brush border of the renal tubular lumen, and/or within luminal cells.

Superoxide dismutase shows saturation of renal clearance as a function of plasma concentration, a phenomenon that is unusual amongst drugs, but familiar to investigators of transport maxima in the kidney with endogenous compounds, as occurs with glucose reabsorption.

9.3 Plasma kinetics and pharmacodynamics

A variety of phenomena familiar to pharmacokineticists is seen with compounds in this category, for example:

- Many quite conventional processes occur, measurement of half-life values, protein binding, growth and decay of plasma concentration curves, renal excretion of metabolites, etc.
- Heparin shows dose-dependent kinetics. This causes difficulty in controlling dosing.
- Trastuzumab also shows dose-dependent kinetics. This is so prominent that it has been turned to advantage in devising optimum dosage regimens for this drug.
- Several of the molecules in this class show bi- or tri-exponential plasma level decay after i.v. bolus or infusion doses. For example, trastuzumab data were fitted to a 'two-compartment linear model' even though its apparent volume of distribution was just 2.95 L (~plasma volume).
- A close correlation between kinetics and effect does not always occur. For example, erythropoietin has a half-life after i.v. injection of 10 h, but the response can be delayed for up to 2–6 weeks. Desmopressin shows a biphasic plasma concentration decay after i.v. dosing, with half-life values of 6.5–9 and 30–117 min, but its antidiuretic effect lasts for 6–20 h. Insulin has a half-life of 5–6 min, and its time-course of effect is controlled

by its pharmaceutical formulation. Somatrophin (HGH) has a half-life of 20–30 min, although its therapeutic value is measured over a period of months or years.
- It would appear that PK-PD models with effect compartments, or invoking principles similar to those applicable with warfarin, are most appropriate with drugs in this class.

9.4 Examples of particular interest

9.4.1 Cholecystokinins

Cholecystokinins are a family of hormones with multiple roles. Individual products of their synthesis are identified by the number of amino acids, for example CCK58 and CCK33. CCK8 (Figure 9.2) is not a drug *per se*, but is of importance as a probe for the study of satiety mechanisms. After intravenous or intranasal pharmacological doses it is to be found almost exclusively in the pyloric region of the stomach wall, where there are stretch receptors that send signals to the brain via the vagus nerve to indicate that the stomach is full and so suppress appetite. This has been explored in human pharmacological investigations by Greenough *et al.* (1998) and others. Cholecystokinin has a half-life measured in minutes when incubated with kidney peptidase preparations. However, it is sufficiently stable *in vivo* for it to be possible to show a clear reduction in meal size in human volunteers given intranasal doses, thus shedding light on satiety mechanisms. There have been attempts to synthesise cholecystokinin analogues that are more stable, and have the same properties, in the search for anti-obesity drugs.

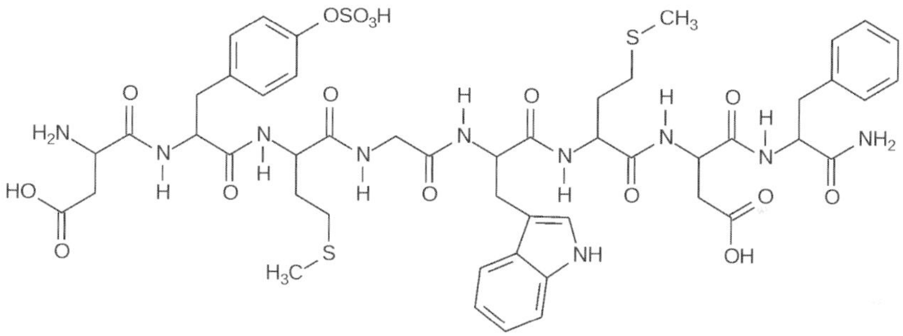

Figure 9.2 Formula of CCK8.

9.4.2 Ciclosporin

Ciclosporin (cyclosporine) is a cyclic polypeptide (Figure 9.3) of fungal origin, which is used as an immunosupressant in transplant medicine. It is lipophilic and hydrophobic. It can be administered intravenously dissolved in a modified castor oil/ethanol mixture, or orally in a soft capsule when it has systemic availability of 20–50%. Soft capsules containing a micro-emulsion formulation have 10% greater availability. It has an apparent volume of distribution of 1.3 L kg^{-1}, and approximately 30 known metabolites. It is a cytochrome P-450 substrate, and so is prone to the drug interactions common with small molecule drugs. Its half-life is 5–6 h. It is mostly excreted as metabolites in bile and hence faeces. Its therapeutic use requires close monitoring of concentrations in plasma.

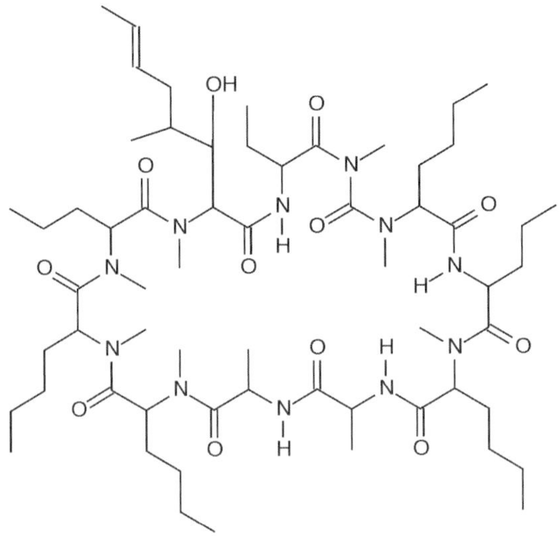

Figure 9.3 Structure of ciclosporin.

9.4.3 Heparin

Heparin (Figure 9.4) is a naturally occurring negatively charged glycosoaminoglycan, and thus is a polymer of alternating D-glucuronic acid and *N*-acetyl-D-glucosamine residues. It is produced in the body in mast cells, where it is stored in granules along with histamine. It is used as an anticoagulant, and it is obtained for medical purposes from biological sources as a by-product of the meat industry. Different batches are not consistent with each other. Its molecular weight varies from 4,000 to 30,000 depending on the degree of breakdown of its mucopolysaccharide side chains. The low molecular weight fractions and high molecular weight fractions have effects at different points in the blood-clotting cascade.

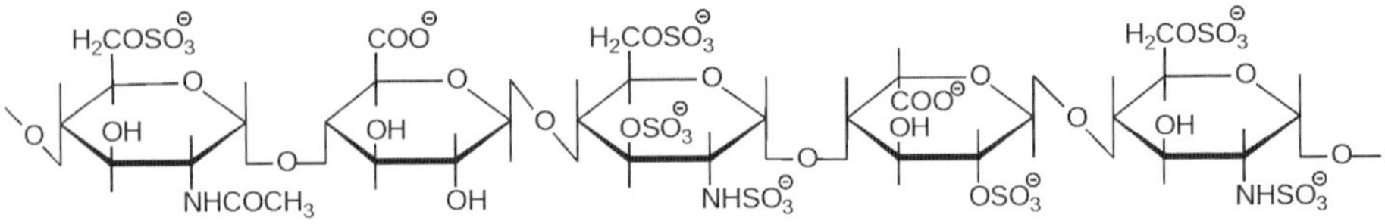

Figure 9.4 Representation of heparin. Heparin consists of repeating disaccharide units which may be sulfated to varying degrees.

Heparin is standardized and dosed in internationally recognized potency units, not in mass units, and even plasma concentration measurements use these units. It is given by i.v. injection for a rapid effect, or by subcutaneous injection, when the onset time is 1–2 h. Even with i.v. injection there is a short delay in onset time, and the effect lasts longer than would be expected from the plasma concentration–decay kinetics. This has led to suggestions that 'activation' is needed. The apparent volume of distribution is 0.058 L kg^{-1}. There is a short phase of fast decline of plasma concentrations after a bolus injection, with a half-life of approximately 5 min, followed by a slower phase with a half-life in the range 1–5 h. There has been extensive discussion about whether there is or is not dose-dependent decline of plasma concentrations but the evidence favours dose dependence. The half-lives of the terminal phase of concentration decline in plasma are generally accepted to be 1, 2.5 and 5 h after doses of 100, 400 and 800 units kg^{-1} respectively, showing marked non-linearity in the elimination kinetics of the drug.

Part of the confusion over the kinetics of heparin has undoubtedly been caused by the methods of analysis. While direct chemical assay is desirable, several 'pharmacokinetic' studies have really been studies of the time course of effect, using various methods to assess the clotting time. The thromboplastin time (APTT) assay is probably the most accepted. In this assay:

- oxalated plasma + a partial thromboplastin time reagent + a surface activator are incubated for 3–5 min
- calcium chloride is added
- the coagulation time is measured.

A plot of *APTT* versus heparin concentration may be linear or log–linear, depending on circumstances that are not always fully understood. Thus the apparent non-linearity in kinetics may have been because of a lack of appreciation of the non-linearity of the assays used. However, there is broad acceptance now that a log–linear model is applicable to the effect of heparin, thus:

$$APTT = APTT_0 \exp(mC) \tag{9.1}$$

Taking logarithms gives:

$$\ln(APTT) = mC + \ln(APTT_0) \tag{9.2}$$

where m is the slope of the graph of ln(*APTT*) versus concentration, C.

Heparin plasma concentrations are considered to be best described by a Michaelis–Menten type equation. During a constant rate infusion the rate of change in concentration is:

$$\frac{\mathrm{d}C}{\mathrm{d}t} = \frac{R_0}{V} - \frac{V_{max}C}{Km + C} \tag{9.3}$$

where R_0 is the rate of heparin infusion (units h^{-1}), V_{max} is the maximal rate of heparin elimination (units h^{-1}) and Km is the heparin concentration at half-maximal velocity (units mL^{-1}). The steady-state concentration is given by rearranging Equation 4.36. However, because this is non-linear kinetics, clearance decreases with increasing concentrations according to Equation 4.47. Combining to two equations gives:

$$C^{ss} = \frac{R_0}{CL_{ss}} = \frac{R_0\left(Km + C^{ss}\right)}{VV_{max}} \tag{9.4}$$

where CL_{ss} is the clearance at steady state. This approach aids the understanding of the relationship between time after the dose and concentration in plasma, and when considered together with the equations for *APTT* above, the effect.

A scatter diagram of *APTT* versus plasma concentration of heparin in a population of patients is truly a scatter diagram, showing no meaningful between-patient correlation. Heparin is a complex drug to understand, and a difficult drug to manage in the clinic.

9.4.4 Adalimumab and trastuzumab

These two compounds are the best known examples of monoclonal antibodies devised by the biotechnology industry. Adalimumab is used for autoimmune diseases, including Crohn's disease, ulcerative colitis and rheumatoid arthritis, and trastuzumab is for the specific treatment of HER-2 positive breast cancer (Tokuda *et al.*, 1999; McKeage & Perry, 2002). Trastuzumab shows marked non-linearity in its elimination kinetics, with the decrease in clearance being accompanied by a corresponding increase in half-life (Figure 9.5), with little change in apparent volume of distribution. As the aim of therapy is to maintain trough concentrations in plasma of 20 μg L^{-1} over a 25–30 week period, it has been possible to replace the originally proposed dosing regimen of a 4 mg kg^{-1} loading dose followed by 2 mg kg^{-1} weekly for about 6 months, with an 8 mg kg^{-1} loading dose followed by 6 mg kg^{-1} every 3 weeks for the same duration. Because the drug must be given intravenously in an outpatient clinic setting, this has resulted in a much more convenient regimen, with the number of clinic visits reduced by two-thirds. The half-life of trastuzumab is similar to that of the endogenous IgGI immunoglobulin that constitutes the backbone of the drug. Of particular interest is the fact that trastuzumab forms a complex with the extracellular domain of HER-2 and the clearance of the complex is greater than that of the drug, demonstrating a rational approach to removing an unwanted residue from the body.

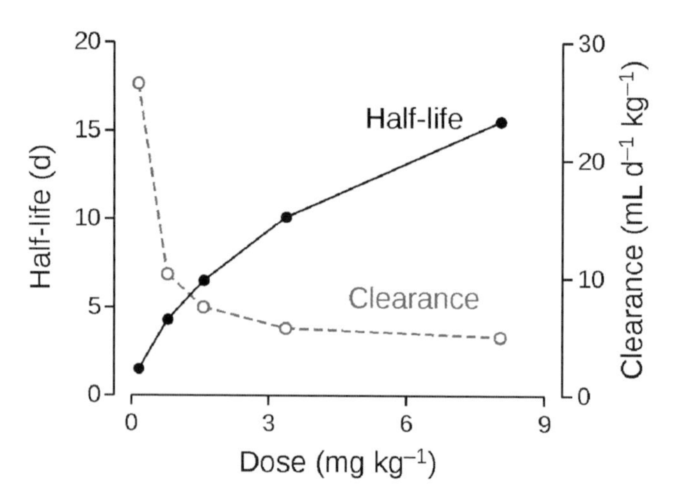

Figure 9.5 Half-life and clearance of trastuzumab as a function of dose.

9.4.5 Erythropoietin

This compound is typical of the protein drugs formed using recombinant DNA techniques to imitate the normal body constituent. It consists of 193 amino acids residues and has a molecular weight of approximately 30,000 g mol^{-1}. It is heavily glycosylated, and is used to treat the anaemia of chronic renal disease. There is always a baseline erythropoietin concentration in plasma, so therapy adds drug to the endogenous amount. The recombinant product is epoetin alfa, and it can be given i.v. or subcutaneously.

Its half-life is 10 h, although its t_{max} is often said to be between 5 and 24 h; the response can be delayed for 2–6 weeks in some patients, and success in treatment is measured over long periods of time.

9.4.6 Vasopressin and desmopressin

Vasopressin, antiduretic hormone (ADH), is a natural nonapeptide (Figure 9.6) that acts in the body to control urine production in the process of homeostasis. Desmopressin is a synthetic peptide analogue. Both compounds are used in pharmacological doses to control the excessive water loss of diabetes insipidus. Vasopressin has been used in this way for many years, and was, apart from cocaine, the first drug to be administered intranasally, having been formulated as a powder that was known as 'pitressin snuff.' This route of administration, while being the only feasible method for this drug, does provide a measure of control of dosing for what is a very difficult drug to manage.

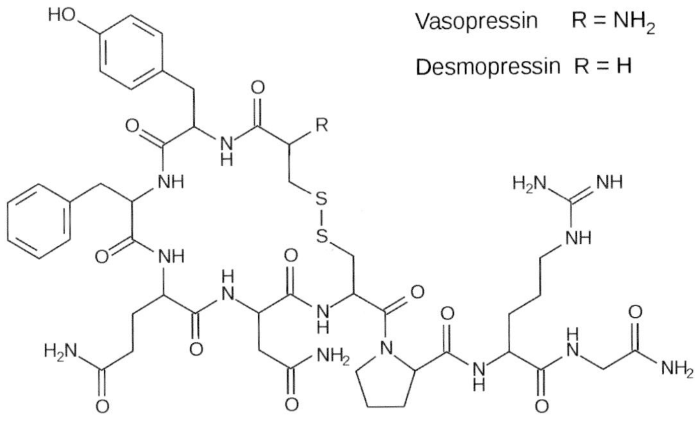

Figure 9.6 Structures of vasopressin and desmopressin.

9.4.7 Mipomersen

This drug is an antisense oligonucleotide used to treat familial hypercholesterolaemia, which has a prevalence of 1 in 500 people. It is the second of its type to be approved, the first being fomiversen, which is used as a treatment for cytomegalovirus retinitis.

The principle of antisense technology is to identify a region on the target mRNA and to develop a complimentary strand of nucleic acid to bind to that region, thereby preventing the mRNA from adopting the single-stranded configuration that is needed for protein synthesis to take place. In the case of mipomersen the protein is apolipoprotein B. The 'sense' sequence of bases is that on the mRNA while the complementary sequence is referred to as the 'antisense' sequence so, for example, an antisense oligonucleotide, 3′-UUCCAG-5′, would block a sense sequence of 5′-AAGGUC-3′, effectively turning off the targeted gene.

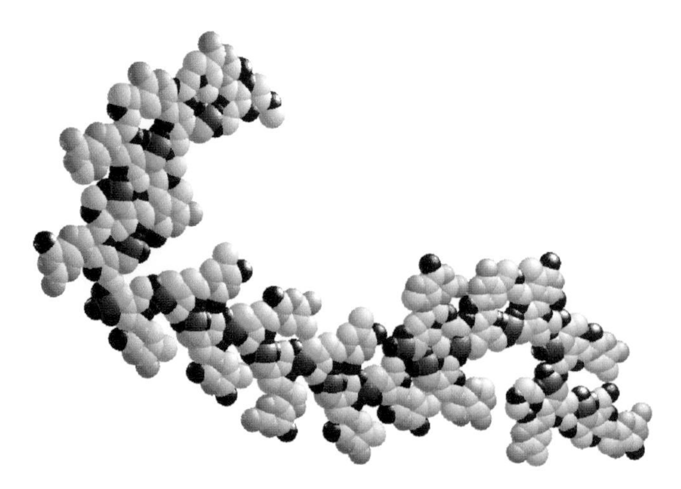

Figure 9.7 *Space-filled model of mipomersen showing the phosphorothioate linkages. Grey, carbon; blue, nitrogen; red, oxygen; purple, phosphorous; yellow, sulfur.*

For therapeutic use, antisense oligonucleotides have to be modified to prevent rapid destruction by intracellular nucleases and to lessen toxicity thought to be caused by the individual nucleotides that are produced (Dias & Stein, 2002). In mipomersen ($M_r = 7594.8$) the nucleotides are connected by phosphorothioate linkages (Figure 9.7) rather than the phosphodiester linkages of native RNA and DNA. This increases both stability and endocytotic uptake into cells. The molecule is further protected from degradation by using five molecules of 2′-*O*-methoxyethyl-modified ribose, rather than 2′-deoxyribose, at the terminal ends. The stability of mipomersen is such that it need only be injected weekly. The drug accumulates selectively in the liver, which is its site of action.

9.4.8 Cocaine hydrolases

At least two synthetic variants of human butyrylcholinesterase have been made that are relatively stable in body fluids (Chen *et al.*, 2016; Cohen-Barak *et al.*, 2015). The enzymes are claimed to increase the rate of hydrolysis of cocaine and are currently undergoing clinical trials to determine their potential value as antidotes to cocaine. Cocaine (Figure 3.3) is a major cause of medical emergencies, causing dangerous cardiovascular effects and acute convulsions if used to excess. The artificial enzymes have been modified so that they have long half-lives after dosing by injection, the therapeutic concept being that any cocaine taken while the user has a significant concentration of the enzyme in his or her blood would be very rapidly metabolized to ecgonine methyl ester and ecgonine, thus preventing it from causing acute effects or reinforcing dependence. Rapid hydrolysis also prevents cocaine from being metabolized to norcocaine, which shares many of the properties of the parent compound. Clearly, the use of such a treatment in attempting to reverse dependence requires a willing patient. Quite apart from such considerations, this work may turn out to lead to the first ever use of a drug-metabolizing enzyme as a therapeutic product.

Summary

The last 30 years have seen the birth of a brave new world of scientifically conceived highly specific 'biological' drugs. The examples quoted here are just the leading edge of what is to come in the search for specificity in drug delivery and effect, in freedom from unwanted pharmacological effects and in what is starting to be called 'personalized medicine'.

9.5 Further reading

Estes JW. Clinical pharmacokinetics of heparin. *Clin Pharmacokinet* 1980; 5: 204–20.

Fahr A. Cyclosporin clinical pharmacokinetics. *Clin Pharmacokinet* 1993; 24: 472–95.

Wills RJ. Clinical pharmacokinetics of interferons. *Clin Pharmacokinet* 1990; 19: 390–9.

9.6 References

Chen X, Xue L, Hou S, Jin Z, Zhang T, Zheng F, Zhan CG. Long-acting cocaine hydrolase for addiction therapy. *Proc Natl Acad Sci U S A* 2016; 113: 422–7.

Cohen-Barak O, Wildeman J, van de Wetering J, Hettinga J, Schuilenga-Hut P, Gross A, Clark S, Bassan M, Gilgun-Sherki Y, Mendzelevski B, Spiegelstein O. Safety, pharmacokinetics, and pharmacodynamics of TV-1380, a novel mutated butyrylcholinesterase treatment for cocaine addiction, after single and multiple intramuscular injections in healthy subjects. *J Clin Pharmacol* 2015; 55: 573–83.

Dias N, Stein CA. Antisense oligonucleotides: basic concepts and mechanisms. *Mol Cancer Ther* 2002; 1: 347–55.

FDA. Information on biosimilars. 2016. http://www.fda.gov/Drugs/DevelopmentApprovalProcess/HowDrugsareDevelopedandApproved/ApprovalApplications/TherapeuticBiologicApplications/Biosimilars/ (accessed 29 September, 2016).

Greenough A, Cole G, Lewis J, Lockton A, Blundell J. Untangling the effects of hunger, anxiety, and nausea on energy intake during intravenous cholecystokinin octapeptide (CCK-8) infusion. *Physiol Behav* 1998; 65: 303–10.

Lopes MA, Abrahim BA, Cabral LM, Rodrigues CR, Seica RM, de Baptista Veiga FJ, Ribeiro AJ. Intestinal absorption of insulin nanoparticles: contribution of M cells. *Nanomedicine* 2014; 10: 1139–51.

McKeage K, Perry CM. Trastuzumab: a review of its use in the treatment of metastatic breast cancer overexpressing HER2. *Drugs* 2002; 62: 209–43.

Tokuda Y, Watanabe T, Omuro Y, Ando M, Katsumata N, Okumura A, Ohta M, Fujii H, Sasaki Y, Niwa T, Tajima T. Dose escalation and pharmacokinetic study of a humanized anti-HER2 monoclonal antibody in patients with HER2/neu-overexpressing metastatic breast cancer. *Br J Cancer* 1999; 81: 1419–25.

10

Pharmacogenetics and Pharmacogenomics

Learning objectives

By the end of the chapter the reader should be able to:

- explain the terms 'pharmacogenetics' and 'pharmacogenomics'
- describe the approaches to phenotyping
- discuss the influence of acetylator status on the toxicity of isoniazid
- discuss the effects of 'mutant' cytochrome P-450 genes on drug therapy
- explain how genetic differences in OATP1B1 can lead to increased exposure to statins but reduce the therapeutic response to other drugs
- discuss the role of personalized medicine in prescribing appropriate doses of drugs.

10.1 Introduction

Genetic differences in drug response may be due to differences in pharmacodynamics (different receptor populations) or in drug disposition (differences in drug-metabolizing enzyme and transporters). When genetic differences are due to a single gene mutation and the incidence is >1% then such differences may be detectable in population studies as a bi- or tri-modal distribution. There is interest in such polymorphisms because metabolism not only inactivates or activates drugs, but also carcinogens and procarcinogens, and much

Introduction to Drug Disposition and Pharmacokinetics, First Edition. Stephen H. Curry and Robin Whelpton.

Companion website: www.wiley.com/go/curryandwhelpton/IDDP

of the recent literature is devoted to assessing the role of genetics as a risk factor in cancer. The term 'pharmacogenetics' is usually applied to the study of drug interactions with a relatively restricted number of genes, whereas pharmacogenomics aims to study the effect of the entire complement of genes (i.e. the genome) on drug action. As this is a rapidly developing field, anything written one year is likely to be superseded a few years later. Thus, this chapter will use selected examples to illustrate the principles involved.

10.1.1 Terminology

In mammalian cells *chromosomes* are thread-like structures in the nucleus comprising DNA and associated proteins. Typically there are two sets of chromosomes, arranged in pairs (*diploid*), one set being inherited from each parent. *Genes* are sequences of nucleic acids located on regions (*loci*) of chromosomes that define the characteristics or traits of the organism. They can be considered as the basic units of heredity. Different forms of a gene are known as *alleles* and it is possible to inherit the same alleles, in which case the individual is referred to as a *homozygote* and said to be *homozygous*, or when the alleles are different, a *heterozygote*. The genetic makeup of an organism is known as the *genotype*. *Phenotype* refers to the physical characteristics exhibited and these can be influenced by inherited and environmental factors. The phenotype in heterozygotes will be largely determined by the interaction of the different alleles, which can often be referred to as *dominant* or *recessive*. In the simplest case, a dominant allele will produce the same phenotype as that when both dominant alleles are present. Dominant alleles may be denoted R and recessive ones r, so that heterozygotes are Rr whilst homozygotes are either RR or rr. The different phenotypes are referred to as being *polymorphic* (having different forms). Clinically it is the phenotype that is important but knowledge of the genotype may help to explain the phenomenon.

By convention genes (e.g. *CYP2C19*) and alleles (e.g. *CYP2C19*1*) are written in italics and in capitals when referring to human genes, whilst the gene products (enzymes, transporters etc.) are written in the standard font (e.g. CYP2C19*1 or CYP2C19.1). The term *wild-type* may be encountered. It was introduced to describe the form of allele found in nature, that is, it was considered to be the 'standard' or 'normal' allele, others being mutant alleles. However, most genes exist in a variety of forms, the frequency of which varies depending on the geographic range of the species and, in the case of humans, the extent to which populations have migrated and interbred.

Often alleles occur because of a *single-nucleotide polymorphism* (SNP), which can give rise to a protein in which one amino acid is substituted for another. In the case of enzymes this may result in reduced activity or no activity. The site(s) of the SNPs may be identified, for example in *CYP1B1*3* cytosine is replaced by guanine, C432G, which produces an enzyme where leucine at 432 is replaced by valine, Leu432Val. Occasionally SNPs lead to enzymes with increased activity. A *null allele* is one that either produces no protein or a protein that lacks any function.

10.2 Methods for the study of pharmacogenetics

10.2.1 Studies in twins

An obvious way to investigate whether a phenomenon is genetically related is to investigate it in twins. The major influence of genetic control on drug disposition has been demonstrated by studying the half-lives of several drugs, including phenazone,

Table 10.1 Paired plasma half-life values (days) for the decline in phenylbutazone in seven pairs of identical and non-identical twins (Vesell & Page, 1968)

	1	2	3	4	5	6	7
Identical twins	2.8	2.6	2.8	4.0	3.9	1.9	3.2
	2.8	2.6	2.9	4.0	4.1	2.1	2.9
Non-identical twins	7.3	2.9	2.6	1.9	2.1	2.3	2.8
	3.6	3.0	2.3	2.1	1.2	3.3	3.5

dicoumarol, phenylbutazone and nortriptyline, in identical (monovular) and fraternal (biovular) twins. The similarity in values for identical twins can be striking, as with the study by Vesell and Page (1968), who investigated the elimination half-life of phenylbutazone after oral doses (Table 10.1).

10.2.2 Phenotyping and genotyping

Early observations of the influence of genetics on drug disposition arose because subjects could be classified according to their phenotype, for example they were either 'slow' or 'fast' acetylators of isoniazid (see below). Today phenotyping may be carried out systematically, using drugs as 'probes' to determine a subject's metabolizer status. Mixtures of drugs have been developed and some bear the name of the institution in which they were developed, for example the Pittsburgh, Indiana and Karolinska Cocktails (Table 10.2). These drugs may also be used when investigating the effects of age, sex and drug interactions on enzyme activity. The tests may simply determine the concentration of the test drug in plasma or urine at a defined time after it has been administered, or specific metabolites may be measured. Alternatively, serial samples may be collected so that *AUC* and oral clearance values can be calculated. The erythromycin breath test involves administering [^{14}C-methyl]-erythromycin and collecting breath (in a balloon) for measurement of $^{14}CO_2$, which reflects the degree of demethylation. Intravenous and oral administration of midazolam has been used to differentiate between hepatic and gut wall (+ hepatic) metabolism by CYP3A4. Additionally, inhibitors may be given in an attempt to confirm the identity of the enzyme involved. A limitation to this approach is the lack of specificity of some substrates and inhibitors for some enzymes and transporters. This is a particular problem with CYP3A4, CYP3A5 and P-gp, which have similar substrate specificities.

Polymerase chain reactions (PCR), originally with restriction fragment length polymorphism (RFLP), are used to sequence genes and so identify particular genotypes and variant alleles. Furthermore, transfection of cDNA into organisms (e.g. *E. coli*) or cell lines allows production of recombinant enzymes which can be sequenced to identify changes in amino acid composition and enzyme activity, expressed as Km and V_{max} values for the substrates of interest. This can lead to confusion as a 'mutant' enzyme may be more active per weight of protein than the wild-type when tested *in vitro*, but if the variant allele results in much less enzyme being expressed, then the *in vivo* activity may be reduced.

10.3 *N*-Acetyltransferase

There are two major forms of arylamine *N*-acetyltransferase. Substrates of type 1 include *p*-aminobenzoic acid, *p*-aminosalicylic acid and endogenous *p*-aminobenzylglutamate, whereas a number of primary aromatic amine and hydrazine drugs are acetylated by

Table 10.2 *Examples of substances used to phenotype individuals for drug-metabolizing activity*

Enzyme	Probe	Measurement	Sample/time
CYP1A2	Caffeine[*]	Caffeine/1,7-dimethylxanthine	Plasma/8 h
	Caffeine[†]	1,7-Dimethylxanthine/caffeine	Serum/6 h
	Caffeine[‡]		Plasma/4 h
CYP2B6	Bupropion[**]	Hydroxylation	
CYP2C8	Amodiaquine[**]	Desethylation	
CYP2C9	Losartan[‡]	5-Carboxylic acid metabolite (E-3174)	Urine/8 h
	Tolbutamide[†]		Serum/serial samples
CYP2C19	Mephenytoin[*]	4-Hydroxymephenytoin	Urine/8 h
	Omeprazole[‡]	5-Hydroxyomeprazole	Plasma/3.5 h
CYP2D6	Debrisoquine[*,‡]	4′-Hydroxydebrisoquine/(4′-hydroxydebrisoquine + debrisoquine)	Urine/8 h
	Dextromethorphan[†]	Dextrorphan	Urine/serial samples
CYP2E1	Chlorzoxazone[*]	6-Hydroxychlorzoxazone/chlorzoxazone	Plasma/4 h
CYP3A4	Dapsone[*]	Dapsone hydroxylamine/(hydroxylamine + dapsone)	Urine/8 h
(hepatic) + (hepatic + intestinal wall)	Midazolam (i.v.)[†]		Serum/serial samples
	Midazolam (p.o)[†]	1′-Hydroxymidazolam	
	Cortisol	6-β-Hydroxycortisol/cortisol	Urine
	[^{14}C]-Erythromycin	[^{14}C]-CO_2	Breath/20 min
CYP3A4/5	Quinine[‡]	3-Hydroxyquinine	Plasma/16 h
	Dextromethorphan	3-Methoxymorphinan	
CYP3A5	Midazolam[**]	1′-Hydroxymidazolam	
NAT2	Dapsone[*]	Monacetyldapsone	Plasma/8 h
	Sulfadimidine	Acetylsulfadimidine	Plasma/8 h

* Pittsburg cocktail (Frye *et al.*, 1997).
† Indiana cocktail (Wang *et al.*, 2001).
‡ Karolinska cocktail (Christensen *et al.*, 2003).
** O'Donnell *et al.* (2007).

N-acetyltransferase type 2 (NAT2), including the examples of Figure 3.14. Over 30 variants of the *NAT2* gene have been identified. *NAT2*4* is considered to be the 'wild-type' allele.

10.3.1 Isoniazid

The bimodal nature of metabolism of the anti-tuberculosis drug isoniazid was known in the early 1950s but it was the classic experiments of Evans *et al.* (1960) that demonstrated the genetic nature of the polymorphism. Plasma isoniazid concentrations were shown to be bimodally distributed when 483 subjects were given identical doses. A subset of results from 267 members of 53 families confirmed the hereditary nature of the phenomenon. Subjects with plasma concentrations of <2 $mg\,L^{-1}$ were referred to as 'rapid inactivators' while those with lower concentrations were classed as 'slow-inactivators' (Figure 10.1).

The rapid allele (R) is dominant, and so only homozygotes (rr) are slow acetylators. The distribution of fast to slow acetylators is approximately 50:50 in Caucasian and African Americans, so the gene frequency of the slow gene(s) must be ~75%. Inuits and Japanese are primarily fast acetylators (95%) but some Mediterranean Jews are mainly slow (20% fast). Fast acetylators may require higher doses of isoniazid. Slow acetylators may develop a peripheral neuropathy due to the imine (Schiff's base) formed between isoniazid and pyridoxal, which depletes the vitamin. Rapid acetylators, on the other hand, are prone to hepatotoxicity, which is probably caused by *N*-acetylhydrazine that is released from the acetyl metabolite (Chapter 13).

It is sometimes possible to identify heterozygotes, but not from histograms of the type shown in Figure 10.1. For example, systemic clearance, rather than plasma concentration, was shown to correlate with the number of *NAT2*4* alleles (Kinzig-Schippers, *et al.*, 2005).

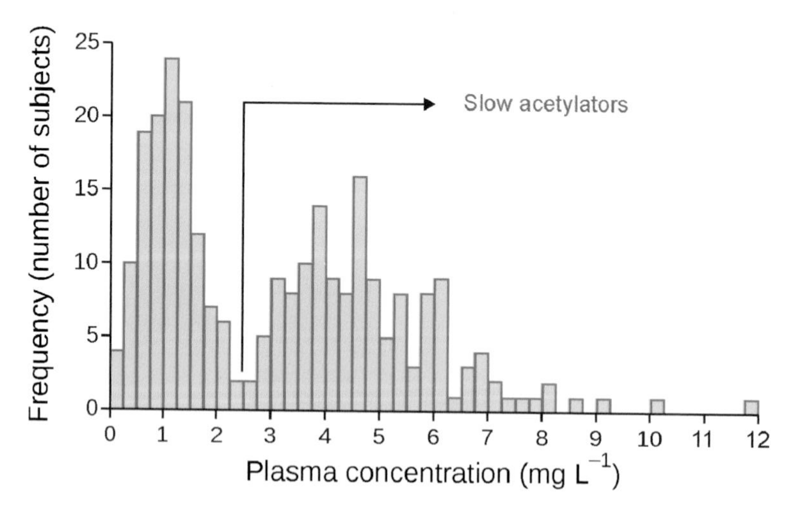

Figure 10.1 *Frequency distribution for plasma isoniazid concentrations 6 h after oral doses of 9.7 $mg\,kg^{-1}$ in 267 members of 53 families. Slow acetylators have the higher isoniazid concentrations. Redrawn with permission from Evans* et al. *(1960).*

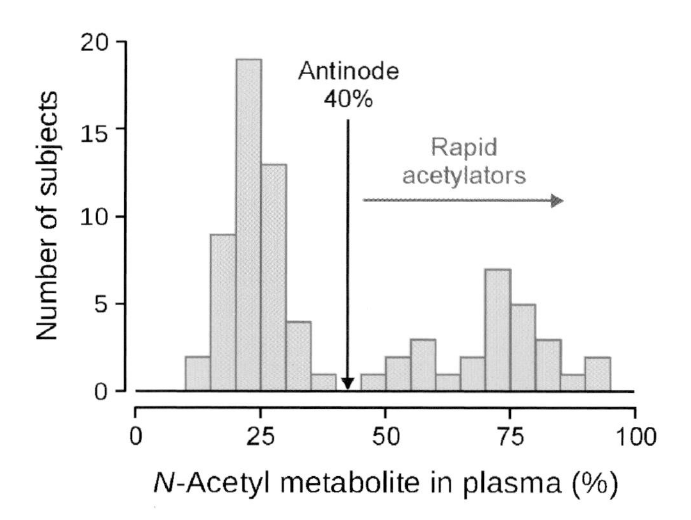

Figure 10.2 *Distribution for the percentage of N-acetyl metabolite in plasma 6 h after a test dose of sulfadimidine (0.5 g, p.o.) in undergraduate medical students. Rapid acetylators have higher concentrations of the acetyl metabolite. Drawn from the data of Whelpton* et al. *(1981).*

10.3.2 Sulfonamides

The acetylation of the primary amine groups of several sulfonamides, including sulfadimidine (sulfamethazine) shows the same polymorphism as that of isoniazid. Because sulfadimidine concentrations could be measured using a relatively simple colorimetric assay, this drug has been used to test for acetylator status. Following a test dose of 0.5 g, the proportion of N^4-acetylsulfadimide in plasma was bimodally distributed, with an antinode at ~40% (Figure 10.2). Thus, those with >40% acetyl metabolite in plasma are fast acetylators. With test doses of 2 g the antinode was ~25%, (Rao *et al.*, 1970) presumably showing that at the higher dose the absorption or metabolism is becoming saturated. As with isoniazid, it has been shown that sulfadimidine elimination half-lives can be assigned to three distinct groups, reflecting the three phenotypes.

10.3.3 Other drugs

Other substrates of NAT2 include hydralazine, phenelzine and dapsone. The acetyl metabolites are considered inactive when compared to the parent drug. Generally, fast acetylators may not respond adequately to treatment whilst slow acetylators are more prone to adverse effects. Procainamide is unusual in that the acetyl metabolite has similar pharmacological properties to the parent drug and is marketed as acecainide.

10.4 Plasma cholinesterase

Several genotypes for plasma cholinesterase (pseudocholinesterase) have been discovered. Approximately 94% of the population are homozygous for the 'normal' allele and are designated EuEu. Of the atypical alleles, the one coding for a dibucaine-resistant form of the enzyme (Ea) is probably the most important clinically. EaEa homozygotes show prolonged paralysis when given the muscle relaxants suxamethonium (succinylcholine) and mivacurium, and may be sensitive to other drugs, including procaine, cocaine, pilocarpine and donepezil. Fluoride-resistant (Ef) and silent (Es) alleles have been identified.

10.4.1 Suxamethonium

In normal subjects the duration of action of suxamethonium is approximately 5–10 min. Most of the injected dose is hydrolysed so that only some 5–10% reaches the acetylcholine receptors of the motor endplate. Drug that diffuses from the receptors is hydrolysed by the normally functioning enzyme. Approximately 1 in 3000 people remains paralysed for an unusually long period following the drug (Figure 10.3). Should this occur during an operation then mechanical ventilation must be continued until the patient can breathe normally.

As well as taking a family history, patients can be tested using standard cholinesterase assays in the presence of a standard concentration of the local anaesthetic dibucaine, which inhibits the normal enzyme activity by 80%. The enzyme from EaEa homozygotes is only inhibited by 20%. These values are known as the dibucaine number. Heterozygotes (EuEa) have dibucaine numbers of ~60, and although these individuals may have a longer duration of apnoea, it rarely lasts for more than 1 h and is not considered clinically important. A more serious situation arises in homozgyotes carrying the Es gene, who have no pseudocholinesterase activity and the duration of apnoea may be over 8 h. The frequency of this polymorphism is 1:100,000.

Demonstrating the polymorphic hydrolysis of other drugs is complicated by the fact that they may also be substrates for the many other esterases that exist in plasma.

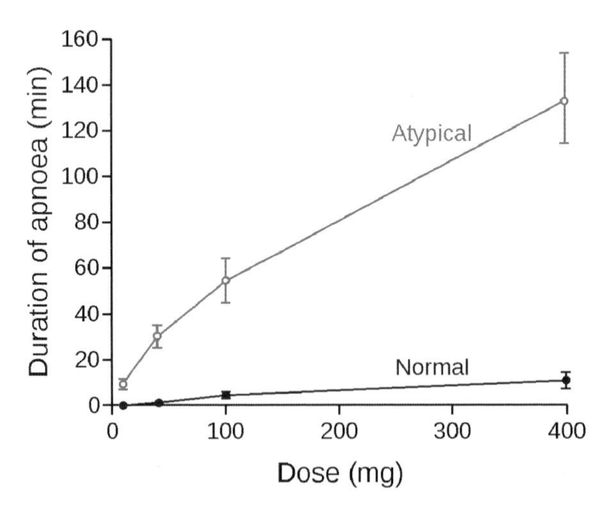

Figure 10.3 Duration of apnoea in adult male patients with normal and atypical cholinesterase. From Kalow & Gunn (1957), with permission.

10.5 Cytochrome P450 polymorphisms

It is probable that polymorphisms exist for all the drug-metabolizing cytochromes and several important clinical differences have been demonstrated for CYP2D6, CYP2C9 and CYP2C19. Individuals with two functioning genes are referred to as extensive metabolizers (EMs) and those with no, or only one, functioning gene, are classed as poor (PMs) and intermediate (IMs) metabolizers, respectively. People with more than two functioning genes are ultrarapid metabolizers (UMs).

10.5.1 Cytochrome 2D6

One of the first observations of a polymorphism in microsomal drug-metabolizing enzymes was the exaggerated response to the obsolete antihypertensive, debrisoquine. The metabolism to the 4-hydroxy metabolite is catalysed by CYP2D6 and has been used to phenotype poor metabolizers (Table 10.2). Urine is collected for 8 h following a test dose of debrisoquine (10 mg, p.o.) and the metabolic ratio (*MR*) calculated from:

$$MR = \frac{\%\text{ of dose as debrisoquine}}{\%\text{ of dose as 4-hydroxydebrisoquine}} \tag{10.1}$$

Poor metabolizers are defined as those having a log(*MR*) value >1.1 (i.e. a ratio >12.5). At the same time as polymorphisms in debrisoquine metabolism were being investigated similar patterns were demonstrated for sparteine and nortriptyline. Subsequently it was shown that the differences were due to different *CYP2D6* alleles. A large number of drugs are substrates for CYP2D6, including several β-blockers, neuroleptics and SSRIs (Table 3.1). It has been claimed that PMs obtain no pain relief from codeine as they are unable to metabolize it to morphine, while UMs show an exaggerated response to codeine. Tamoxifen is another drug that relies on CYP2D6 for its activation and PMs do not respond as well as EMs do to this drug.

The distribution of PMs varies amongst different ethnic groups. In Europeans it is ~7%, which means that some 20–30 million Europeans have no CYP2D6 enzymes. However it has also been estimated that 15–20 million Europeans have multiple copies of the gene. Approximately 2% of Swedish and 7% of Spanish people are UMs. The figure may be as high as 29% for Ethiopians. Many of these patients fail to respond to standard doses of CYP2D6 substrates and may be classed as non-responders. Figure 10.4 shows plasma nortriptyline concentrations in subjects carrying different numbers of functional genes (0–13) after a single dose. The differences in concentrations will be even more marked after multiple dosing to steady-state concentrations.

10.5.2 Cytochrome 2C9

It has been estimated that CYP2C9 catalyses approximately 10% of cytochrome P450-mediated drug metabolism. *CYP2C9*2* and *CYP2C9*3* alleles arise from SNPs and the enzymes have been estimated to confer 70% and 10% of the intrinsic clearance of the wild-type enzyme (CYP2C9*1), respectively. Approximately 35% of Caucasians have at least one *2 or *3 allele. Although rare, *CYP2C9*6* is a null allele, conferring no enzyme activity.

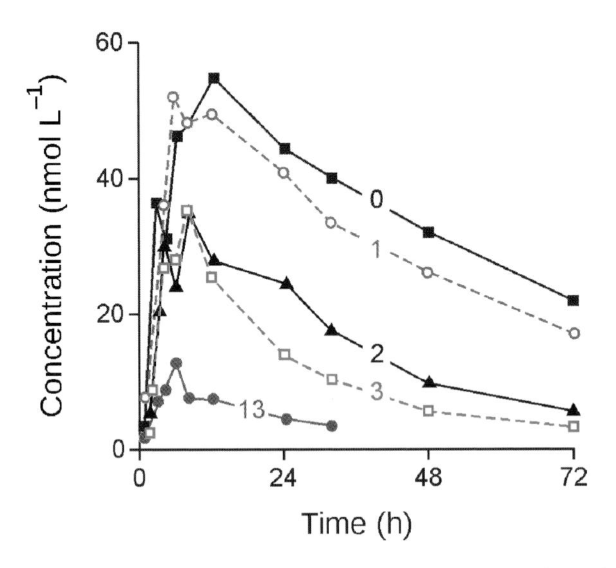

Figure 10.4 *Mean plasma concentration of nortriptyline after a single oral dose in subjects with varying numbers (0–13) of* CYP2D6 *genes, as indicated on the lines. There were five subjects in each group apart from* n = *1 for the subject with 13 genes. Drawn with permission from Dalén* et al. *(1998).*

Table 10.3 *Examples of CYP2C9 substrates*

Analgesics	Anticonulsants	Oral hypoglycaemics	NSAIDs
Paracetamol Phenacetin Amidopyrine	Phenytoin Phenobarbital Valproate	Tolbutamide Glibenclamide Glipizide	*S*-Naproxen Diclofenac Celecoxib
Oral anticoagulants	**Angiotensin II antagonists**	**SSRIs**	**Others**
Warfarin Dicoumarol	Losartan Candesartan	Sertraline Venlafaxine	Fluvastatin Sildenafil

Substrates of this enzyme include phenytoin, tolbutamide, valproate and warfarin (Table 10.3). In the past there have been reports of unusually long half-lives of some of these drugs in a small number of patients. Adverse drug reactions to phenytoin have been ascribed to patients having defective *CYP2C9* alleles, particularly in *3/*3 diplotypes, who comprise ~0.4% of Caucasians.

10.5.3 Cytochrome 2C19

Polymorphic 4′-hydroxylation of *S*-mephenytoin was reported in the 1980s and it has since been shown that the enzyme responsible is CYP2C19. This enzyme catalyses the metabolism of several frequently prescribed drugs. It catalyses hydroxylation of the proton pump inhibitor omeprazole, and the demethylation of diazepam, imipramine and citalopram. The incidence of poor metabolizers is higher in East Asian populations (13–23%) compared with Caucasians (2–5%). *CYP2C19*1* is the wild-type allele whilst *CYP2C19*2* and *CYP2C19*3* are considered defective mutants that result in reduced enzyme activity. Thus, following a single dose of omeprazole the mean *AUC* value for *2/*2 homozygotes was nearly 10 times that measured for *1/*1 homozygotes. The *AUC* for heterozygotes was only twice that of the EM subjects.

The degree to which the pH of gastric contents was increased and the success of ulcer treatment with omeprazole was highly dependent on phenotype, with 100% success in the poor metabolizers, but only 25% in EMs. It has been suggested that CYP2C19 genotyping is cost effective in predicting response to omeprazole and amoxicillin in the treatment of *Helicobacter pylori* infection and peptic ulcer.

10.5.4 Cytochromes 3A4/5

The *CYP3* alleles are clustered on chromosome 7 along with the *MDR1* gene that encodes P-glycoprotein (P-gp). As CYP3A4/5 have been estimated to metabolize some 50% of commonly used drugs and have important roles in first-pass metabolism, polymorphisms could have major effects on bioavailability, and hence pharmacological/toxicological activity. However, because of the similar substrate/inhibitor specificities of CYP3A4/5 and P-gp it is not always possible to ascertain whether individual differences in bioavailability are due to polymorphisms in the enzymes or the transporter. The situation is further complicated by the fact that high levels of these enzymes are expressed in enterocytes and hepatocytes. The enzymes can accommodate both small drugs, such as midazolam, and larger substrates, including ciclosporin. *CYP3A4* alleles have been identified but their frequencies are low.

Higher levels of CYP3A5 are expressed in Africans and saquinavir *AUC*s were 34% lower in 'CYP3A5 producers'. Two alleles arising from SNPs, *CYP3A5*3* and *CYP3A5*6*, have relatively high frequencies and some subjects may not have any functioning CYP3A5. Dosage adjustments of immunosupressants, ciclosporin and tacrolimus may be required for some individuals, but the effect of polymorphism in *MDR1* may be a contributory factor (Section 10.9).

10.5.5 Other cytochrome P450 polymorphisms

Polymorphisms in *CYP2B6* have been identified, of which the variant *CYP2B6*6* appears to be the most important. The frequency of homozygotes for this allele is 3% in Caucasians and 20% in African Americans, and these individuals have higher plasma concentrations and increased adverse reactions with the anti-HIV drug, efavirenz. Cyclophosphamide is a prodrug with complicated activation and inactivation pathways, some of which are non-enzymatic. The first step, oxidation to 4-hydroxycyclophosphamide, is catalysed by CYP2B6 and individuals with the variant *CYP2B6*6* produce a protein that metabolizes the drug at a faster rate. However, the amount of enzyme *expressed* by *6/*6 homozygotes was considerably less than those with the wild-type allele. Polymorphism in nicotine oxidation has been attributed to an inactive variant allele of *CYP2A6*. An alternative explanation is that in some Asian subjects a *CYP2A6* gene may be deleted and may be responsible for reduced nicotine metabolism. Some individuals may have multiple copies of *CYP2A6*.

10.6 Alcohol dehydrogenase and acetaldehyde dehydrogenase

Alcohol dehydrogenase (ADH) is a dimeric enzyme made up of six separate subunits, encoded by three genes, ADH_1, ADH_2 and ADH_3. Many combinations of isoenzymes exist, leading to different rates of metabolism amongst white, black African and Asian populations.

Aldehyde dehydrogenase (ALDH) is a mitochondrial enzyme. Some Asians have ALDH different from that of Caucasians and about 50% of Asians (principally Chinese) have inactive ALDH, leading to flushing and other unpleasant effects when these individuals consume ethanol. These effects are similar to those seen with disulfiram (Section 3.2.2.1).

10.7 Thiopurine methyltransferase

Phenotyping or genotyping of thiopurine methyltransferase (TPMT) is used to guide treatment with azathioprine to avoid life-threatening acute toxicity (Lennard *et al.*, 1989). The incidence of very low TPMT activities is relatively high (1:300), whilst 11% of subjects have intermediate activity. 6-Mercaptopurine, derived from azathioprine, is normally metabolized via one of three pathways: (i) methylation by TPMT, (ii) oxidation by xanthine oxidase or (iii) by hypoxanthine phosphoribosyltransferase to active thiopurine metabolites, including 6-thioguanine nucleotides. Patients with low TPMT activity have unusually high levels of 6-thioguanidine incorporated into DNA, which is, in part, responsible for azathioprine toxicity.

10.8 Phase 2 enzymes

Because of the historical importance of NAT2 polymorphism and the fact that it provides a simple, but clear, example of the issues involved, this phase 2 enzyme was discussed earlier. Other phase 2 enzymes, the UDP-glucuronosyltransferases (UGT), sulfotransferases (SULT) and glutathione transferases (GST), are superfamilies, much as the cytochrome P450s are a superfamily.

10.8.1 UDP-glucuronosyltransferases

The two major classes of *UGT* genes are *UGT1* and *UGT2*, which produce at least 18 enzymes. UGT1A1 catalyses the glucuronidation of bilirubin. The wild-type allele is *UGT1A1*1*, but a common mutant, *UGT1A1*28*, leads to a mild form of hyperbilirubinaemia, known as Gilbert's syndrome, which occurs in 5–10% of the population. As a consequence sufferers may be prone to the adverse effects of drugs or metabolites metabolized by UGT1A1. This has been shown to be the case with the anti-cancer drug irinotecan, which is metabolized to the active, and potentially toxic, metabolite known as SN-38. Glucuronidation and inactivation of this compound is catalysed by UGT1A1 and those with Gilbert's syndrome are more prone to neuropenia and diarrhoea. In 94% of people with Gilbert's syndrome two other UGTs are affected, one of which is thought to catalyse the glucuronidation of paracetamol.

3-Glucuronidation of morphine is catalysed by several UGTs, but only UGT2B7 has been shown to catalyse both 3- and 6-glucuronidation. Some studies of the allelic variant *UGT2B7*2* have shown that the rate of glucuronidation is greater in carriers of this allele, however other studies have failed to substantiate this claim, and whether this genotype has any clinical significance is equivocal. Of course the situation is complicated by the fact that the 3-glucuronide is inactive whereas the 6-glucuronide is analgesic.

10.8.2 Sulfotransferases

The most widely studied sulfotransferase is SULT1A1, also known as 'thermostable' or phenol SULT because it catalyses sulfation of a large number of endogenous and exogenous phenols, including paracetamol and the 4-hydroxy active metabolite of tamoxifen. SULT1A1 and 1A3 are highly expressed in platelets, thereby facilitating study. A common variant is *SULT1A1*2*, in which G638A substitution results in Arg213His in the allozyme. This enzyme is less thermally stable, has reduced activity compared to SULT1A1*1 and a shorter half-life. *SULT1A1*3* arises from a A667G SNP, leading to a Met223Val substitution in the enzyme.

Ethnic variations have been described. The frequency of the *1A1*1* allele is 0.914 in Chinese but only 0.656 and 0.477 in Caucasians and African Americans, respectively. The frequency of the *1A1*2* allele is 0.332 and 0.294 in the latter groups and African Americans have an incidence of 0.229 of the *1A1*3* allele. Much of the research into the functional effects these alleles has concerned sulfation of flavanoids, which may protect against cancer, 17-β-oestradiol, which has been implicated in breast cancer, and 4-hydroxytamoxifen. It has been suggested that women carrying the *1A1*2* or *1A1*3* alleles sulfate the active metabolite of tamoxifen to a lesser extent and so have higher exposure. However, they also have reduced sulfation of oestradiol. Despite the role of sulfation in the metabolism of paracetamol, there appears to be no definitive observation on the impact of SULT variants.

10.8.3 Glutathione transferases

Human GST families are designated by uppercase Greek letters, for example alpha (A), mu (M), pi (P) and theta (T). Polymorphisms are known in the *GSTP1* gene, and deletions occur frequently with *GSTM1* so that 50% of the population are homozygous for the null allele. The frequency of deletions with *GSTT1* is ethnically determined, ~20% of Europeans and ~60% of Orientals and Africans being *GSTT1*0/*0*. High levels of GSTT1 are found in erythrocytes. Thus, it would appear that there is potential for phenotypic polymorphism in GSH conjugation, however most studies have concentrated on the role of these alleles in the development of cancers.

With regard to the effects on drug disposition, the clearance of busulfan, an alkylating agent used for treating chronic myelogenous leukaemia, was significantly lower in *GSTA1*A/*B* heterozygotes than in those with homogenous *GSTA1*A/*A*. The plasma concentrations were correspondingly higher in the heterogeneous group. Ekhart *et al.* (2009) investigated the effects of pharmacogenetics on the oxidation and conjugation of the alkylating anti-cancer drug thiotepa, which is metabolized to tepa. The conclusion was that patients who were homozygous for a *GSTP1* variant allele had greater exposure to the two compounds.

10.9 Transporters

The widespread distribution of multifunctional transport proteins results in these being of prime importance in pharmacogenetics. Interest is primarily in those genes that encode P-gp, multidrug-resistance-associated protein, organic anion transport polypeptides and organic cationic transporters. Some of these transporters are located in the basolateral membranes whilst others are in the luminal membrane and may work in concert to eliminate drugs. The complexity of these combinations can be appreciated from Figure 10.5.

The *MDR1* gene which encodes for P-gp is highly polymorphic. One variant, C3435T, has been shown to result in half the normal P-gp expression in the duodenum and this was associated with higher digoxin concentrations in these subjects. Some 24% of the subjects were homozygous for this SNP. Similarly, the frequency in children being treated for HIV was C/C 44%, C/T 46% and T/T 10%. The children that were heterozygous had higher plasma nelfinavir concentrations and lower oral clearance (although it is not clear whether the clearance values were corrected for differences in bioavailability). Children in this C/T group responded more quickly to treatment than the others. On the other hand, Kim *et al.* (2001) found conflicting results with fexofenadine but pointed out that additional SNPs were possible (C1236T and G677T) and Chowbay *et al.* (2003) showed that haplotypes where all three positions were substituted with thiamine (T-T-T) had increased *AUC* and C_{max} values for ciclosporin.

One of the members of the organic anion transporter polypeptide family, OATP1B1, is thought to be the most polymorphic. It is highly expressed in hepatocytes and the effect of genomic variants on the uptake of drugs has been reviewed by Niemi *et al.* (2011). Two

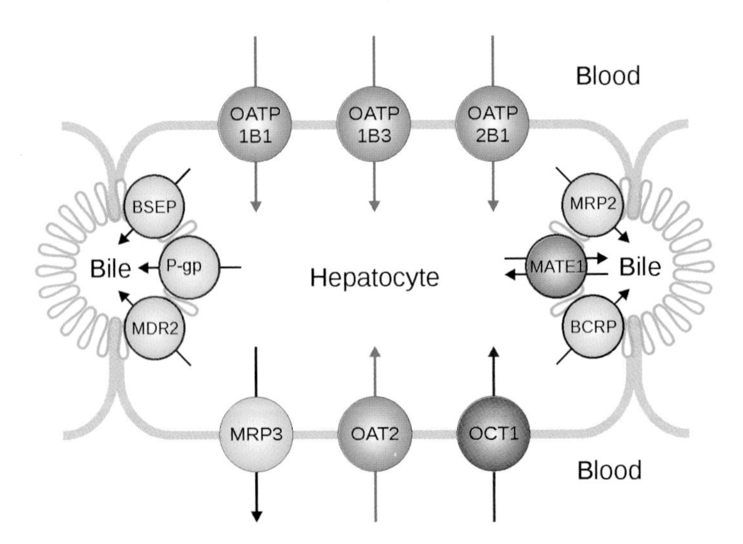

Figure 10.5 Simplified diagram of hepatic transporters. Organic anion transport polypeptides (OATP1B1, OATP1B2 and OAT2B1), organic anion transporter 2 (OAT2) and organic cation transporter 1 (OCT1) facilitate the uptake of solutes into hepatocytes. Multidrug-resistance-associated protein 3 (MRP3) is an efflux pump. The hepatic canalicular efflux proteins include multidrug resistance proteins (P-gp and MDR2), MRP2, breast cancer resistance protein (BCRP) and bile salt export pump (BSEP), which utilize ATP (not shown), and multidrug and toxin extrusion protein 1 (MATE1).

variants, OATP1B1*5 and OATP1B1*15, are associated with reduced activity, whereas some studies have shown OATP1B1*1B to have increased activity. Substrates include drugs with low therapeutic indices, including the anti-cancer agents, irinotecan and methotrexate, and several statins. One of the first examples of a functional effect was the clearance of pravastatin. In subjects with the *OATP1B1*15* variant the total and non-renal clearances were lower than those subjects without this allele. The one subject who was homozygous **15/*15* had the lowest *CL* value and the largest *AUC* (Nishizato *et al.*, 2003). Different statins are affected to varying degrees as evidenced by increases in *AUC* (Figure 10.6). This is clinically important because increased exposure increases the risk of drug-induced myopathy. Furthermore, those statins that are prodrugs have to be taken up by the liver prior to metabolism, for example simvastatin is metabolized to the acid, simvastatin acid, and defective transporters may lead to reduction of activity of these prodrugs.

The risk of GI toxicity observed when i.v. methotrexate is given to children with acute lymphoblastic leukaemia is increased by *OATP1B1* variants that are associated with increased transport activity. This has been explained as enhanced hepatic uptake and biliary clearance of methotrexate increasing the intestinal exposure. Subjects with an SNP that conferred reduced activity experienced less toxicity.

OCT1 is widely distributed and it has been estimated that it is inactive in some 9% of Caucasians. The significance of genetic variants has been reviewed by Arimany-Nardi *et al.* (2015). *OCT1*5* and **6* are associated with reduced, or no, uptake of several drugs, including morphine and sulpiride. The active metabolite of tramadol is internalized in cells and defective alleles result in higher plasma concentrations and increased CNS effects. The antiretroviral drug lamivudine normally enters its target cells via OCT1 and its uptake may be reduced in subjects with variant transporter proteins.

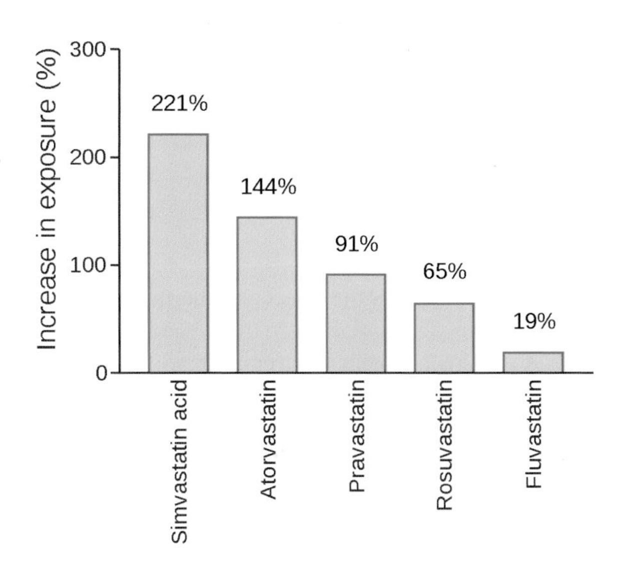

Figure 10.6 *Increase in exposure (as assessed by AUC) in subjects with OATP1B1*15/*15 variant genes when compared with 'normal' genes. Repeated studies in 32 volunteers. After Niemi* et al. *(2011), with permission.*

OCT2 is the major organic cation transporter in the basolateral membrane of renal proximal tubular cells and metformin is a suitable probe as 98% of a dose is eliminated via the kidneys. The renal clearance (CL_R) ranges from ~400 to 600 mL min^{-1}, indicating tubular secretion. There is little plasma protein binding so creatinine clearance, CL_{cr}, has been used to estimate filtration, enabling the contribution from secretion, CL_{sec}, to be estimated:

$$CL_{sec} = CL_R - CL_{cr} \tag{10.2}$$

At least 28 variants of OCT2 have been reported. A study in Chinese subjects to investigate the functional effects of a common variant (G808T) showed the clearance of metformin was lowest in T/T homozygotes, while T/G heterzygotes had intermediate clearance values. Cimetidine, a specific inhibitor, reduced the metformin *AUC* in G/G homozygotes but had little effect on the *AUC* measured in T/T subjects. In a study in Caucasians, low-activity OCT1 was associated with reduced metformin clearance.

An investigation of the role of mutations in OCT genes, in the disposition of the loop diuretic torsemide, revealed that mutations in OCT4 (located in the luminal membrane of proximal tubular cells, see Figure 3.18) rather than OCT1 or OCT3 variants (found in the basolateral membrane) reduced the renal clearance of this drug.

10.10 Ethnicity

In the above, frequent references have been made to ethnic diversity in the distribution of variant genes encoding both drug-metabolizing enzymes and transporters. Knowledge of such differences is important in both the development and clinical use of drugs. It was also pointed out in the introduction to this chapter that migration and interbreeding are responsible for regional distribution of genetic differences. Tables and maps have been created to illustrate the geographic distribution of genes. One such map is that showing the distribution of *SLCO1B1*, the gene that encodes for OAPT1B1 (Figure 10.7).

10.11 Pharmacodynamic differences

Naturally, this chapter concentrates on the effects of pharmacogenetics on drug disposition but it is worth remembering that variations in drug response may be due, wholly or in part, to differences in drug targets. Genetic variants in β-adrenoceptors have been long known. Differences in sensitivity to warfarin also arise because of mutations in the *VKORC1* gene, which encodes for vitamin K epoxide reductase. African Americans are relatively resistant to warfarin. The situation may be exacerbated in those who have low CYP2C9 activity. Ten common polymorphisms have been identified in the *VKORC1* gene, of which haplotypes 1 and 2, collectively known as type A, are associated with a low dose requirement of warfarin. Individuals with haplotypes 7-9, which are collectively known as type B, require higher doses of the anticoagulant. Thus, patients with the combination of A/A VKORC1 haplotypes and CYP2C9*2 or *3 isoforms of the drug-metabolizing enzymes require the lowest doses (Figure 10.8).

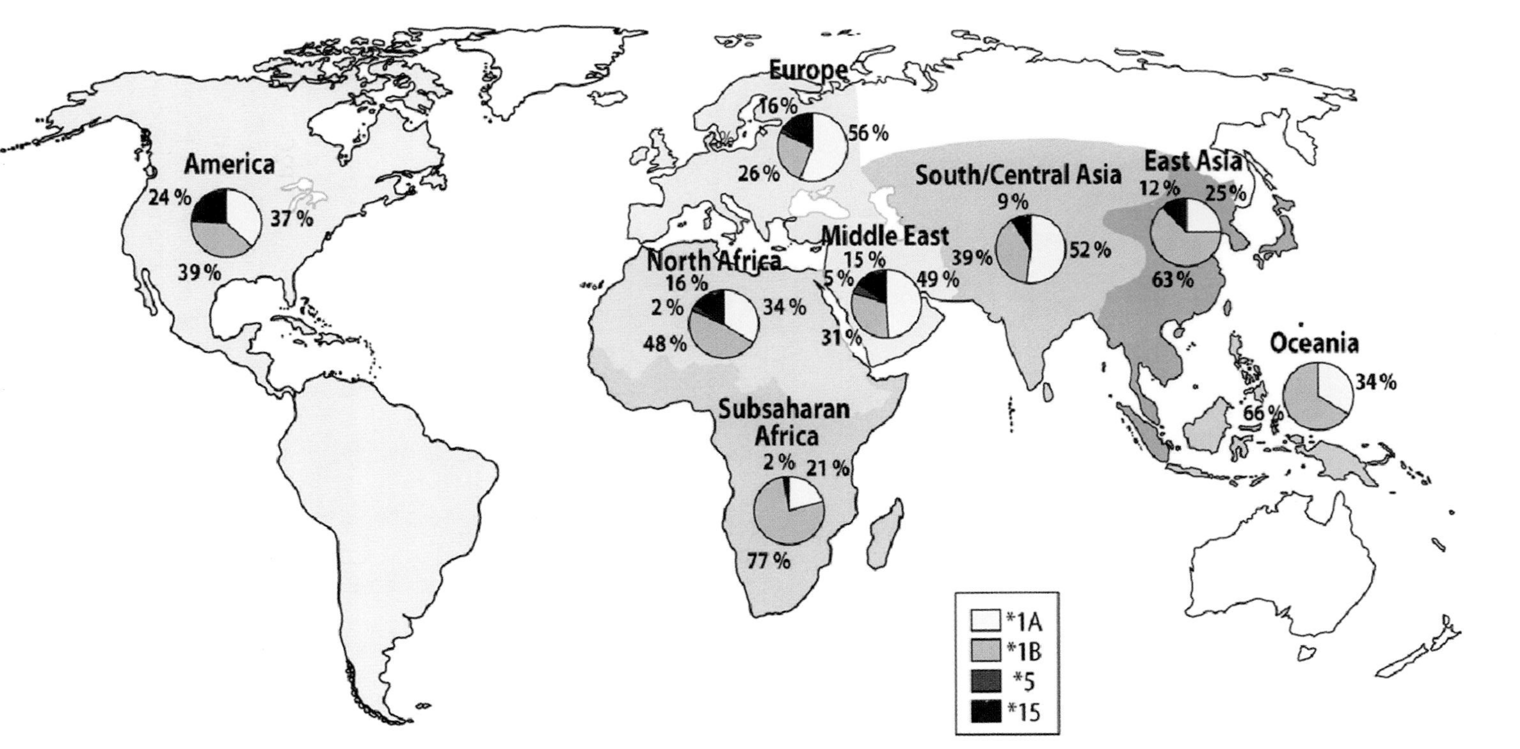

Figure 10.7 *Global distribution of selected SLCO1B1 haplotypes as shown by the key. Reproduced with permission from Niemi* et al. *(2011).*

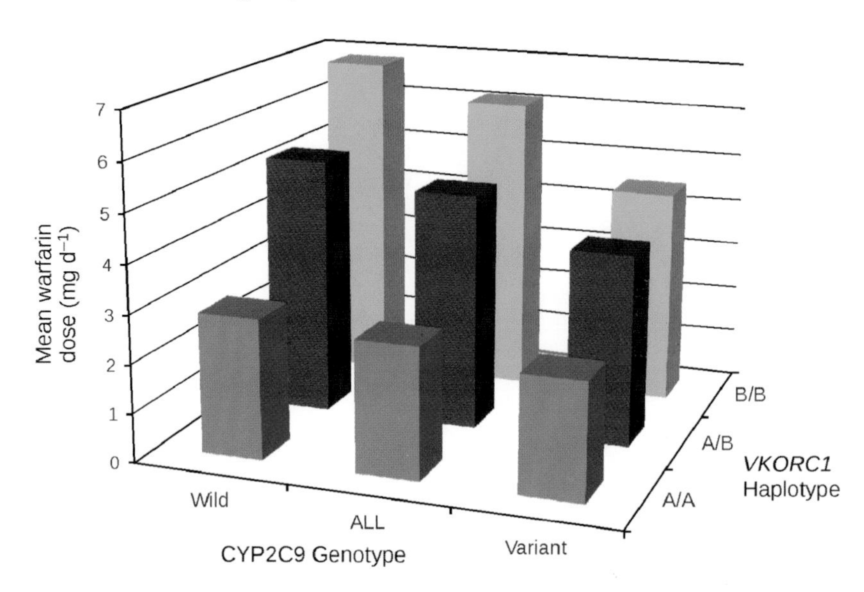

Figure 10.8 *Effect of genotype on warfarin dosing. Individuals with variant CYP2C9*2/*3 and VKORC1 haplotypes A/A (see text for details) require the lowest dose. Drawn from the data of Rieder* et al. *(2005).*

A similar situation exists for phenytoin. Severe toxicity was observed in patients who were homozygous for the *CYP2C9*3* gene as a result of reduced metabolism. The dose requirement was even further reduced in subjects with a variant *SCN1A* gene, which encodes for voltage-regulated sodium channels – the target for phenytoin.

10.12 Personalized medicine

Pharmacogenomics offers the promise of being able to tailor the appropriate dose of a medicine and possibly avoid certain drug combinations based on the genetic profile of the individual patient – 'personalized medicine' as it has become known. The concept is not particularly new, Rao *et al.* (1970) used sulfadimidine acetylor status as guide for weekly (rather than daily) dosing of isoniazid and dibucaine number has been used to screen those who may be susceptible to the prolonged effects suxamethonium. It is usual to phenotype or genotype patients for TPMT status before starting treatment with azathioprine to avoid life-threatening acute toxicity (Section 10.7).

More recently, interest has been in predicting CYP2D6 and CYP2C9 activity, and several genotyping kits have been approved by the FDA, including a chip-based system (AmpliChip) from Roche Molecular Diagnostics. The system can identify 33 *CYP2D6* and three *CYP2C9* alleles. Of course it is then necessary to predict how the genotype affects the enzyme activity. It has been suggested that starting doses of nortriptyline should be 10, 100 and 500 mg d^{-1} for PMs, IMs and EMs, and UMs, respectively (Xie & Frueh, 2005).

Genotypying is not inexpensive, and currently at least one medical insurance company in the USA will not pay for the service. However, one must assume that those companies that do, believe the approach is cost effective. At present it is expensive for some but profitable for others.

Summary

This chapter has dealt with what is an ever-expanding subject as more and more genetic variants are being discovered, in both enzymes and transporters. This knowledge is useful in explaining what were once considered idiosyncratic differences in drug response.

10.13 Further reading

Guillemette C. Pharmacogenomics of human UDP-glucuronosyltransferase enzymes. *Pharmacogenomics J* 2003; 3: 136–58.

Ingelman-Sundberg M. Pharmacogenetics of cytochrome P450 and its applications in drug therapy: the past, present and future. *Trends Pharmacol Sci* 2004; 25: 193–200.

König J, Müller F, Fromm MF. Transporters and drug–drug interactions: Important determinants of drug disposition and effects. *Pharmacol Rev* 2013; 65: 944–66.

10.14 References

Arimany-Nardi C, Koepsell H, Pastor-Anglada M. Role of SLC22A1 polymorphic variants in drug disposition, therapeutic responses, and drug–drug interactions *Pharmacogenom J* 2015; 1515 473–87.

Chowbay B, Cumaraswamy S, Cheung YB, Zhou Q, Lee EJ. Genetic polymorphisms in MDR1 and CYP3A4 genes in Asians and the influence of MDR1 haplotypes on cyclosporin disposition in heart transplant recipients. *Pharmacogenetics* 2003; 13: 89–95.

Christensen M, Andersson K, Dalen P, Mirghani RA, Muirhead GJ, Nordmark A, Tybring G, Wahlberg A, Yasar U, Bertilsson L. The Karolinska cocktail for phenotyping of five human cytochrome P450 enzymes. *Clin Pharmacol Ther* 2003; 73: 517–28.

Dalén P, Dahl ML, Bernal Ruiz ML, Nordin J, Bertilsson L. 10-Hydroxylation of nortriptyline in white persons with 0, 1, 2, 3, and 13 functional CYP2D6 genes. *Clin Pharmacol Ther* 1998; 63: 444–52.

Ekhart C, Doodeman VD, Rodenhuis S, Smits PH, Beijnen JH, Huitema AD. Polymorphisms of drug-metabolizing enzymes (GST, CYP2B6 and CYP3A) affect the pharmacokinetics of thiotepa and tepa. *Br J Clin Pharmacol* 2009; 67: 50–60.

Evans DAP, Manley KA, McKusick VA. Genetic control of isoniazid metabolism in man. *Br Med J* 1960; 2: 485–91.

Frye RF, Matzke GR, Adedoyin A, Porter JA, Branch RA. Validation of the five-drug 'Pittsburgh cocktail' approach for assessment of selective regulation of drug-metabolizing enzymes. *Clin Pharmacol Ther* 1997; 62: 365–76.

Kalow W, Gunn DR. The relation between dose of succinylcholine and duration of apnea in man. *J Pharmacol Exp Ther* 1957; 120: 203–14.

Kim RB, Leake BF, Choo EF, Dresser GK, Kubba SV, Schwarz UI, Taylor A, Xie HG, McKinsey J, Zhou S, Lan LB, Schuetz JD, Schuetz EG, Wilkinson GR. Identification of functionally variant MDR1 alleles among European Americans and African Americans. *Clin Pharmacol Ther* 2001; 70: 189–99.

Kinzig-Schippers M, Tomalik-Scharte D, Jetter A, Scheidel B, Jakob V, Rodamer M, Cascorbi I, Doroshyenko O, Sorgel F, Fuhr U. Should we use N-acetyltransferase type 2 genotyping to personalize isoniazid doses? *Antimicrob Agents Chemother* 2005; 49: 1733–8.

Lennard L, Van Loon JA, Weinshilboum RM. Pharmacogenetics of acute azathioprine toxicity: relationship to thiopurine methyltransferase genetic polymorphism. *Clin Pharmacol Ther* 1989; 46: 149–54.

Niemi M, Pasanen MK, Neuvonen PJ. Organic anion transporting polypeptide 1B1:A genetically polymorphic transporter of major importance for hepatic drug uptake. *Pharmacol Rev* 2011; 63: 157–181.

Nishizato Y, Ieiri I, Suzuki H, Kimura M, Kawabata K, Hirota T, Takane H, Irie S, Kusuhara H, Urasaki Y, Urae A, Higuchi S, Otsubo K, Sugiyama Y. Polymorphisms of OATP-C (SLC21A6) and OAT3 (SLC22A8) genes: consequences for pravastatin pharmacokinetics. *Clin Pharmacol Ther* 2003; 73: 554–65.

O'Donnell CJ, Grime K, Courtney P, Slee D, Riley RJ. The development of a cocktail CYP2B6, CYP2C8, and CYP3A5 inhibition assay and a preliminary assessment of utility in a drug discovery setting. *Drug Metab Dispos* 2007; 35: 381–5.

Rao KVN, Mitchison DA, Nair NGK, Prema K, Tripathy, SP. Sulphadimidine acetylation test for classification of patients as slow or rapid inactivators of isoniazid. *Br Med J* 1970, 3(5721): 495–7.

Rieder MJ, Reiner AP, Gage BF, Nickerson DA, Eby CS, McLeod HL, Blough DK, Thummel KE, Veenstra DL, Rettie AE. Effect of VKORC1 haplotypes on transcriptional regulation and warfarin dose. *N Engl J Med* 2005; 352: 2285–93.

Vesell ES, Page JG. Genetic control of drug levels in man: phenylbutazone. *Science* 1968; 159: 1479–80.

Wang Z, Gorski JC, Hamman MA, Huang SM, Lesko LJ, Hall SD. The effects of St John's wort (*Hypericum perforatum*) on human cytochrome P450 activity. *Clin Pharmacol Ther* 2001; 70: 317–26.

Whelpton R, Watkins G, Curry SH. Bratton-Marshall and liquid-chromatographic methods compared for determination of sulfamethazine acetylator status. *Clin Chem* 1981; 27: 1911–4.

Xie H-G, Frueh FW. Pharmacogenomics steps towards personalized medicine. *Personalized Medicine* 2005; 2: 325–337.

11

Additional Factors Affecting Plasma Concentrations

Learning objectives

By the end of the chapter the reader should be able to:

- list the factors that affect plasma concentrations, and hence the effects of drugs
- describe the basic differences between males and females that might be expected to result in changes in pharmacokinetics
- discuss the problems encountered when trying to demonstrate differences in drug disposition in males and females
- explain how physiological properties change in obesity and how these changes influence drug disposition
- describe how time of day may be important with regard to drug dosing and when plasma samples are taken for analysis
- explain how posture affects pharmacokinetic parameters
- compare the advantages and disadvantages of using fixed-weight dosing of medicines.

11.1 Introduction

During drug development, the drug disposition and pharmacokinetics of the proposed new chemical entity (NCE) that may one day become a new medicine will be investigated using animal models and healthy human volunteers, often relatively young men in the

Introduction to Drug Disposition and Pharmacokinetics, First Edition. Stephen H. Curry and Robin Whelpton.

Companion website: www.wiley.com/go/curryandwhelpton/IDDP

Box 11.1 Factors that affect plasma concentrations

- Those associated with disposition and fate, such as the route of administration (Chapter 2) and rate of elimination (Chapter 3).
- Pharmaceutical factors such as tablet and capsule properties.
- Individual variation such as weight, sex, pregnancy, genetics (Chapter 10) and age (Chapter 12).
- Lifestyle, including food and diet, smoking and alcohol consumption (Chapter 13), exercise and sleeping patterns.
- Pathological state, especially diseases of the organs involved in disposition and fate (Chapter 12).
- Drug interactions (Chapter 13).

first instance. Animals are usually from a tightly bred colony of the same strain and study results often show little variability between individual animals. However, the patient population is much more diverse. A stroll around a busy high street or market will highlight many of the characteristics that lead to much greater variation in pharmacokinetic parameters and consequently variations in plasma concentrations and clinical response. There will be men and women, some thin, others fat and some short whilst others are tall. There will be children and the elderly. Some people may be smoking and others at bars eating and drinking. Some of the women may be pregnant. Various ethnicities may be apparent. Thus, we have our list of some of the factors that might affect plasma concentrations in a particular individual (Box 11.1). Other factors may not be so obvious, such as posture and exercise, illness and even the time of day.

Many of the factors listed in Box 11.1 are interrelated. Diet and exercise will affect weight. Diet, smoking and alcohol may affect enzyme activity and lead to drug interactions. With regard to sex, differences between male and female will include differences in weight and weight distribution as well as hormonal differences. Additionally there is the influence of the menstrual cycle: prepubescent girls and postmenopausal women could be considered under age rather than sex. The incidence of disease increases with aging. Clearly the issues are complex and, not surprisingly, studies often report conflicting results.

In considering the factors that affect plasma concentrations of drugs following single doses, the plasma concentration–time relationships discussed at the beginning of Chapter 4 will apply. The variability between what we might call 'normal' and 'special' populations will be manifest in differences in absorption, including bioavailability, volume of distribution and clearance. Special populations, including regulatory aspects and clinical trials, have been reviewed by Grimsrud *et al.* (2015). It will be helpful to recall the following with regard to plasma concentrations:

- A change in the *extent* of absorption will change the concentrations at all times, leading to a larger or a smaller area under the concentration–time curve (AUC).
- A change in the *rate* of absorption, with no changes in the extent, will lead to changes in one direction at early time points, with changes in the opposite direction at later time points, and no change in AUC.

- A change in the rates of metabolism and/or excretion will lead to changes in the concentrations in the same direction at all points, and to changes in *AUC*.
- Changes in tissue localization will lead to changes in concentrations in the opposite direction, e.g. increased tissue binding leading to reduced plasma concentrations.
- Any one factor can change one or more of the pharmacokinetic properties of the drug.

With regard to multiple dosing, it is essential to refer to the standard pattern shown in Figure 4.2. With oral dosing, in addition to the average concentration within each dosage period and the time to reach that average, and the pharmacokinetic pseudo steady state that it represents, it is necessary to consider the fluctuation between peaks and troughs. Thus factors affecting the rate and extent of absorption of a drug in different ways, and also factors affecting elimination, will cause complex changes in the time to reach a plateau and in the height of that plateau.

Some of the differences that have been observed within these special populations occur in the early, pre-distributional phase so recalling the principles of multicompartment models, as discussed in Chapter 5, including the effect of redistribution on the duration of action of drugs such as thiopental, will be helpful.

This chapter briefly considers those factors listed in Box 11.1 that are not considered in other chapters.

11.2 Pharmaceutical factors

In the late 1960s and early 1970s, publications appeared showing the effects of tablet formulation on pharmacokinetics. The observations were mainly altered bioavailabilities as a result of differences in the form and size of the crystals of the active ingredient or the effects of excipients (i.e. additives to aid the formulation process, improve stability, solubility and dissolution, and the like). A classic example of the effect of using difference excipients is that of the antiepileptic drug phenytoin, which has a low therapeutic index. A number of patients showed signs of toxicity when one of the excipients, calcium sulfate, was replaced by lactose. The lactose was more easily wetted, resulting in faster dissolution of the tablet and hence higher plasma concentrations (Figure 11.1).

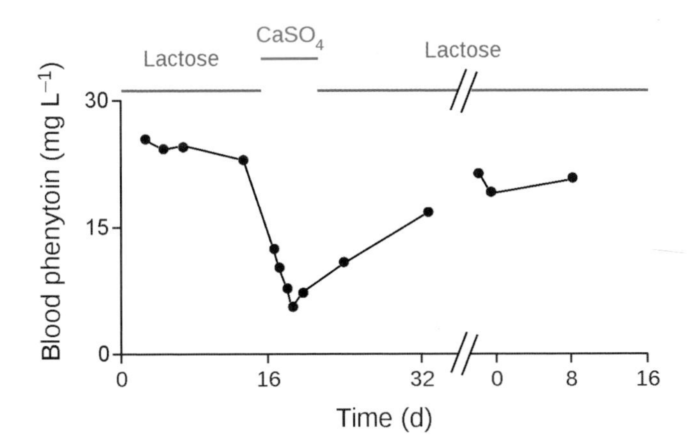

Figure 11.1 Influence of lactose and calcium sulfate as excepients on the concentrations of phenytoin in blood in a patient taking 400 mg per day. Redrawn with permission from Tyrer et al. *(1970).*

Table 11.1 Examples of the factors leading to inequivalence in drug formulations

Drug	Mechanism	References
Chloramphenicol	Polymorphic crystals	Aguiar *et al.* (1967)
Digoxin	Formulation produced in UK resulted in two-thirds higher plasma levels than older or other formulations	Fraser *et al.* (1972) Shaw *et al.* (1974)
Phenacetin	Particle size: *AUC* after oral doses greater as particle size decreased	Prescott *et al.* (1970)
Tolbutamide	Formulation: demonstrated different bioavailability from two preparations	Varley (1968)
Triamterene	Tablets versus capsules: reduced bioavailability with capsules	Tannenbaum *et al.* (1968)

Other factors were shown to influence the bioavailabilities of what were thought to be equivalent preparations (Table 11.1). The problems highlighted led to the requirement for bioequivalence studies (Section 4.5.3.3), not only to compare products from different pharmaceutical companies but when a manufacturer changed a formulation. For drugs such as ciclosporin, where it is necessary to maintain the plasma concentrations within a very small range, to avoid toxicity at high concentrations and organ rejection at sub-therapeutic concentrations it is advisable not to change the product for that of another manufacturer.

11.3 Sex

There has been for many years a general belief that females are more sensitive to drugs than males. This has been based on observations with alcohol, on extrapolations from knowledge of glomerular filtration rates in males and females, on early studies that showed that male rats tend to metabolize drugs relatively rapidly, and on clinical observations, to some extent made by anaesthetists. Table 11.2 shows some of the physiological similarities

Table 11.2 Selected physiological properties in men and women

Parameter	Male	Female
Height (cm)	177	163
Weight (kg)	73	60
Body surface area (m^2)	1.9	1.66
BMI ($kg\,m^{-2}$)*	23.3	21.8
Total body water (L)	42.0	29.0
Extracellular fluid (L)	18.2	11.6
Fat (kg)	14.6	18.0
Cardiac output ($L\,min^{-1}$)	6.5	5.9
Cardiac output ($L\,min^{-1}\,kg^{-1}$)	0.089	0.098
Tissue blood flow ($L\,min^{-1}$ kg of tissue^{-1}):		
liver	0.94	1.07
kidneys	4.05	4.08
fat	0.022	0.016

*Body mass index, see Equation 11.1.

and differences between men and women. The table is based on those by Mattison (2013) and Soldin & Mattison (2009) but we have scaled some of the parameters to body weight and organ weight. Although there are obvious differences in height and weight, differences in other parameters are smaller when scaled for weight; indeed cardiac output was calculated to be slightly higher in women than men. The one striking difference is that women have a higher proportion of fat than men (see Figure 12.1) and this is reflected in the lower perfusion rate (litres per minute per gram of tissue) of fat in women. The blood flows to the liver and kidneys are very similar when the cardiac output and organ size are taken into consideration.

11.3.1 Absorption and bioavailability

Gastrointestinal transit time is longer in women (mean 91.7 h) than in men (44.8 h), as is gastric emptying time. This can be expected to cause delays in the absorption of drugs, resulting in the same *AUC*, but longer lag times and lower rate constants of absorption, and hence lower C_{max} and later t_{max} values. Following oral doses of levofloxacin and losartan, there were no differences in t_{max}, but C_{max} and *AUC* were significantly greater in woman until these parameters were normalized for weight, when there were no differences (Carrasco-Portugal & Flores-Murrieta, 2011). In a similar study with fluconazole, t_{max} was shorter in women (2.17 h) than in men (5.05 h), but again there were no differences in C_{max} and *AUC* after weight normalization. Relatively fast absorption of oral salicylate has been shown in females, and a population study with mizolastine, an orally administered antihistamine drug, demonstrated relatively slow absorption in males. Absorption of ferrous sulfate has been shown to be faster in prepubertal girls than in boys.

Gastric alcohol dehydrogenase levels are relatively low in females. This leads to lesser pre-systemic metabolic losses, and so to relatively high blood alcohol levels. This accounts, at least in part, for the observations of sensitivity differences with this drug. In contrast, intestinal concentrations of CYP3A4 do not show a similar, or indeed, any consistent pattern. However, men have high GI concentrations of the efflux transporter P-gp. A difference between metabolism in males and females has been claimed for verapamil (Krecic-Shepard *et al.*, 2000). The oral bioavailability (*F*) was low in both sexes, but higher in women, indicative of greater pre-systemic elimination in men. This could be as a result of greater first-pass metabolism or efflux. After intravenous infusion, the systemic clearance (*CL*) of verapamil was greater in women. Thus, the difference in oral bioavailability may be due to greater efflux from enterocytes in men. Of course the oral clearance (*CL*/*F*) was higher in men because the *AUC* was lower, but there is no reason to suppose that systemic clearances were different after oral doses. This illustrates the limitation of comparing oral clearances when there may be changes in *F* (Section 4.4.2.1).

11.3.2 Distribution

Important differences between males and females are that men have higher volumes of total body water whereas women have a larger proportion of fat (Table 11.2). Thus drugs which are distributed in total body water (fluconazole, metronidazole, ethanol) have higher plasma concentrations in women. A 2-min i.v. infusion of vecuronium, a skeletal muscle relaxant, produced a more rapid onset in women, which can be explained by differences in distribution. The plasma concentration–time data were fitted to a three-compartmental model, with central compartment volumes of 0.04 and 0.05 L kg^{-1}, and V_{ss} values of 0.17 and 0.20 L kg^{-1} in women and men, respectively (Xue *et al.*, 1998). These values approximate to weight-adjusted plasma and extracellular fluid volumes, respectively (Table 2.1). Thus, as might be expected for a quaternary ammonium compound, vecuronium is distributed in ECF. The rapid onset seen in women is probably due, in part, to higher perfusion of skeletal muscle.

Some lipophilic drugs, diazepam and nitrazepam for example, show relatively high apparent volumes of distribution in females, as might be expected given their higher body fat content. For many other drugs it is not possible to generalize as to how sex affects distribution.

11.3.2.1 Plasma protein binding

Men and women have similar serum albumin concentrations but α_1-acid glycoprotein (AAG), which binds basic and neutral drugs, is lower in women. The expression of AAG is under hormonal control and is further reduced in pregnancy and by oral contraceptives. Where differences between the sexes have been observed, for example chlordiazepoxide and warfarin, protein binding is higher in males. The non-bound fractions of diazepam in males, females and females taking oral contraceptives were 1.46%, 1.67% and 1.99%, respectively. The corresponding values for non-bound lidocaine were 32%, 34% and 37%, respectively. The reduced binding in women will tend to increase the apparent volumes of distribution of these drugs.

11.3.3 Metabolism

Many studies have looked for sex-related differences in metabolism by the various cytochrome P450 isoenzymes and a pattern is beginning to emerge, but there are exceptions that are difficult to explain. Metabolism of some substrates of CYP1A2, clozapine and olanzapine for example, is faster in men. The activity of this enzyme is further reduced in women taking oral contraceptives, or who are pregnant. Men metabolize chlorzoxazone, a CYP2E1 substrate, faster than women. *In vitro* studies have mostly shown relatively high CYP3A4 concentrations in female tissues, and erythromycin and tamoxifen have been shown to be metabolized more rapidly *in vivo* in women. The expression of CYP3A4 is thought to be under hormonal control, but that does not mean necessarily that levels vary during the menstrual cycle. A clear sex difference in the metabolism of midazolam, a marker for CYP3A4, has not been demonstrated and the clearance of zolpidem is slower in women, even though 60% of the metabolism is believed to be catalyzed by CYP3A4.

Women are said to be less sensitive to the effects of propofol and recover from anesthesia more quickly than men. Believing this to be because women metabolize propofol more quickly than men, Choong *et al.* (2013) measured the concentrations of the drug and four of its metabolites (Figure 11.2) during intravenous infusions.

Propofol may be conjugated directly to propofol glucuronide, a reaction which is believed to be catalysed by an isoform of uridine diphosphate (UDP)-glucuronosyltransferase, UGT1A9. Oxidation to 4-hydroxypropofol is primarily catalysed by CYP2B6 with subsequent glucuronidation to 1- and 4-hydroxypropofol glucuronides. Small quantities of sulfate conjugates are also formed (not shown in Figure 11.2). *AUC* values, corrected for weight and total dose delivered, were calculated for propofol and each metabolite (Figure 11.3).

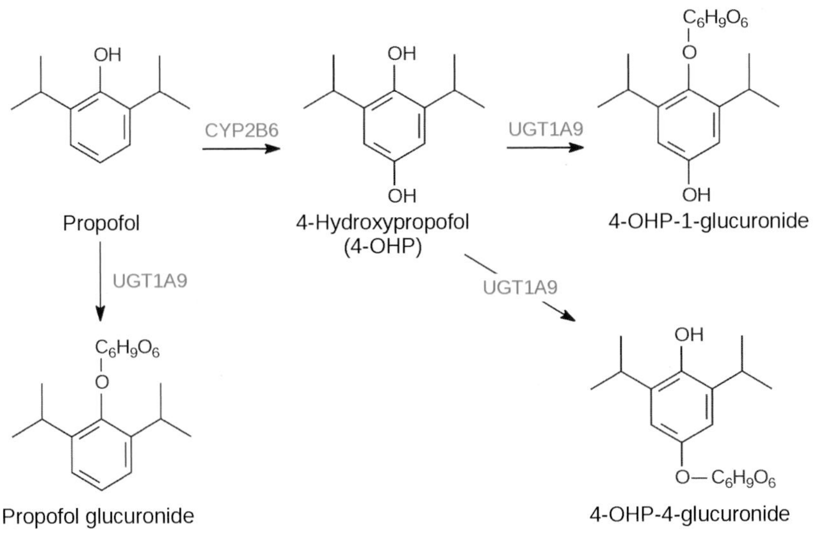

$C_6H_9O_6$
OH OH O
CYP2B6 UGT1A9
OH OH
Propofol 4-Hydroxypropofol (4-OHP) 4-OHP-1-glucuronide
UGT1A9 UGT1A9
$C_6H_9O_6$ OH
O
$O-C_6H_9O_6$
Propofol glucuronide 4-OHP-4-glucuronide

Figure 11.2 *Selected metabolic pathways for propofol. Direct conjugation produces propofol glucuronide. After phase 1 oxidation to 4-hydroxypropofol, 1- or 4-glucuronides are formed.*

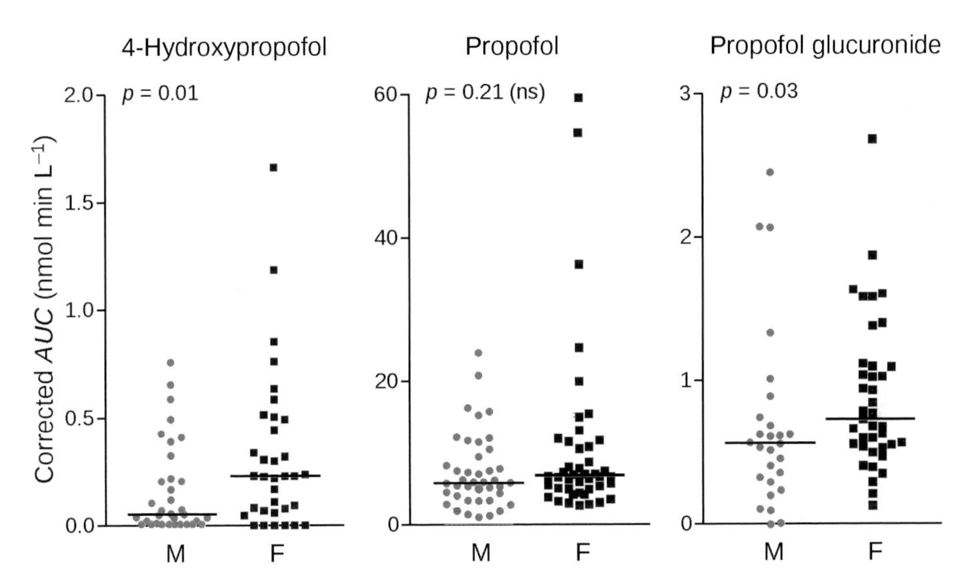

Figure 11.3 *Comparison of areas under the curve, corrected for dose and weight, in male and female patients. Centre: Propofol showed no significant differences in AUC. Left: females had significantly higher AUCs of 4-hydroxypropofol. Right: females had significantly higher AUCs of propofol glucuronide (see Figure 11.2 for structures). Redrawn after Choong* et al. *(2013), with permission.*

There was no significant difference between men and women with regard to the corrected *AUC* for the parent drug. However, *AUC* values for 4-hydroxypropofol and propofol glucuronide were significantly higher in women than in men (Figure 11.3). This was taken as indicating that women both oxidized and conjugated propofol faster than men. The differences were even greater for the metabolites that underwent both pathways: 4-hydroxypropofol-1-glucuronide ($p \leq 0.0001$) and 4-hydroxypropofol-4-glucuronide ($p = 0.001$). The results are interesting because other studies have shown increased mRNA, protein and enzyme activity for CYP2B6 in women when compared with men. Furthermore, there appears to be an ethnic component as the activity in Hispanic females is some 3.6 times higher than in Caucasian females. With regard to glucuronidation, there are examples of activity being greater in men, but other or additional UGT isoforms were involved: oxazepam (UGT2B15), paracetamol (UGT1A9, UGT1A1, UGT1A6), propranolol (UGT1A9, UGT2B4, UGT2B7). Thus, it may be the *relative* activities of particular isozymes that determine whether or not a sexual dimorphism is observed.

11.3.4 Excretion

The glomerular filtration rate in men is about 10% higher than in females after normalization for weight. The renal clearances of several drugs (aminoglycosides, cephalosporins, vancomycin) are lower in women. Some evidence of sex differences in active tubular secretion of drugs comes from studies in rats and mice. Activity of the OAT2 transporter is under hormonal control. The renal clearances of *p*-aminohippuric acid and furosemide, which are substrates for renal tubular transporters, were some 54% and 56% less in female rats when compared with males. Similarly, amantadine, which is a substrate for an organic cation transporter, shows faster clearance in males.

11.3.5 Effects

Links between sex differences in pharmacokinetics and effect have been sought with prednisolone, for which there are pharmacokinetic differences but no pharmacodynamic differences. In contrast, vecuronium shows a relatively high effect in females because of differences in the apparent volumes of distribution (Section 11.3.2). In regard to centrally-acting drugs, a limited number of studies has confirmed that there are cases of relatively high pharmacological sensitivity in females. Undoubtedly, part of the sex difference in alcohol response is the result of pharmacokinetic influences (greater bioavailability and a smaller volume of distribution) but there seems to be a pharmacodynamic contribution to this. With morphine, there is a relatively narrow therapeutic index in females, with a 60% higher incidence of nausea and vomiting associated with a comparable analgesic effect to that in males. With diazepam and some antidepressants there is evidence of relatively high pharmacodynamic sensitivity in females.

Because of their smaller size and differences in clearance and distribution it would seem that fixed-dose prescribing results in women have greater drug exposure than men. This may explain the greater reporting of adverse drug effects in women. Because of their sensitivity to zolpidem and the increased incidence of 'hangover effects' the FDA has recommended lower doses (50%) for females.

11.4 Pregnancy

Pregnancy leads to a wide variety of anatomical, physiological and biochemical changes, and all of them have the capacity to modify the pharmacokinetic properties of drugs. While there has always been a tendency to discourage the use of medication during pregnancy because of the risk of teratogenic effects on the foetus, many patients have to continue with chronic medication, such as with antiepileptics, antiasthmatics and antidepressants, during pregnancy. There is also an ongoing need for acute treatments, such as with anti-infective agents, during pregnancy, and it has been estimated that pregnant women receive an average of 1.3 prescriptions per clinic visit.

11.4.1 Physiological and biochemical changes

The cardiovascular system shows profound changes in pregnancy. Cardiac output, heart rate and stroke volume increase, and peripheral resistance and blood pressure (except in abnormal situations) decrease. Plasma volume can also increase. Total hepatic blood flow can increase by over 50% above non-pregnant rates, especially in the third trimester. Renal blood flow and glomerular filtration rate also increase by as much as 50% as the result of renal vasodilatation. Thus changes in drug absorption, tissue distribution, metabolism and excretion can all be proposed as likely. However, the pharmacokinetics of the majority of drugs remains to be studied in this condition (Hodge & Tracy, 2007). Renal excretion has been investigated the most. For example, in one study the renal clearance of atenolol, a drug commonly studied for its renal elimination because of its near dependence on the kidney for its removal from the body, was 12% above the postpartum level during the third trimester. Similarly, the renal clearance of digoxin, which is 80% excreted unchanged, increased by 21% and the clearance of lithium doubled during pregnancy.

The isoforms of the cytochrome P450 system show variable changes during pregnancy as do several phase 2 enzymes (Table 11.3).

11.4.2 Hormonal effects

The significance of hormonal changes *per se* is not clear. Studies in pregnancy are in short supply, but much can be learned from studies during oral contraceptive use. There is considerable evidence of a component of hormonal control over the activity of CYP1A2, the activity of which is reduced among women taking oral contraceptives, although studies have shown no correlation between either oestrogen or progesterone levels and the activity of this isoform. An analogous situation exists with CYP2A6, the activity of which is

Table 11.3 Examples of changes in enzyme activity during pregnancy.

Enzyme	Drug example	Metabolism	Notes
CYP1A2	Theophylline	Decreased	Reduce dose
CYP2C19	Proguanil*	Decreased	Increase dose
CYP2D6	Fluoxetine	Increased	Increase dose
CYP2C9	Phenytoin	Increased	Adjust dose as necessary
CYP2A6	Nicotine	Increased	
CYP3A4	Methadone	Increased	
UGT1A1	Paracetamol	Increased	
UGT1A4	Lamotrigine	Decreased	
UGT2B7	Lorazepam	Increased	
NAT2	Caffeine	Decreased	

*Prodrug converted to cycloguanil.

increased during oral contraceptive use. The metabolism of omeprazole (CYP2C19) is decreased during oral contraceptive use, while that of dextromethorphan (CYP2D6) is apparently unchanged. Nifedipine and midazolam (CYP3A4) show decreased metabolism in users of oral contraceptives, and it is theorized that this may be due to inactivation of the CYP3A4 enzyme by oestrogen. Similar incomplete information is to be found for the conjugating enzymes.

11.4.3 Transporters

Studies of transporters, especially P-gp, multidrug-resistance-associated protein (MRP), and breast-cancer-resistance protein (BCRP), have been mostly restricted to the role of transporters in the placenta. There is some evidence that hormonal changes may induce or inhibit the expression of transporter proteins affecting intestinal uptake and efflux, and renal excretion and reabsorption. For example, *in vitro* accumulation studies with digoxin appear to indicate potential for changes in the disposition of this drug in the intestine and kidney during pregnancy, although details remain unclear. In the placenta, although approximately 20 transporter proteins have been identified, few have been linked to xenobiotic transport. For example, P-gp is expressed on the maternal side of the placenta, and the use of knock-out mice has shown that the experimental teratogen avermectin is at least in part prevented from reaching the foetus by this efflux protein. Also, measurement of foetal levels of digoxin, paclitaxel and saquinavir, all P-gp substrates, has been used to demonstrate a similar exclusion. Measurement of P-gp levels in the human placenta has been considered to provide evidence of similar exclusions in women. P-gp expression is relatively high in the earlier stages of pregnancy, the time when the foetus is most susceptible to teratogenic damage. Analogous, but less complete, information has been obtained for MRP and BCRP.

11.4.4 The foetus

Some of the enzymes involved in drug metabolism are present as early as the sixth week of gestation, although appreciable levels are mostly not reached until after birth. CYP3A7 is the predominant enzyme in the foetus, declining in activity after the end of the first week from birth. Data from *in vitro* studies support the belief that this enzyme has a role in detoxifying certain endogenous compounds, notably dehydroepiandrosterone-3-sulfate, and also the potentially toxic metabolites of retinoic acid. Enzymes detected as present also include CYP2C9, CYP2C19 and CYP2D6, and it can be presumed that these enzymes are present to metabolize exogenous molecules that fail to be excluded by transporters such as P-gp. There is relatively little expression of UGT activity in the foetus, although *in vitro* studies have shown measurable glucuronidation of morphine in cells from foetal livers (Section 12.2.4.2).

11.5 Weight and obesity

Drug doses are mostly given in multiples of unit doses, as represented by a number of tablets or capsules, and while higher doses may be given to heavier people, exact proportionality in dosing is rarely attempted. Within a group of individuals, changes in dose on a mg kg^{-1} basis mostly lead to proportionate changes in concentrations in plasma and tissues, and it might be reasonable to suppose that changes in weights of individuals will lead to proportionate changes in drug concentrations. During drug development, human pharmacokinetic studies are, generally speaking, conducted in subjects and patients in the 'normal' range with regard to their weight and age. Rarely does a pharmacokinetic protocol involve tight control of the weight of the subjects, even less frequently are objective measures of weight used as selection criteria for inclusion in a study. However, the increasing rise in the numbers of people with obesity is leading to increasing numbers of studies on the effects of obesity on pharmacokinetic parameters such as volume of distribution and clearance. Any differences in these parameters may be important because loading doses are usually calculated on the basis of V and maintenance doses based on CL (Section 4.5.2).

Body mass index (BMI) provides a convenient and useful indicator of obesity.

$$BMI = \frac{\text{weight in kilograms}}{(\text{height in metres})^2} \tag{11.1}$$

Individuals with BMI values of 25–30 kg m^{-2} are considered to be overweight, and those with values >30 are designated as obese. Obesity is further classified as moderate (BMI 30–35), severe (35–40) or morbid (>40). Obviously, it is possible to have a very muscular body without excess fat, and thus have a relatively high BMI, as this index does not differentiate adipose tissue and muscle mass. In fact BMI (Equation 11.1) is linearly related to weight (Figure 11.4(a)). Nevertheless, generally speaking, a high BMI indicates an excessive body content of fat.

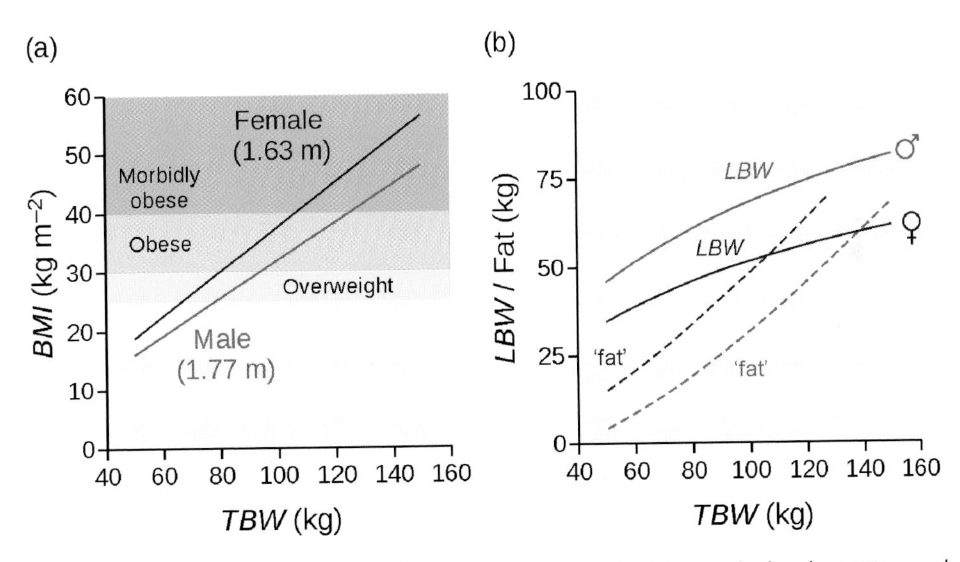

Figure 11.4 (a) BMI as a function of total body weight (TBW) in a male, height 1.77 m, and a female, height 1.63 m. (b) Lean body weight (LBW) as a function of TBW for the same hypothetical male and female subjects as in (a), calculated using Equations 11.2 and 11.3. The difference between TBW and LBW gives an approximate estimate of the amount of fat, shown by the broken lines. Note the much greater proportion of fat in females.

As weight increases, the sizes of organs increase, including those of the heart and the liver, and cardiac output increases. The blood vessels are wider and the blood volume greater in the obese. Lean body weight (*LBW*) increases but not as rapidly as the amount of fat, so total body weight, *TBW* (not to be confused with total body water), and *BMI* are poor indicators of the proportion of lean to fat mass. Several formulae have been suggested for calculating *LBW*, some of which could produce negative values at extreme weights. The one below does not (Janmahasatian *et al.*, 2005):

For males

$$LBW\left(\text{kg}\right)=\frac{9270\times TBW\left(\text{kg}\right)}{6680+216\times BMI\left(\text{kg}\,\text{m}^{-2}\right)} \tag{11.2}$$

and for females:

$$LBW\left(\text{kg}\right)=\frac{9270\times TBW\left(\text{kg}\right)}{8780+244\times BMI\left(\text{kg}\,\text{m}^{-2}\right)} \tag{11.3}$$

The change in the proportion of *LBW* as *TBW* increases can be seen in Figure 11.4(b), which shows calculated values for a standard/average man and woman. The difference between *TBW* and *LBW* gives an approximate weight for the amount of fat, the proportion of which increases as *TBW* increases. *LBW* includes a small amount of fat, chiefly as lipid membranes.

Ideal body weight (*IBW*) is a concept derived from data collected by the Metropolitan Life Insurance Company of New York, which produced tables relating weight to mortality data. It is an estimate of desirable weight corrected for sex, height and frame size. Some studies have investigated *IBW* or percent of *IBW* (*%IBW*) as a scaling factor for drug dosage. At one time, individuals who were 20% above their *IBW* were classed as obese.

11.5.1 Effects of obesity on pharmacokinetics

When attempting to predict or explain changes in drug disposition it should be remembered that in obesity there are increases in the:

- size of the heart
- blood volume
- diameter of blood vessels.

Thus, there is an increase in cardiac output with increased hepatic and splanchnic blood flows.

11.5.1.1 Absorption

Generally speaking, no consistent or important influences of obesity on drug absorption have been identified, but this may relate to the study design. Knibbe *et al.* (2015) pointed out that it is not possible to determine oral bioavailability without including an intravenous study. For example, the systemic clearance (*CL*) of propranolol after i.v. dosing was the same in obese and control subjects, but oral clearance (*CL/F*) was less in the obese, indicating

Table 11.4 *A representative selection of examples of apparent volumes of distribution in obesity*

Drug	Therapeutic group	V (L)*		V (L kg^{-1})	
		Normal	Obese	Normal	Obese
Ciprofloxacin	Anti-infective	219	269†	3.08	2.46†
Ifosfamide	Anticancer	33.7	42.8†	0.53	0.55
Carbamazepine	Anticonvulsant	69.7	98.4†	0.96	0.87†
Propofol	Sedative/anaesthetic	13.0	17.9	2.09	1.8
Dexfenfluramine	Appetite suppressant	668.7	969.7†	11.3	10.2
Propranolol	β-Blocker	180.0	180.0	3.1	2.4

* V_{ss} for propranolol.
† Significantly different.

greater bioavailability. For several other drugs, including midazolam, triazodone, ciclosporin and dexfenfluramine, there were differences in the extent or rate of absorption.

11.5.1.2 *Apparent volume of distribution*

One would expect the volume of distribution to increase as *TBW* increases, that is, lower plasma concentrations for a given dose in a larger individual, and indeed this is the case. However, for many drugs, *V* is often less in the obese after normalization to body weight (Table 11.4), presumably indicative that there is greater sequestration in lean body mass than adipose tissue. Thus, it would be expected that lipophilic drugs would have higher volumes of distribution per kilogram of *TBW*. This has been demonstrated for thiopental and diazepam. The volume of distribution of digoxin decreased in the obese when calculated as litres per kilogram, which is to be expected as this drug is largely bound to skeletal muscle (Section 2.4.1).

Increases in plasma protein binding would be expected to reduce the apparent volume of distribution (Section 2.5.1). There appears to be little change in protein binding for drugs that primarily bind to serum albumin. However, binding of drugs to α_1-acid glycoprotein has been shown to increase and decrease in obesity. Some studies have shown an increase in AAG concentrations, sometimes accompanied by increases in binding (e.g. propranolol) and sometimes not. Elevated serum free fatty acids in obesity may displace bound drug from plasma protein binding sites, thereby increasing the apparent volume of distribution.

11.5.1.3 *Clearance*

Several studies have shown that clearance increases in obesity. This is thought to be as a result of increased cardiac output to the eliminating organs, with less blood going to the increased stores of fat. There was a significant increase in the clearance of ciprofloxacin, but this did not reduce the elimination half-life (Table 11.5). The situation is complicated with regard to hepatic clearance because the livers of obese people show fatty infiltration, and clearance can be lower, higher or no different in obesity. Few investigations have attempted to assess enzyme activity in obesity, although studies with phenazone (antipyrine) have suggested that there are no specific effects on hepatic intrinsic clearance. The clearance

Table 11.5 A representative selection of examples of clearance and half-life in obesity

Drug	*CL* (L h^{-1})		*Half-life* (h)	
	Normal	Obese	Normal	Obese
Ciprofloxacin	44.6	53.8*	4.0	4.2
Ifosfamide	4.33	4.56	4.9	6.4
Carbamazepine	1.38	1.19	31.0	59.4*
Propofol	1.70	1.46	4.1	4.05
Dexfenfluramine	37.3	43.9	13.5	17.8
Propranolol	41.6	46.2	3.4	3.9

*Significantly different.

of chlorzoxazone, a marker for CYP2E1 activity, and production of its 6-hydroxy metabolite, increased in obesity. However, the *N*-desmethylation of erythromycin, a marker for CYP3A4 activity (Section 10.2.2), showed a negative correlation with *%IBW*.

The glucuronidation of oxazepam and lorazepam has been shown to increase in obesity. The clearance of paracetamol (acetaminophen) was enhanced, an observation supported by a study which showed increased excretion of paracetamol glucuronide but not sulfate in obese Zucker rats, a strain of rat that occurred spontaneously in 1961 and which has since been bred as a model for obesity.

There are conflicting reports on the effect of obesity on renal clearance. Glomerular filtration, as assessed by creatinine clearance, has been shown to decrease, increase or remain unchanged. It was thought that the renal clearance of aminoglycoside antibiotics was increased in obesity, but a more recent study showed that there was no change in the clearance of vancomycin until patients' weights were 30% greater than their *IBW*. The renal clearances of some drugs, including ciprofloxacin (Table 11.5), that are both filtered and secreted, are elevated in obesity.

11.5.2 Dose adjustment in obesity

The requirements for dose adjustment based on a patient's weight will be different when the drug is to be given orally over a period of time from when it is to be administered as a single acute dose. With multiple dosing, the plasma concentrations will rise towards steady-state concentrations (Figure 4.2) with a high degree of equilibration between tissues and plasma. Under these circumstances the dose can be adjusted as indicated by the clinical response, an approach sometimes referred to as 'titrating' the dose. However, the size of an acute dose has to be decided before it is administered. This presents a particular problem for (i) drugs that are injected intravenously and (ii) those that have a small therapeutic window. Such drugs include i.v. general anaesthetics and some anticancer drugs.

The volume of distribution and elimination half-life of thiopental is increased in the obese. Because cardiac output is an important determinant of the early distribution kinetics, the onset may be sooner, but the duration of action less, because of redistribution (Section 2.4.3.1). Cardiac output has been correlated with *LBW* and so it is has been suggested that the induction dose should be scaled to *LBW* in the obese. Similarly, peak concentrations of propofol are influenced by cardiac output so rapid infusions for induction should be based on *LBW*. However, because the apparent volume of distribution and clearance of propofol correlate with *TBW*, this parameter should be used to calculate the maintenance infusion rate (Ingrande & Lemmens, 2010). It has been recommended that suxamethonium (succinylcholine) should be dosed according to *TBW* because the amount of plasma cholinesterase is greater in the obese, but that vecuronium dosing should be based on *IBW* to avoid overdose in the obese.

11.6 Food, diet and nutrition

The influence of food on the absorption of orally administered drugs was described in Section 2.3.1.4. Food and drug interactions are discussed in Chapter 13. Similarly, because the effects of smoking and alcohol consumption on plasma drug concentrations may be classed as interactions, these are also considered in Chapter 13.

The presence of food in the GI tract, particularly the stomach, provides *adsorbing* surfaces to which drugs can adhere, in competition with sites of *absorption*, and prevents free access of drug molecules to the absorbing surface by reducing the efficiency of mixing. Food stimulates gastric acid secretion, affecting the ionization of weak electrolytes and decomposing some drugs. Changes in gastric emptying time change the rate of movement of drugs from the stomach to the intestine.

It is thus not surprising that food can markedly reduce the rate, and sometimes also the extent, of absorption. This may be an advantage if a relatively low, prolonged effect is wanted, or if, as with some of the non-steroidal anti-inflammatory drugs (NSAIDs), reduction in GI irritation and bleeding is required. It can be a nuisance if the highest possible peak concentration or the earliest possible effect is required, or if bioavailability (Chapter 4) is reduced, so prescribing instructions sometimes include advice on timing of doses in relation to food, depending on the clinical need.

Food in the stomach both stimulates and delays gastric emptying. In the fasting state, the stomach experiences periodic waves of peristalsis, which cause any accumulated fluid to pass through the pyloric sphincter. When solid or semi-solid food enters an empty stomach, the pyloric sphincter closes or remains closed. A slurry, called 'chyme', is created by the combined effect of stomach acid and the grinding effect of stomach muscle, so that virtually no solids enter the intestine. Once the slurry is formed, the pyloric sphincter opens to allow the chyme through. Fatty food, especially, slows gastric emptying. Drugs may shorten (metoclopramide) or lengthen (propantheline) the gastric emptying time.

11.6.1 Diet and nutrition

Compounds in the diet may interact with drugs to alter plasma concentrations, as discussed previously. However, nutrition can have effects on drug disposition. Campbell & Hayes (1974) reviewed the effects of nutrition on the drug-metabolizing enzymes in animals. Unsurprisingly, starvation reduced enzyme activity, but more specifically activity was reduced by:

- sugar intake (glucose, sucrose and fructose)
- fat-free diets
- reduced protein
- vitamin deficiency, particularly B, C and E, and, to a lesser extent, A
- calcium, magnesium and iron deficiencies.

Early human studies supported these observations. Phenazone (antipyrine) and theophylline metabolism increased when subjects were on a low-carbohydrate/high-protein diet, leading to reductions in half-lives: phenazone from 16.2 to 9.5 h and theophylline from 8.1 to 5.2 h. A change to a high-carbohydrate/low-protein diet reversed these effects.

More recent data support the above conclusions. Long-term consumption of a high-protein diet has been shown to increase the clearance of propranolol and phenazone. A high-carbohydrate diet has been shown to reduce the clearance of theophylline, and a high-fat diet has been shown to increase the clearance of ciclosporin after i.v. injection and the fraction absorbed after oral dosing. Restriction of calorific intake has been shown to reduce the clearance of aminopyrine, as has intravenous nutrition with glucose. The changes that occur with alterations in dietary protein intake are supported by studies in patients on total parenteral nutrition (TPN). They can be detected within a few days of dietary changes.

11.6.1.1 Dietary protein and gastric emptying

Consumption of a high-protein meal increased the systemic availability of propranolol by as much as 70% within 5 min, the effect lasting for about 30 min. This occurred with both i.v. and oral (immediate-release) but not controlled-release dosage forms. The explanation was that the meal caused a dramatic increase in hepatic blood flow, raising the portal vein concentrations of propranolol after both i.v. and oral doses to concentrations that partially saturate the enzymes responsible for pre-systemic metabolism. Similar observations have been made with several other high extraction ratio drugs, notably oral metoprolol, labetalol and intravenous lidocaine. Any meal might be expected to increase hepatic blood flow but the effect seems to be particularly evident after one rich in protein.

11.7 Time of day

The dosing of drugs ‘three times a day with meals’ was at one time deeply ingrained in the practice of medicine. Sometimes this was modified to ‘before meals’ or ‘after meals’, depending on whether food was thought to modify the absorption or, as in the case of aspirin, to protect from GI upset. The frequency of dosing was unimportant in comparison with the need to provide a reminder that it was time for a dose, and so increase compliance. Zero compliance results in zero efficacy!

Some drugs, such as antacids, proton pump inhibitors and anti-emetics, may be taken before meals for obvious reasons, but the introduction of drugs with pharmacokinetic properties that permit less frequent administration means that most drugs are given once or twice a day. A morning dose is generally suitable for a once-a-day regimen, when absorption is generally faster (Section 11.7.1). Sleep-inducing drugs are generally taken in the evening, and it may be recommended that drugs with sedative properties, some antihistamines and neuroleptics for example, are taken at night.

Cholesterol synthesis is greater at night and at least two studies have shown that the LDL-cholesterol lowering effect of simvastatin is greater when taken at night. However, the situation is complex: statins vary in first-pass effects, which can in turn be affected by diurnal rhythms. They also vary in that some of them are prodrugs and have different affinities for transport proteins. Simvastatin has a short half-life but the effect of time of the dose should be less with statins with longer half-lives. A study with atorvastatin showed no differences between morning and night-time dosing.

11.7.1 Circadian rhythms

Circadian rhythm refers to a cycle in biochemical, physiological and behavioural processes of approximately 24 h. The term is from the Latin *circa* ('around') and *diem* or *dies* ('day'). The formal study of such temporal rhythms is chronobiology and more recently the term 'chronopharmacokinetics' has been coined. Circadian rhythms are internally generated, but they can be reset by external cues, such as daylight, and even by conscious decisions such as eating schedule. A cue that can reset the cycle is referred to as a *zeitgeber*, from the German 'time giver'. Functionally, circadian rhythms facilitate adaptation by the body to environmental changes and homeostasis within a particular 'daily round'.

Diurnal variations occur in all of the physiological processes of importance in pharmacokinetics, such as gastric emptying time (faster in the morning), GI perfusion (faster in the morning), urinary pH (higher in the evenings) and blood flow to tissues (slower at night). Plasma concentrations of albumin and AAG are highest around noon. Consequently, diurnal variations are observable in the absorption, metabolism and excretion of some drugs. However, it should be noted that such variations occur in parallel with, for example, physical activity (Section 11.8) and meals, so it can be difficult to determine the component contributions to pharmacokinetic variation in any particular case.

Shorter gastric emptying times and increased GI perfusion in the morning should favour absorption of orally administered drugs, and in general this is the case (Table 11.6). The kinetics of valproic acid were studied after administration at 8:00 h and 20:00 h after a light breakfast or a more substantial evening meal. Absorption was faster in the morning but the effect was reduced when the meals were switched, showing that some of the effect was related to food. Similarly, increased renal pH in the evening is associated with food and the renal clearance of weak acids would be expected to be greater in the evening (Table 11.6).

Table 11.6 Examples of pharmacokinetic parameters showing time-dependent changes

Drug	C_{max}	t_{max}	*AUC*	Half-life	Comment
Nifedipine	↑ am	↓ am	–	–	Gastric emptying faster in mornings
Propranolol	↑ am	↓ am		–	
Theophylline	↑ am	↓ am	↑ am		
Indometacin				↑ pm	
Ketoprofen				↑ pm	
Diclofenac	↑ am	↔am	↑ am	↔	
Salicylic acid				↓ pm	↑CL_R in evening when pH is higher
Paracetamol				↑ am	
Valproic acid	↑ am	↓ am			Largely influenced by food

11.8 Posture and exercise

The effects of standing, sitting and lying on pharmacokinetics have been described by Queckenberg & Fuhr (2009). Different positons when lying have been described: supine (flat on the back), prone (face down) and lateral (on one side). Sometimes the word supine may be use synonymously with lying, but where possible we differentiate between the two. Changes in posture result in the following:

- On standing after lying, plasma water is transferred to interstitial fluid, resulting in increases in haematocrit (10.8%) and plasma protein concentrations (~20%).
- Lying on the left reduces gastric emptying when compared with other positions.
- Splanchnic–hepatic blood flow is reduced by 37% on standing versus supine.

The plasma concentrations of phenytoin, imipramine and desipramine (as a metabolite of imipramine) all increased by approximately 10% on standing from a supine position. These increases parallel the change in blood volume and increase in plasma protein concentrations. Small molecular weight drugs that are less protein bound may not be as affected because non-bound drug can diffuse into the interstitial space.

Orally administered drugs that are rapidly absorbed are the most likely to be affected by reduced gastric emptying as a result of lying on the left. Generally, t_{max} will be increased and C_{max} reduced. *AUC* may be reduced, particularly if there is normally a degree of saturation of first-pass metabolism. All these effects have been observed for nifedipine, including pharmacodynamic changes in pulse rate and diastolic blood pressure (Figure 11.5).

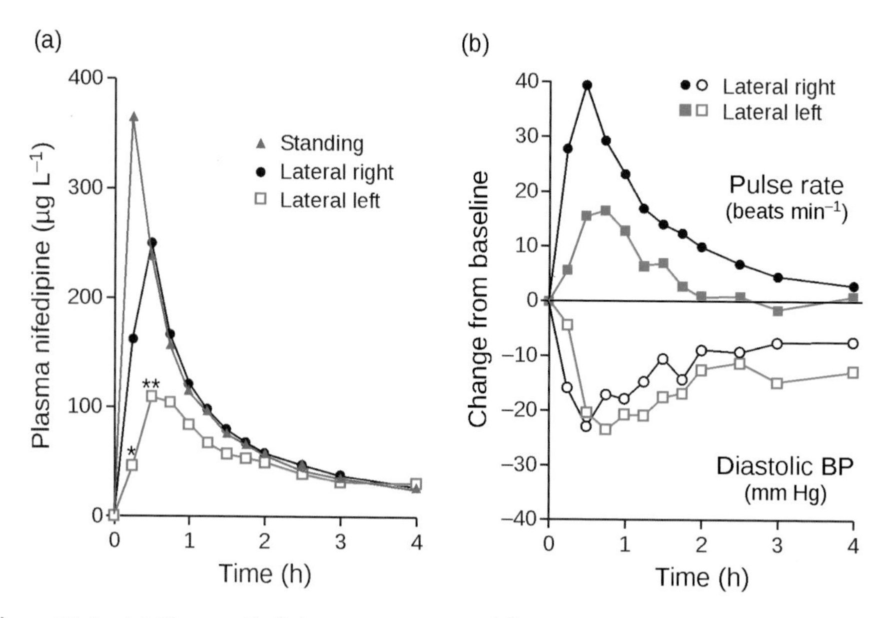

Figure 11.5 (a) Plasma nifedipine concentrations following administration of 20 mg capsules while standing, lying on the right and lying on the left side. *Significantly different from standing; **significantly different from lateral right. (b) Changes from baseline for pulse rate (solid symbols) and diastolic blood pressure (open symbols) for lying on the right (●, ○) and left (■, □). Significant differences in pulse rates between 0.25 and 2.5 h. Redrawn with permission from the data of Renwick et al. (1992).

11.8.1 Exercise

Exercise can be described as 'acute' (relatively short-lived) or 'chronic' (a programme of training continued over a relatively long period of time). These activities obviously involve changes in posture among other body features. The distinction between acute and chronic exercise is important because, for example, during acute exercise blood is shunted to the working organs, primarily the muscles, at the expense of other organs, some of them crucial in drug disposition and pharmacokinetics, such as the GI tract and the liver. In acute exercise, heart rate, cardiac output, systolic blood pressure and pulmonary ventilation all increase, while hepatic blood flow is decreased. In chronic exercise, cardiac output is relatively high, although resting heart rate is lowered, so perfusion of all tissues, including the GI tract and the liver (as with acute exercise), and the brain, kidneys and tissues in which drug molecules are stored such as voluntary muscle and fat deposits, is increased. This is assessed as increased regional blood flow. Also, the blood volume and activity of oxidative enzymes is increased, while fat mass is reduced. Thus there is potential in both acute and chronic exercise for all of the drug disposition sites to operate differently. Maximal oxygen uptake ($VO_{2_{max}}$), expressed as mL min^{-1} or mL min^{-1} kg^{-1}, is commonly used as the best measure of aerobic (cardiorespiratory) fitness, higher values being associated with higher fitness levels. Maximal oxygen uptake is considered to be reached when oxygen consumption shows no further increase with increased exercise workload, and represents the maximal capacity of the system to extract, deliver and utilize oxygen. The effects of exercise on drug disposition and pharmacokinetics have been reviewed by Khazaeinia, *et al.* (2000).

11.8.1.1 Absorption

Gastric emptying is delayed and splanchnic blood flow reduced by 50% when the intensity of acute exercise exceeds 70% of maximum oxygen uptake. Intestinal blood flow can be reduced by as much as 80% in intensive exercise. These effects are primarily a result of increased sympathetic activity. Chronic exercise has the ability to both shorten and lengthen gastric emptying time depending on the type of stomach contents involved. Intestinal transit time shows only minor changes with exercise, and then only with chronic exercise, when it is shortened. The absorption of quinidine, salicylate and sulfadimidine (sulfamethazine) is unaffected by the level of acute exercise. In contrast, sulfamethazole, tetracycline and doxycycline showed increases in C_{max} on acute exercise, but the comparison was with bed rest (as opposed to upright but not exercising), which is a different posture. The *rate* of absorption of digoxin has been shown to be higher during acute exercise, again in contrast with the body in the different posture of the supine position.

At rest, cardiac output to skeletal muscle is approximately 20%, but this can rise to over 80% with maximal exercise. Consequently absorption may be faster from sites that are actively exercising. Subcutaneous injections of insulin were more rapidly absorbed from a leg that was exercising than from the one that was not. It is important for diabetic patients self-injecting with insulin to know and understand the significance, to their own particular needs, of the site of injection in relation to their level of activity. The absorption of atropine injected intramuscularly is increased by moderate exercise, with the rate constant of absorption nearly doubling.

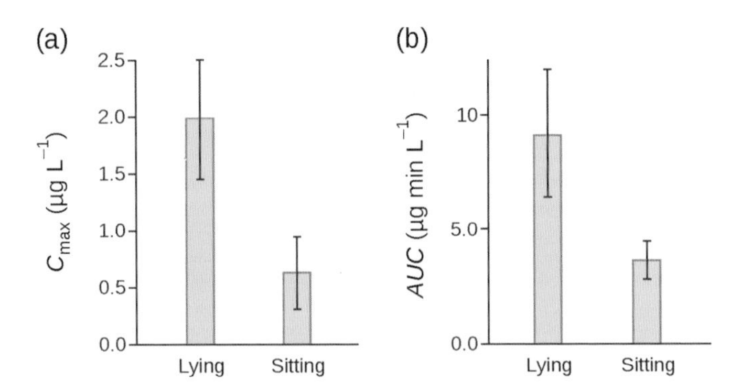

Figure 11.6 (a) Maximum concentration and (b) area under the curve for glyceryl trinitrate in two matched groups of healthy volunteers, one group lying down (face up) and the other sitting. Data are mean ± SEM. Adapted with permission from Curry & Kwon (1985).

Exercise increases skin temperature and can cause vasodilation or vasoconstriction if compensatory mechanisms of the vasomotor centre prevail. In this context, glyceryl trinitrate (GTN, nitroglycerin) concentrations obtained from skin patches have been shown to increase three-fold in exercise, with a weak correlation with skin temperature. However, GTN concentrations in plasma are very sensitive to posture and to hepatic blood flow, and observations attributed to exercise may well be related to changes in the metabolism of the drug in blood vessel walls, caused by changes in cardiac output (Figure 11.6). Similar increases have been observed with nicotine patches.

11.8.1.2 Distribution

Because blood is shunted to the active tissues in acute exercise, changes in apparent volume of distribution are to be expected, and animal studies with atropine, theophylline and phenazone, in which reductions in apparent volume of distribution were observed, support this.

There is about a 10–15% decrease in plasma volume on exercise. The increases in plasma concentrations of propranolol and atenolol may be due in part to release, along with catecholamines, from adrenergic neurones on exercise. There were no changes in plasma protein binding of verapamil and propranolol during acute exercise. However, digoxin concentrations in skeletal muscle have been shown to be increased during exercise, with corresponding decreases in concentrations in erythrocytes. Thus a shunting from red cells to active tissues occurs.

11.8.1.3 Metabolism

Hepatic blood flow is reduced during acute exercise. This has been demonstrated with indocyanine green clearance studies and has obvious implications for drugs with blood flow-dependent clearance, such as lidocaine, whose plasma concentrations are higher during exercise. However, verapamil and propranolol showed no change in clearance, and there is some evidence that the lidocaine observation is more related to posture than to hepatic blood flow controlled by blood flow shunting. Theophylline clearance is reduced in exercise, but there is evidence from phenazone, diazepam and sulfadimidine that there are basically no effects on enzyme activity. Chronic exercise apparently does not cause a long-term increase in liver blood flow, and, generally speaking, few positive effects have been shown on drug metabolism. In one study, a relatively short half-life was observed with aminopyrine in trained athletes, and an analogous observation has been made with phenazone. Prednisolone has also been shown to have a faster elimination rate in trained athletes.

11.8.1.4 Renal excretion

Renal blood flow is reduced as the result of exercise, as is the glomerular filtration rate, shown by creatinine clearance. Plasma protein concentrations can change either way with exercise, because of fluid shunting, and urinary pH is reduced. As a consequence, atenolol renal clearance is reduced, as is that of digoxin, along with active tubular secretion of procainamide.

Summary

This chapter has considered how pharmaceutical formulation, sex, weight, diet, exercise and circadian rhythms can result in changes in plasma concentrations and consequently the effects of drugs. It is necessary to extrapolate from the results obtained in rigidly controlled trials in healthy volunteers to the diverse patient population. These factors should be borne in mind when interpreting pharmacokinetic data, assessing clinical effects, designing clinical trials and particularly when prescribing medicines for special populations. Chapter10 looked at genetic factors whilst Chapter12 will consider the effects of age and disease.

11.9 Further reading

Baraldo M. The influence of circadian rhythms on the kinetics of drugs in humans. *Expert Opin Drug Metab Toxicol* 2008; 4: 175–92.

Cheymol G. Effects of obesity on pharmacokinetics implications for drug therapy. *Clin Pharmacokinet* 2000; 39: 215–31.

Green B, Duffull SB. What is the best size descriptor to use for pharmacokinetic studies in the obese? *Br J Clin Pharmacol* 2004; 58: 119–33.

Harris RZ, Jang GR, Tsunoda S. Dietary effects on drug metabolism and transport. *Clin Pharmacokinet* 2003; 42: 1071–88.

Persky AM, Eddington ND, Derendorf H. A review of the effects of chronic exercise and physical fitness level on resting pharmacokinetics. *Int J Clin Pharmacol Ther* 2003; 41: 504–16.

van Baak MA. Influence of exercise on the pharmacokinetics of drugs. *Clin Pharmacokinet* 1990; 19: 32–43.

11.10 References

Aguiar AJ, Krc J, Jr., Kinkel AW, Samyn JC. Effect of polymorphism on the absorption of chloramphenicol from chloramphenicol palmitate. *J Pharm Sci* 1967; 56: 847–53.

Campbell TC, Hayes JR. Role of nutrition in the drug-metabolizing enzyme system. *Pharmacol Rev* 1974; 26: 171–97.

Carrasco-Portugal M del C, Flores-Murrieta FJ. Gender differences in the pharmacokinetics of oral drugs *Pharmacol Pharm* 2011; 2: 31–41.

Choong E, Loryan I, Lindqvist M, Nordling A, el Bouazzaoui S, van Schaik RH, Johansson I, Jakobsson J, Ingelman-Sundberg M. Sex difference in formation of propofol metabolites: a replication study. *Basic Clin Pharmacol Toxicol* 2013; 113: 126–31.

Curry SH, Kwon HR. Influence of posture on plasma nitroglycerin. *Br J Clin Pharmacol* 1985; 19: 403–4.

Fraser EJ, Leach RH, Poston JW. Bioavailability of digoxin. *Lancet* 1972; 2: 541.

Grimsrud KN, Sherwin CM, Constance JE, Tak C, Zuppa AF, Spigarelli MG, Mihalopoulos NL. Special population considerations and regulatory affairs for clinical research. *Clin Res Regul Aff* 2015; 32: 47–56.

Hodge LS, Tracy TS. Alterations in drug disposition during pregnancy: implications for drug therapy. *Expert Opin Drug Metab Toxicol* 2007; 3: 557–71.

Ingrande J, Lemmens HJ. Dose adjustment of anaesthetics in the morbidly obese. *Br J Anaesth*, 2010; 105 Suppl 1: i16–23.

Janmahasatian S, Duffull SB, Ash S, Ward LC, Byrne NM, Green B. Quantification of lean bodyweight. *Clin Pharmacokinet* 2005; 44: 1051–65.

Khazaeinia, T, Ramsey AA, Tam, YK. The effects of exercise on the pharmacokinetics of drugs. *J Pharm Pharm Sci* 2000. 3: 292–302.

Knibbe CA, Brill MJ, van Rongen A, Diepstraten J, van der Graaf PH, Danhof M. Drug disposition in obesity: toward evidence-based dosing. *Annu Rev Pharmacol Toxicol* 2015; 55: 149–67.

Krecic-Shepard ME, Barnas CR, Slimko J, Jones MP, Schwartz JB. Gender-specific effects on verapamil pharmacokinetics and pharmacodynamics in humans. *J Clin Pharmacol* 2000; 40: 219–30.

Prescott LF, Steel RF, Ferrier WR. The effects of particle size on the absorption of phenacetin in man. A correlation between plasma concentration of phenacetin and effects on the central nervous system. *Clin Pharmacol Ther* 1970; 11: 496–504.

Mattison DR. Pharmacokinetics in real life: Sex and gender differences. *J Popul Ther Clin Pharmacol* 2013; 20: e340–e9.

Queckenberg, C. Fuhr U. Influence of posture on pharmacokinetics. *Eur J Clin Pharmacol* 2009; 65: 109–19.

Renwick AG, Ahsan CH, Challenor VF, Damiels R, Macklin BS, Waller DG, George CF. The influence of posture on the pharmacokinetics of orally administered nifedipine. *Br J Clin Pharmacol* 1992; 34: 332–6.

Shaw TR, Howard MR, Hamer J. Recent changes in biological availability of digoxin. Effect of an alteration in 'Lanoxin' tablets. *Br Heart J* 1974; 36: 85–9.

Soldin OP, Mattison DR. Sex differences in pharmacokinetics and pharmacodynamics. *Clin Pharmacokinet* 2009; 48: 143–57.

Tannenbaum PJ, Rosen E, Flanagan T, Crosley AP, Jr. The influence of dosage form on the activity of a diuretic agent. *Clin Pharmacol Ther* 1968; 9: 598–60.

Tyrer JH, Eadie MJ, Sutherland JM, Hooper WD. Outbreak of anticonvulsant intoxication in an Australian city. *Br Med J* 1970; 4: 271–3.

Varley AB. The generic inequivalence of drugs. *Jama* 1968; 206: 1745–8.

Xue FS, An G, Liao X, Zou Q, Luo LK. The pharmacokinetics of vecuronium in male and female patients. *Anesth Anal* 1998; 86: 1322–7.

12

Effects of Age and Disease on Drug Disposition

Learning objectives

By the end of the chapter the reader should be able to:

- describe the developmental changes that occur from birth to adulthood that might be expected to affect drug disposition
- explain why, on a weight for weight basis, infants metabolize drugs faster than adults
- explain why in heart failure lidocaine plasma concentrations are elevated, but the half-life is unchanged
- describe how plasma protein concentrations change in disease
- explain why in renal failure the intrinsic clearance of propranolol is decreased but the elimination half-life is unchanged
- critique the problems that occur when investigating drug disposition in special populations, such as those of this chapter.

12.1 Introduction

This chapter examines the effect of age and development, and the influence of disease on drug distribution and pharmacokinetics. The two are not unrelated and a major problem is dissociating the effects of age *per se* from the effects of age-related practices and phenomena,

Introduction to Drug Disposition and Pharmacokinetics, First Edition. Stephen H. Curry and Robin Whelpton.

Companion website: www.wiley.com/go/curryandwhelpton/IDDP

such as smoking, caffeine, alcohol and other drug consumption, and dietary and exercise habits. Also, age-related disease is an inevitable complication.

As in previous chapters, the basic principles apply and any changes in the pharmacokinetics will be as a result of changes in rate and extent of absorption, and changes in distribution and/or in clearance. Changes in these variables will be reflected by changes in elimination half-lives and steady-state plasma concentrations. The list of physiological and biochemical factors to be considered include the 'usual suspects':

- GI tract and hepato-portal system
- body composition: fat/lean mass ratio and plasma protein concentrations
- activity of drug-metabolizing enzymes
- kidney function
- cardiac output and blood flow to tissues.

12.2 Age and development

Interest in both the young and the old as special populations is now well established, especially in relation to pharmacokinetics. This reflects a dramatic change since the early days of clinical pharmacology. However, the study of extreme age groups remains hampered by severe limitations. For example, analytical methods suitable for the small blood samples obtainable from children have required specific development, and ethical problems persist for studies in both the very young and very old age groups.

In the young systems are developing and maturing, whereas in the old the tendency is for systems to be deteriorating. There is also the concept of 'chronological' and 'biochemical' age, so that two people of, say, 80 years may have markedly different physiological function. This leads to greater variably in pharmacokinetic parameters in the elderly when compared with the young.

12.2.1 Terminology

One difficulty in age-related studies has been defining when a neonate becomes a child, when a child becomes an adult and when an adult becomes elderly. This terminology has evolved to agreement on the following basic divisions of life's segments for most purposes:

- Premature born before anticipated date of full-term development
- Neonate birth to 1 month
- Infant 1 month to 2 years
- Child 2–12 years
- Adolescent 12–16 years
- Older adult 65 and above.

12.2.2 The gastrointestinal tract

At birth gastric pH is approximately neutral, it rapidly falls to pH 3 within a few hours and then returns to neutral in 7–10 days, reaching adult values at about 3–5 years. This achlorhydria increases the absorption of acid-labile drugs such as penicillin G and erythromycin, and some weak bases. In the newborn gastric emptying is delayed but transit times are shorter in healthy infants and, weight for weight, the area of the GI tract is greater than in adults. Bioavailability, as shown by increased absorption of midazolam, is greater in the young, probably as a result of less first-pass elimination because of fewer drug-metabolizing enzymes and efflux transporters. Biliary excretion and the splanchnic circulation show postnatal development and there is low bile acid secretion in newborns, with implications for enterohepatic circulation involving conjugation and hydrolysis.

In adulthood, gastric pH increases with age so that approximately 35% of those over 60 have achlorhydria. Transit times increase and there is a general delay in absorption of substances as people become older.

12.2.3 Body composition

It should be remembered that babies and children are not simply adults in miniature and nowhere is this more apparent than in the relative proportion of lean tissue to fat. Total body water forms a large proportion of lean body weight (*LBW*) and the two tend to change in parallel (Figure 12.1). When the embryo is just a few rapidly dividing cells it is approximately 90% water, but as the foetus grows the proportion of water falls until at term it is 80–75% of body weight. The percentage of water in premature babies can be as much as 85% of body weight. The proportion of total body water declines rapidly over the first 6 months, partly because of an increase in the amount of fat, which normally doubles by 4–5 months and then rises to adult levels (50–70%), depending on sex and age. The extracellular fluid (ECF) is 40% of body weight in neonates compared with 20% in adults. Adult values are reached by 10–15 years of age. The proportion of intramuscular fluid is especially low at birth. All of this will affect tissue-to-plasma concentration ratios, for both lipid-soluble and lipid-insoluble drugs. Additionally, the relative acidosis of the newborn will affect the tissue distribution of weak electrolytes.

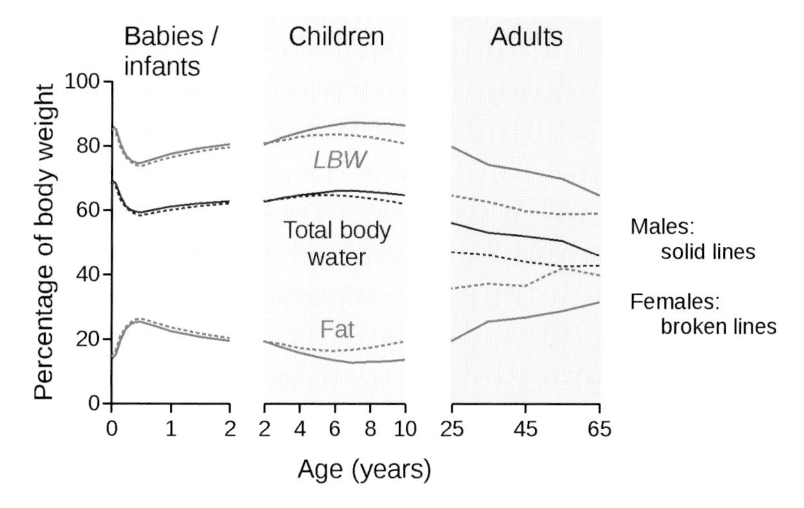

Figure 12.1 Relative distribution of lean body weight (LBW), total body water and fat as a function of age. Compiled from the data of Fomon et al. *(1982) and Chumlea* et al. *(1999).*

12.2.3.1 Protein binding

Protein binding can vary with age; serum albumin concentrations are about 75–80% of adult levels and some studies have shown α_1-acidglycoprotein (AAG) to be about half adult values in the newborn. Furthermore, the presence of foetal albumin which has reduced binding affinity, and bilirubin which can displace bound drugs, may increase the non-bound fraction of some drugs. This difference in binding, combined with a risk of high blood bilirubin concentrations consequent on immaturity of the conjugating systems, compounded by an immature blood–brain barrier, increases the risk of accumulation of bilirubin in the grey matter of the central nervous system. This can cause irreversible neurological damage (kernicterus) in the newborn. The risk may be increased by drugs, for example sulfonamides, which displace bilirubin from those binding sites that are available. Reduced binding to AAG may lead to greater drug toxicity, with local anaesthetics for example, in the young.

Serum albumin concentrations fall with aging, being 20% less in an 80-year-old compared with someone of 20 years. The situation is less clear with AAG, some studies showing an increase in concentration in the elderly. Thus, the effects of reduced protein concentrations should be more pronounced for acid drugs which bind primarily to albumin and the free fractions of salicylate, sulfadiazine and phenylbutazone were higher in a group of subjects aged 72–94 years compared with controls aged 19–40 years. However, the elderly group were also taking one or more additional drugs. It is likely that disease and concomitant use of other drugs have a greater significance on clinical effects in the elderly than protein binding.

The high proportion of total body water, particularly ECF, coupled with low protein binding, results in water-soluble drugs such as gentamicin having large apparent volumes of distribution in neonates (~0.5 L kg^{-1}) compared with adults (0.2–0.3 L kg^{-1}). The resultant low plasma concentrations mean that young children may require higher loading doses of such drugs. However, for lipophilic drugs the apparent volumes of distribution will be lower than adults. As the proportion of fat increases with age so does the apparent volume of lipophilic drugs. In the case of diazepam this leads to an increase in half-life (Klotz *et al.*, 1975). There was no effect of age on clearance so the increase in half-life (Figure 12.2(a))

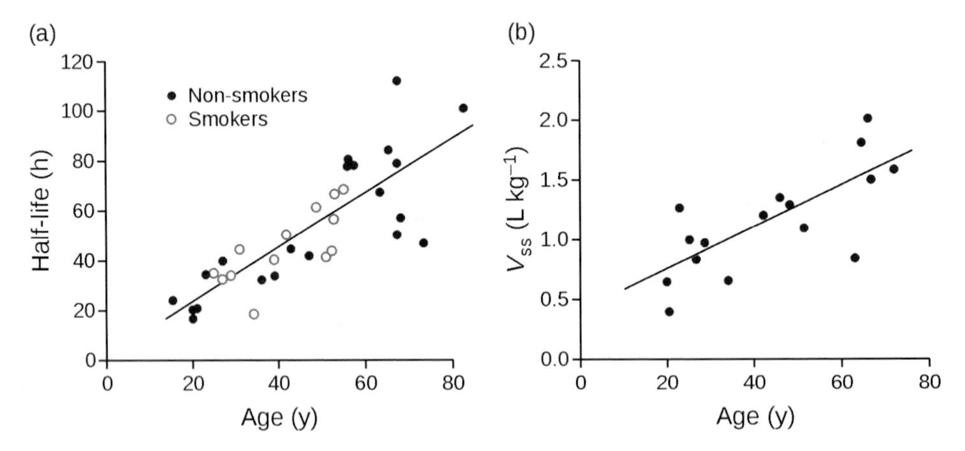

Figure 12.2 *(a) Half-life of diazepam as a function of age. There were no significant differences between smokers and non-smokers. (b) Volume of distribution of diazepam as a function of age. Redrawn from the data of Klotz* et al. *(1975).*

was attributed to the increase in V_{ss} (Figure 12.2(b)). The Klotz study eliminated the effect of smoking on the half-life of diazepam by showing that smoking did not influence the kinetics. Diazepam is metabolized by CYP2C19 and CYP3A4, whereas substances in cigarette smoke induce CYP2E1 (Table 3.1).

12.2.4 Drug-metabolizing enzymes

Most of the drug-metabolizing enzymes found in adults are either absent or have very low activities in the foetus, but there are a few exceptions. CYP3A7, which is sometimes referred to as a foetal form of CYP3A4 because of its similar substrate specificity, and two phase 2 enzymes, SULT1A3 and a glutathione S-transferase, GSTP1, are present. These enzymes are barely measureable in adults. Glucuronidation of morphine has been reported but the activity of UGT2B7 in the foetus is no more than 10–20% of that in adults.

12.2.4.1 Phase 1 metabolism

The activity of CYP3A7 peaks 1–7 days after birth and declines rapidly, being replaced by CYP3A4. The activities of the other drug-metabolizing cytochromes increase after birth but not at the same rate. One of the earliest studies was the metabolism of dextromethorphan by CYP2D6 (Treluyer *et al.*, 1991) showing that activity rises in the first few weeks of life, regardless of gestational age. The rates of metabolism correlated well with measured levels of enzyme protein (Figure 12.3). There was a general rise in messenger RNA, although this was less in adult samples.

Much of the data on enzyme activity and expression are derived from *in vitro* work with liver samples from aborted foetuses, and/or postmortem dissections, biopsies from adult donors undergoing surgery and the like. This approach, although informative, does not

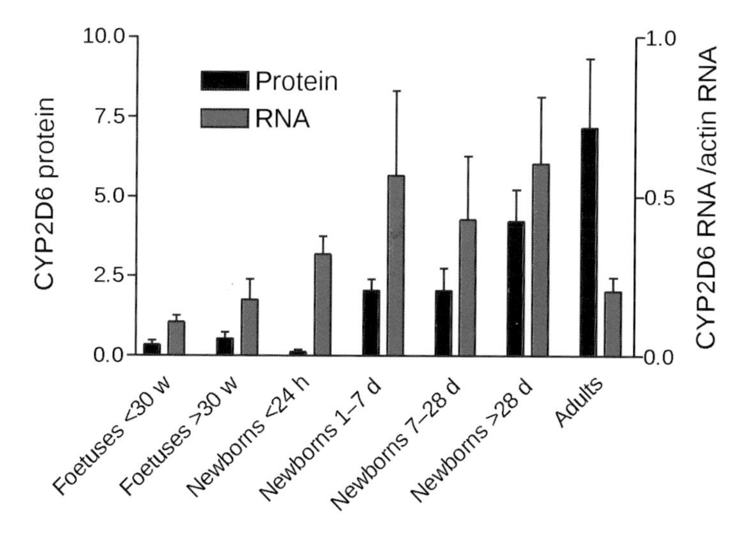

Figure 12.3 *Age-related variation of CYP2D6 protein and CYP2D6 mRNA in human liver. Redrawn with permission after Treluyer* et al. *(1991).*

account for *in vivo* factors such as relative liver size and blood flow. A rough guide to the onset of enzyme activity is:

- days CYP2C9/CYP2D6
- weeks CYP3A4
- months CYP1A2.

When normalized for body weight, the relative size of the liver peaks at about 2 years, when it is approximately twice that of an adult. Thus, the metabolism of some drugs can exceed adult rates when normalized to body weight – a surge in metabolism in toddlers. A typical adult dose of theophylline is $12\,mg\,kg^{-1}\,d^{-1}$ whereas children require ~$22\,mg\,kg^{-1}\,d^{-1}$.

An isoform of flavin-containing monooxygenase, FMO1, is another example of a foetal form. Expression of this enzyme ceases 3 days after birth, but the major adult form (FMO3) is not detectable until 1–2 years.

Esterase activity is low in neonates, and coupled with reduced protein binding may lead to increased toxicity with local anaesthetics.

12.2.4.2 *Phase 2 metabolism*

Examples of phase 2 enzymes and their development in human liver are presented in Table 12.1. As mentioned previously some of these enzymes are detectable in foetal livers and, in general, the activity increases postpartum, reaching adult levels after various amounts of time. UGT1A1 activity increases after birth and because it conjugates bilirubin, amongst other substrates, its development is crucial in limiting the risk of hyperbilirubinaemia in neonates. Because of relatively high sulfotransferase activity and low UGT1A6 activity, the urinary paracetamol sulfate/glucuronide ratio is greater in infants and children when compared with adults. It may take more than 10 years to reach adult levels of activity.

Table 12.1 Development of phase 2 enzymes in human liver

Enzyme	Example substrate	Foetus (% adult values)	Notes
UGT1A1	Bilirubin	–	Rapid increase postpartum Adult activity by 3–6 months
UGT1A3	Bile acids Estrone	30%	
UGT1A6	Paracetamol Propofol	1–10%	>10 years for adult activity
UGT2B7	Morphine	10–20%	Adult activity by 2–6 months
UGT2B17	Androgens	3–8%	11–13% adult activity in neonates
UGT2B?	Chloramphenicol		Grey baby syndrome
SULT1A3	Dopamine	+	Not detected in adult liver
SULT1A1	1-Naphthol	+	
GSTP1	Thiotepa	+	Absent in adults
NAT2	Isoniazid	+	Slow phenotypes initially Adult pattern by 3 years

–, not detected; +, present.

N-Acetyltransferase type 2 (NAT2) has been found in foetal liver but the activity in neonates is low. Rapid acetylator phenotypes (Section 10.3) are not expressed initially so that all newborns would be classed as slow acetylators. The adult pattern may take 2–3 years to develop.

12.2.4.3 Issues assessing the activity of drug-metabolizing enzymes

Apart from the obvious experimental difficulties in obtaining viable tissues from foetuses and neonates, there is the problem of relating *in vitro* and *in vivo* activities. Furthermore, the pattern of expression in one tissue may not be the same as another so, for example, the findings of Table 12.1 should not be applied universally to all tissues. Also there is the issue of substrate specificity: a drug may be metabolized by several enzymes and a given enzyme may exhibit different activities when tested against different substrates. For example, the rate of glucuronidation of serotonin was the same in samples of foetal and adult liver. Although the isoform involved was not identified at the time, serotonin has since been shown to be a substrate for UGT1A6.

A major complication is the degree of enzyme induction (Chapter 13) which affects particular samples, depending on exposure of mother, foetus and baby to environmental chemicals and drugs. Both pre- and postnatal enzyme induction occurs. Expectant mothers may be treated with phenobarbital to induce foetal enzymes and thereby reduce the risk of neonatal hyperbilirubinaemia. The newborns of mothers treated with phenytoin showed the ability to metabolize the drug at rates comparable with those of their mothers.

The rate at which adult activities of CYP1A2 are attained is particularly slow. However, this enzyme is induced by various environmental factors, including smoking, and it is possible that during 'natural development' the activity would peak at a level lower than that measured in an adult population that has been exposed to inducing agents.

12.2.5 Renal function

A relatively small proportion of cardiac output reaches the kidneys in the very young. Also, there is relatively slow glomerular and tubular development, and the loop of Henle is incomplete. Glomerular filtration in human beings is immature at birth, but postnatal development is rapid, reaching adult levels in about a year (Figure 12 4(a)). Kidney mass declines in old age with loss of functioning nephrons and renal blood flow is reduced. In fact there is a gradual decline in GRF, as shown by inulin clearance, from about 20 years (Figure 12.4(b)).

Kidney function can be assessed from creatinine clearance (Section 6.3.1.1). Creatinine distributes throughout total body water, shows little binding or tissue localization, and is eliminated almost entirely by renal excretion. Creatinine clearance, normalized for body weight, is highly depressed in premature newborns and is depressed in the neonate. However, it increases rapidly, reaching a maximum at 6 months, when it is twice the adult rate, in part due to kidney size being relatively larger in infants. From age 20 onwards creatinine clearance (mL min^{-1}) is given by the following formulae:

Males:

$$CL_{\text{creatinine}} = \frac{(140 - age) \times W}{70} \quad (12.1)$$

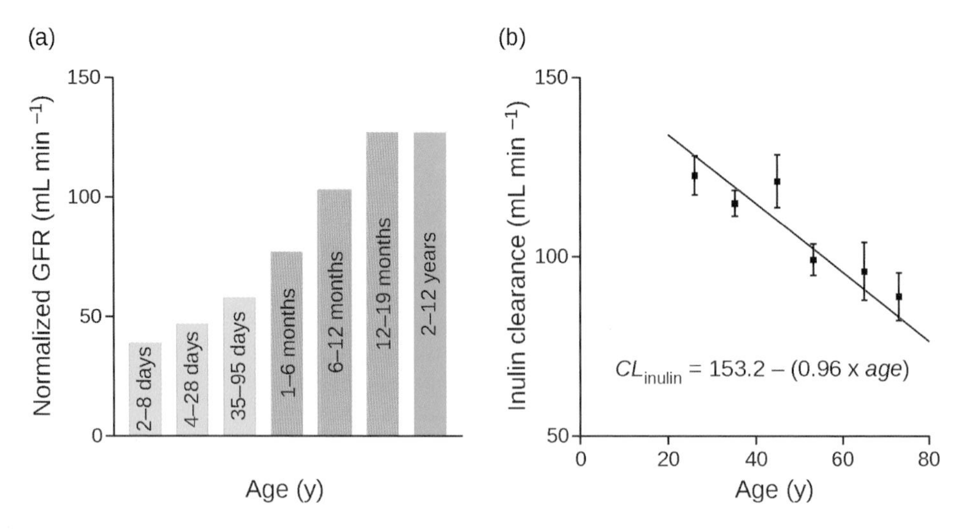

Figure 12.4 *(a) GFR in neonates and children. Values have been normalized to a standard body surface area of 1.73 m². (b) Inulin clearance, mean ± SEM in adults normalized to 1.73 m². Drawn from the data of Davies & Shock (1950).*

Females:

$$CL_{creatinine} = \frac{(140 - age) \times W}{85} \quad (12.2)$$

where *age* is in years and weight, *W*, in kg.

In practice it is not uncommon for serum creatinine measurements to be used to estimate creatinine clearance, thereby avoiding the need for urine collection. These estimates are prone to error, actually leading to underestimates of renal clearance as the serum concentration of creatinine reflects both its production and elimination. Creatinine is produced from creatine, a constituent of skeletal muscle. In the elderly serum creatinine may not be elevated because, although renal clearance may be low, the elderly have lower muscle mass and so produce less creatinine; an important consideration when accessing renal function from *serum* creatinine.

Elimination half-lives are often very elevated in newborns and neonates but quickly fall over a few days as the elimination processes develop. This is illustrated by the half-life of ampicillin in neonates and infants, which is 4.0 h in 2–7-day-olds, 2.8 h in 8–14-day-olds and declines to 1.6 h by 2 months (Axline *et al.*, 1967).

12.2.6 Dose adjustments for different age groups

Over the years a number of different methods have been devised for calculating the doses of drugs to be given to children. These doses are usually a fraction of the adult dose, dependent on the age and weight of the child, with the full adult dose being introduced at puberty. Another dosing scalar is body surface area (*BSA*). Several equations for calculating *BSA* have been proposed, including that of Dubois and Dubois in 1916:

$$BSA = W^{0.425} \times H^{0.725} \times 71.84 \times 10^{-3} \quad (12.3)$$

where *BSA* is in square metres, *W* is body weight in kilograms and *H* is height in centimetres. Different formulae may be used for different age ranges.

Clearly, surface area increases with increases in weight and height. In fact, the increase in surface area with age is linear, and there are good correlations between surface area and cardiac output, renal blood flow, and glomerular filtration rate (GFR). These correlations are better than those with body weight. The concept is that therapeutic precision would improve if doses were related to surface area rather than weight. In 1928, based on very little data, $1.73\,m^2$ was declared the *BSA* of the standard or average 25-year-old (Heaf, 2007) and it has been used ever since, particularly to normalize GFR. Thus, the values of GFR and inulin clearance in Figure 12.4 were normalized to $1.73\,m^2$ by the original authors.

For a drug that is primarily eliminated via the kidney an infant dose may be calculated from:

$$dose_{Infant} = dose_{Adult} \times \frac{BSA_{Infant}}{BSA_{Adult}} \quad (12.4)$$

However, normalization to *BSA* produces estimates of *GFR* that are too low in young children and an alternative formula has been suggested for children under 2 years:

$$dose_{Infant} = dose_{Adult} \times \frac{GFR_{Infant}}{GFR_{Adult}} \quad (12.5)$$

where *GFR* is in $mL\ min^{-1}$, that is, *not* normalized to $1.73\,m^2$.

The use of *BSA* in calculating doses has been positively discouraged by authorities such as the American Academy of Pediatrics, at least in part because it can lead to prescribing of drugs in ways different from the FDA approval for those drugs. However, some literature still recommends the use of *weight* for premature and full term neonates, and *BSA* for children.

As with any prescribing it is important to understand the properties of the drug in question. The maxim 'start low and increase the dose as necessary' has been suggested for dosing in the elderly. Because the half-lives of several drugs may be longer, steady-state concentrations will be higher and, importantly, the times to reach steady state will be longer, so sufficient intervals must be left before assessing the clinical response.

12.2.7 Metabolic and pharmacodynamic phenomena

It is to be expected that drug responses might vary with age as the result of pharmacodynamic sensitivity or metabolic differences. Children appear to tolerate high doses of paracetamol better than adults, possibly because of increased ability to sulfate this drug, higher glutathione transferase activity to conjugate the toxic metabolite and because the activity of CYP2E1 (which produces the toxic metabolite) is less than in adults, particularly those adults exposed to ethanol (Chapter 13). Children are less prone to the nephrotoxic effects of gentamicin, possibly because they do not excrete as much and so do not concentrate it in the kidney as adults do. The incidence of isoniazid hepatotoxicity (Chapter 13) is less, possibly reflecting the reduced acetylation in the young.

Both the young and the elderly are prone to exaggerated effects of centrally acting drugs. This may be, in part, because of an underdeveloped blood-brain barrier or, in the case of the elderly, one that is deteriorating. The susceptibility of neonates to morphine may arise from reduced metabolism, increased transfer into the CNS or the higher expression of μ-opioid receptors, which occurs in the young. The elderly are susceptible to the sedative effects of drugs such as diazepam and nitrazepam, and may suffer confusion and ataxia, leading to falls. These effects seem to be greater than would be predicted simply from an increase in elimination half-life.

12.3 Effects of disease on drug disposition

Drug responses are affected by disease states because of changes in both pharmacokinetics and pharmacodynamics. This is especially apparent with diseases that affect drug disposition and pharmacokinetics. Clearly, diseases of the liver and kidney would be expected to affect the clearance and, hence the half-lives, of drugs that are eliminated via these organs. It is not unreasonable to expect diseases of the GI tract to affect the rate and extent of drug absorption. However, the influence of the cardiovascular system should not be forgotten because tissue perfusion and delivery of drug to the liver and kidneys play major roles in drug disposition.

12.3.1 Gastrointestinal disorders and drug absorption

The effect of GI pathology on drug absorption is complex. For example, changes in pH do not necessarily affect absorption because of the relationship between pH, site of absorption, gastric emptying and so on. Similarly, in coeliac disease multiple changes occur, including increased rate of gastric emptying, increased gastric acid secretion and prolonged reduction in pH after eating, reduced surface area for absorption, increased permeability of the gut wall, and deceased local enzyme concentrations. These factors interact to either accelerate or decelerate absorption, and it is not surprising that a mixed pattern of changes has been reported. Additionally there are acute bacterial infections, such as salmonella and shigella, food-borne infections that cause diarrhoea.

12.3.1.1 Inflammatory conditions of the intestines and coeliac disease

Inflammatory bowel conditions such as Crohn's disease and ulcerative colitis are characterized by abdominal cramps and diarrhoea. Crohn's disease affects the full thickness of the intestinal wall, most commonly occurring in the lower part of the small intestine and in the large intestine. In contrast, ulcerative colitis affects only the large intestine, and does not affect the full thickness of the bowel wall. Inflammatory diseases have the potential to change the surface area available for absorption, the thickness of the intestinal wall and therefore the distance over which diffusion takes place, intestinal pH, mucosal enzymes that metabolize drugs, intestinal microflora, gastric emptying and peristalsis, and transporters that control inward and outward movement of nutrients and drugs. It is not surprising therefore that a variety of different observations has been made with various drugs in these conditions. For example, in Crohn's disease, the absorption of clindamycin and propranolol has been shown to be increased in extent, while that of many

Table 12.2 Some examples of the influences of GI pathology on drug absorption

Condition	Drug(s) affected	Nature of effect
Prolongation of gastric emptying	L-dopa	More than usual destruction by gastric acid
Achlorhydria	Aspirin and cefalexin	Impaired absorption because of pH effects
Gastric stasis/pyloric stenosis	Paracetamol	Impaired absorption
Shigella gastroenteritis/ fever	Ampicillin and iron	Impaired/reduced absorption
Malabsorption syndromes	Tetracycline and digoxin	Absorption reduced
Biliary tract disease	Cefalexin	Reduced bile secretion affects solubilization prior to absorption
Coeliac disease	Many drugs	Examples exist of both decreased and increased absorption (see text)
Crohn's disease	Many drugs	Examples exist of both decrease and increased absorption (see text)
Diarrhoea in children	Ampicillin/ co-trimoxazole	Impaired absorption/no change
Gastroenteritis in children	Sulfonamides	Decreased renal clearance resulting from relative acidosis

other drugs is decreased. The expression of CYP3A4 and P-gp levels was significantly higher in biopsy samples from a group of children with Crohn's disease when compared with controls. These differences could account for decreased bioavailabilities. An *in vivo* study with radiolabelled prednisolone in adults showed reduced bioavailability in Crohn's patients. In the case of paracetamol (acetaminophen), the mean rate constant of absorption was not reduced in Crohn's disease, the conclusion being that any pharmacokinetic differences were related to slower gastric emptying. Similarly, absorption of trimethoprim, methyldopa and lincomycin has been shown to be reduced while that of sulfamethoxazole was increased. Enhanced absorption of macromolecules has been observed in Crohn's disease (e.g. horseradish peroxidase) using biopsy samples and *in vitro* methods of study. Examples of some of the observations made with regard to GI absorption are summarized in Table 12.2.

12.3.2 Congestive heart failure

In congestive heart failure (CHF) the cardiac output is reduced so that insufficient quantities of oxygen and nutrients are delivered to the tissues for their normal functioning. Associated with CHF are atrial fibrillation and flutter. The reduced tissue perfusion leads to oedema. Digoxin may be used to control some cardiac dysrhythmias and it also has a diuretic effect because the improved cardiac efficiency increases renal perfusion. Furosemide may be used orally and/or intravenously, particularly after myocardial infarction, to treat pulmonary oedema. Lidocaine infusions are used to treat ventricular arrhythmia.

In CHF the perfusion of the organs that are involved in absorption, tissue distribution, metabolism and excretion is reduced as a result of reduced cardiac output. Redistribution of cardiac output can cause splanchnic vasoconstriction, and reduction of hepatic and renal blood flows would be expected to reduce the clearance of some drugs. The volumes of distribution of some dugs have been reported to decrease in CHF, possibly because of reduced tissue perfusion, and the situation is further complicated by the presence of oedema.

12.3.2.1 Oedema

It is particularly difficult to determine whether any pharmacokinetic changes in oedematous subjects are caused by the underlying disease or by the oedema itself. For example, furosemide has been well studied in this context, as it is the drug of choice for rapid reversal of oedema in both CHF and renal failure. The rate of absorption of oral doses has been shown to be decreased in oedema of CHF, although with no change in bioavailability, C_{max} values were generally decreased although paradoxically the time to achieve C_{max} was described as 'reduced or unchanged'. Some of the changes in drug absorption may be due to changes in splanchnic blood flow but also to changes in the GI tract, including oedema of mucosal cells.

12.3.2.2 Clearance and apparent volume of distribution

Generally, systemic clearance is reduced in CHF but changes in volume of distribution may increase or decrease. With lidocaine, for example, the apparent volume of distribution was reduced from 1.7 to 1.0 L kg^{-1} but half-lives in the patient group (1.8 h) and controls (2.2 h) were not significantly different, resulting in high plasma concentrations in CHF (Figure 12.5).

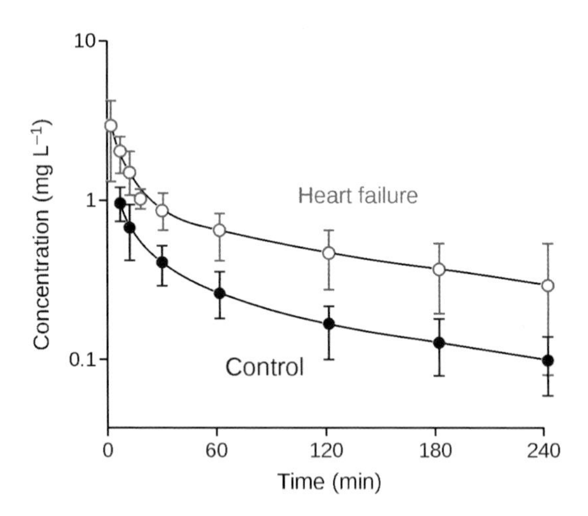

Figure 12.5 Lidocaine plasma concentrations (mean ± SD) in seven patients with heart failure and in control subjects after i.v. bolus dose of 50 mg. Adapted with permission from Thomson et al. (1971).

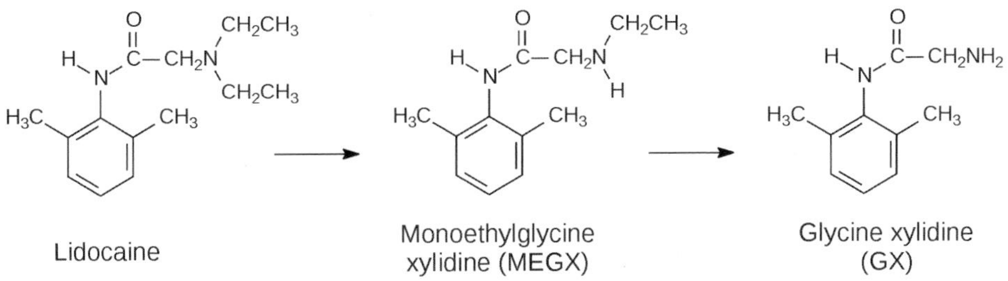

Lidocaine

Monoethylglycine xylidine (MEGX)

Glycine xylidine (GX)

Figure 12.6 Desethylation of lidocaine to the less active monoethyl metabolite (MEGX), which accumulates in plasma, and to the inactive desdiethylmetabolite (GX), which is normally excreted via the kidneys.

This is in keeping with reduced clearance as a result of reduced hepatic blood flow, something which may be generally applicable to drugs with flow-limited clearance. In the example of Figure 12.5, clearance in the patient group was almost half that in the control group of subjects. When lidocaine is used to control ventricular arrhythmia, it is often given as an intravenous bolus loading dose and a maintenance infusion. The example above nicely illustrates the need for using V to calculate the loading dose and CL to decide an appropriate rate of infusion, rather than the half-life, which was the same in both groups (Figure 12.5).

Lidocaine is *N*-desethylated to an active metabolite (MEGX) that accumulates in the plasma in CHF, leading to CNS toxicity. MEGX is not normally cleared by the kidney but is further desalkylated to GX, which is eliminated (40–60%) via the kidney (Figure 12.6). It would appear that MEGX metabolism is reduced in CHF patients and this, along with a decrease in apparent volume of distribution, could explain the increase in plasma concentrations.

Aminophenazone (aminopyrine) shows a different pattern from lidocaine. The metabolically labile methyl group can be isotopically labelled (^{14}C or ^{13}C) so that labelled carbon dioxide and/or formaldehyde can be measured in exhaled breath, in much the same way as for the erythromycin breath test (Section 10.2.2). In the patients, 2.6% of the dose was recovered in this way, compared with 5.6% in healthy volunteers (Hepner *et al.*, 1978). The total body clearance, reflecting the reduced metabolism, was 29.7 $mL\,min^{-1}$ in CHF patients compared with 125.1 $mL\,min^{-1}$ in the control group. However, unlike lidocaine, the apparent volume of distribution was greater (63.6 L) in the patients compared with the controls (43.6 L). This may be explained by the fact that the apparent volume of distribution in lidocaine is greater than total body water, so there must be some tissue binding, whereas aminophenazone is distributed in a volume equivalent to total body water. The proportion of extracellular fluid is much greater in CHF patients, so the increase in V can be explained, at least in part, by the fact that CHF patients are oedematous. Aminophenazone is 15–20% bound to plasma proteins, so a decrease in binding could also contribute to the increase in V observed in CHF.

The apparent oral clearance (CL/F) of the anti-arrhythmic, mexiletine (mean ± SD) was reduced in the CHF patients ($0.264 \pm 0.093\,L\,h^{-1}\,kg^{-1}$) compared with control values of ($0.393 \pm 0.082\,L\,h^{-1}\,kg^{-1}$). The reasons for the difference could be as a result of differences in hepatic and/or renal activity towards the drug, resulting from reduced blood flow and/or enzyme activity differences and/or differences in bioavailability. Such are the vagaries of oral clearance when so many variables may be changing.

12.3.3 Liver disease

Liver disease can affect:

- *liver blood flow*, creating the potential for effects on drugs with high extraction ratios such as propranolol, lidocaine and indocyanine green (which is used diagnostically to assess liver function)
- *hepatic intrinsic clearance*, creating the potential for effects on drugs such as metoprolol and phenazone
- *plasma enzymes* such as pseudocholinesterase, creating the potential for effects on such drugs as suxamethonium (succinylcholine) and cocaine
- *plasma protein binding*, because the binding proteins are synthesized in the liver, creating the potential for effects on warfarin and naproxen.

Some drugs, for example phenytoin, diazepam and tolbutamide, may be affected by changes in both intrinsic clearance and protein binding.

12.3.3.1 Pathophysiology of liver disease

When discussing liver disease it will be helpful to know the following:

- *Cirrhosis* is destruction of normal liver tissue leaving non-functioning scar tissue surrounding areas of functioning liver tissue. Many conditions lead to cirrhosis, including alcoholic liver damage and chronic hepatitis.
- *Hepatitis* is inflammation of the liver. It can be caused by viruses and/or chemicals, and can be acute or chronic.
- *Jaundice* causes a yellow discolouration of the skin and whites of the eyes, resulting from an increase in bilirubin in blood. One source of jaundice is cholestasis, i.e. reduction or stoppage of bile flow (*obstructive jaundice*).
- *Ascites* is fluid in the peritoneal cavity, associated with liver disease, especially cirrhosis.
- *Hepatic encephalopathy* refers to neuropsychiatric abnormalities such as personality changes, intellectual impairment and a depressed level of consciousness in patients with cirrhosis. The cause is multifactorial and ammonia, which arises from increased urea concentrations (uraemia), is one of the substances that have been implicated.

As liver disease progresses there are gross distortions to the structure and bypasses (*shunts*) may develop to change the liver blood flow. Collateral vessels may develop so that blood is shunted between the portal and systemic circulation. Hepatorenal syndrome is the development of life-threatening renal failure, probably as a result of changes in renal blood flow.

Two of the theories that have been postulated to explain the changes in clearance as a result of liver disease are:

- *sick-cell theory*: reduced intrinsic hepatic clearance is a result of damaged hepatocytes
- *intact-cell theory*: there is a reduced number of hepatocytes but these function normally and are normally perfused, and the clearance of high-extraction drugs is reduced because of intrahepatic shunts.

12.3.3.2 Assessing and investigating liver disease

The standard alanine transaminase and aspartate transaminase (ALT/AST) clinical laboratory tests that are used to test the general health of a broad range of patients may give a first indication of liver disease. Other markers include increased serum bilirubin, reduced albumin and increased prothrombin time and the presence of encephalopathy, and ascites. These form the basis of the Child–Pugh score, which ranks and scores each in terms of the severity on a scale of 1–3, and the sum is used to assess the severity:

- Group A (mild): 5–6
- Group B (moderate): 7–9
- Group C (severe): 10–15

The MELD test considers serum bilirubin and creatinine concentrations, prothrombin time and the cause of the disease. It is used in the end-stages of the disease and is effective in predicting 3-month mortality in patients awaiting liver transplants.

Indocyanine green, which is almost entirely eliminated through the bile, has a relatively high extraction ratio (0.5–0.8). It is administered intravenously to test hepatic blood flow and intrinsic clearance. It has been suggested that sorbitol is better. Galactose concentrations measured at a single time (1 h after intravenous administration of 0.5 $g\,kg^{-1}$) correlate well with a wide variety of indicators of liver function (Hu *et al.*, 1995). Imaging techniques (ultrasound and PET) have been used for measuring liver blood flows.

Drug cocktails (see Section 10.2) have been used to assess the *in vivo* activities of various CYP450 isoforms, and drugs such as lorazepam used to investigate glucuronidation. Biopsy samples have been used to study enzyme expression and activity *in vitro*.

12.3.3.3 Absorption

The oedema which results from portal hypertension leads to changes in the GI tract (Section 12.3.2.1) and this may delay the absorption of some drugs. However, portocaval shunts (i.e. from the portal vein to the inferior vena cava) allow orally administered drugs to bypass the liver and so avoid first-pass metabolism. Therefore in cirrhosis, the oral bioavailability is substantially increased for drugs that have a moderate to high hepatic extraction ratio. Examples range from verapamil, with a 1.6-fold increase, to chlormethiazole, with an 11.6-fold increase, and include pethidine, morphine, nifedipine, midazolam and most of the β-adrenoceptor antagonists.

12.3.3.4 Protein binding, drug distribution and clearance

The concentrations of several plasma proteins are reduced in liver disease. The mean AAG concentration in patients with cirrhosis was 195 $mg\,L^{-1}$ compared with 674 $mg\,L^{-1}$ in control subjects. This resulted in the free fraction of erythromycin rising from ~20% in plasma from controls to almost 80% in the patient with the lowest AAG concentration. Similarly, the fall in serum albumin, which is a marker for liver disease, results in reduced binding, particularly of acidic drugs. Increased plasma concentrations of bilirubin (and other substances) may decrease protein binding further by displacing drugs from their binding sites. Assuming there is little change in tissue binding, decreased plasma protein binding will increase the apparent volume of distribution and this is generally the case in patients with cirrhosis. Systemic clearance is reduced as a result of reduced hepatic clearance and this,

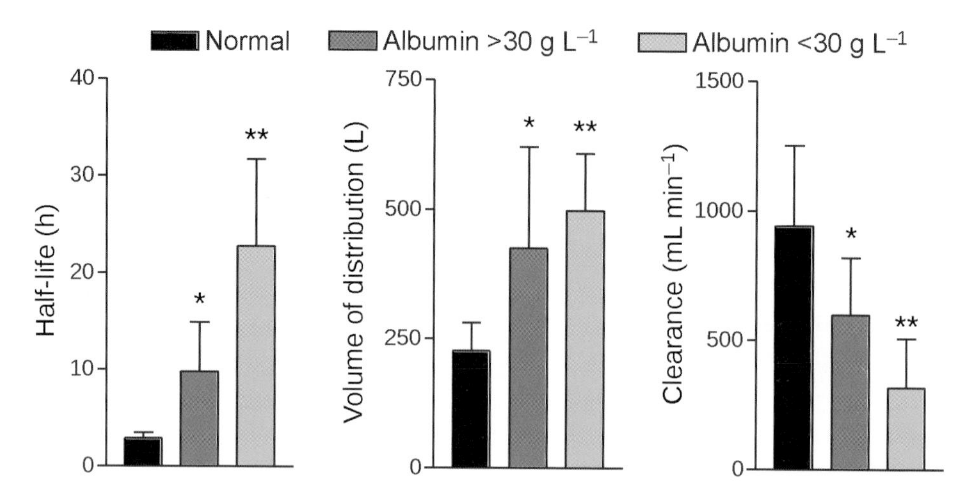

Figure 12.7 *Half-life, apparent volume of distribution and clearance after i.v. injection of 40 mg (+)-propranolol in normal subjects, patients with serum albumin >30 g L^{-1} and patients with serum albumin <30 g L^{-1}. Data are mean + SD. * $p<0.05$, ** $p<0.001$ relative to normal subjects. Redrawn with permission from Branch* et al. *(1976).*

coupled with the increase in the volume of distribution, can lead to marked increases in the elimination half-life. A typical pattern of these changes is shown for (+)-propranolol in Figure 12.7. Patients with albumin concentrations less than 30 g L^{-1} were considered as having more severe disease, and were classed separately from patients with albumin >30 g L^{-1}. Being basic, propranolol normally binds to AAG rather than albumin and there was a good correlation between the percentage of propranolol bound in plasma and the apparent volume of distribution ($r^2 = 0.81$).

Propranolol is a high hepatic extraction drug and protein binding would not be expected to affect the hepatic clearance (Equation 7.15), so the decrease in clearance must be as a result of reduced flow to the hepatocytes or reductions in intrinsic clearance. For drugs of intermediate extraction, with high protein binding, then Equation 7.19 is more applicable. Table 12.3 shows the kinetics of a selection of drugs with varying degrees of plasma

Table 12.3 *Pharmacokinetic parameters in normal subjects (N) and patients with cirrhosis (LD) after intravenous administration*

Drug	Half-life (h)		V_{ss} (L kg^{-1})		*CL* (mL min^{-1} kg^{-1})		Bound (%)		Reference
	N	LD	N	LD	N	LD	N	LD	
Theophylline	6.7	25.6	0.79	0.51	1.03	0.70	55.6	36.8	Piafsky *et al.* (1977)
Diazepam	46.6	105.6	1.13	1.74	0.39	0.21	97.3	95.3	Klotz *et al.* (1975)
Morphine	1.7	4.2	4.0*	4.1*	28	11.4	–	–	Halsselstom *et al.* (1990)
Thiopental	8.8	11.9	2.3	3.5	3.9	4.4	85.5	74.8	Pandele *et al.* (1983)
Propranolol†	2.9	19.0	3.46*	8.09*	14.3	6.49	87.3	82.3	Branch *et al.* (1976)

* V_{area}
† Calculated from a subset of control and cirrhosis patients.

protein binding in normal controls and patients with cirrhosis. The data were taken from studies in which the drugs were administered intravenously. In every case there is a clear prolongation of the half-life. Thiopental is the only example in Table 11.4 where *CL* was not reduced. However, the *non-bound intrinsic clearance* was reduced from (mean ± *SD*) 28.3 ± 9.0 to 18.2 ± 10.5 mL min^{-1} kg^{-1}. The volume of distribution of the non-bound fraction of thiopental was similar in controls and patients: 16.9 ± 4.6 and 14.0 ± 6.9 L kg $^{-1}$, respectively.

The plasma protein binding of morphine was not determined in the study by Hasselstrom *et al.* (1990), but it is normally low (20–30%), so any change in the non-bound fraction would be expected to be low. This is in keeping with there being no change in V_{ss} (Table 12.3).

12.3.3.5 Drug-metabolizing enzymes and transporters

One of the earliest examples of reduced enzyme activity in liver disease must be suxamethonium (succinylcholine). It is metabolized by plasma (pseudo) cholinesterases, which are synthesized in the liver. This hydrolysis reaction can be reproduced *in vitro*. Figure 12.8 shows the metabolism of suxamethonium *in vitro* by pseudocholinesterase enzymes from the plasma of patients with impaired hepatic function, including impaired synthesis of plasma pseudocholinesterase enzymes.

The existence of different isoforms of P-450, and their differential ability to metabolize their various substrates, may explain some of the differences observed between the clearances of different drugs. In many cases the liver appears be able to metabolize drugs until the very last stages of the disease. Frye and his colleagues (2006) administered caffeine, mephenytoin, debrisoquine and chlorzoxazone to measure the activities of CYP1A2, CYP2C19, CYP2D6 and CYP2E1, respectively, in patients with liver disease. CYP2C19 activity was lost early in the process of decay of liver function and CYP1A2 activity fell

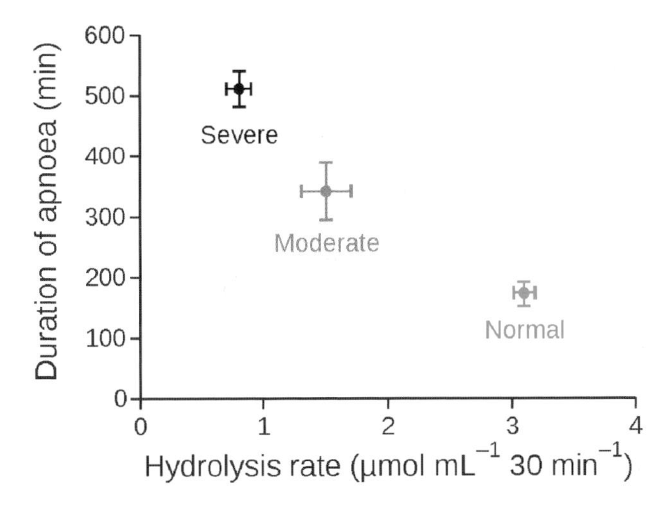

Figure 12.8 Relationship between enzymatic hydrolysis of suxamethonium and duration of apnoea after 0.6 mg kg^{-1} in patients with severe and moderate liver disease and normal livers. Data are mean ± SEM. After Birch et al. *(1956) and Foldes* et al. *(1956).*

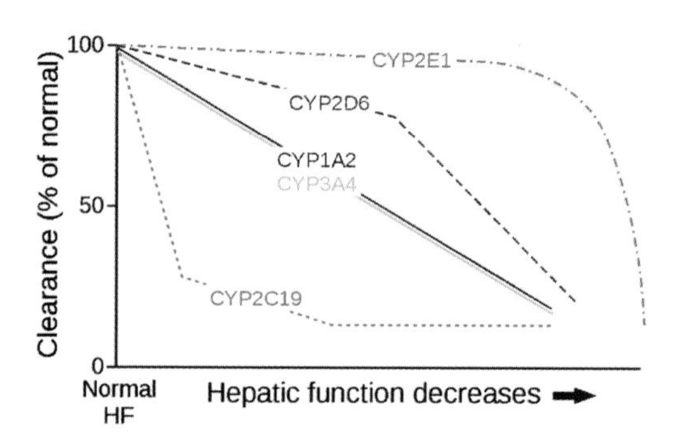

Figure 12.9 Sequential progressive model of hepatic dysfunction: implications for clearance of drugs predominantly metabolized by individual CYP pathways in liver. From Branch (1998), with permission.

in an almost linear fashion with increasing severity of the disease. CYP2D6 did not fall as quickly while CYP2E1 was conserved until the hepatorenal syndrome was established. The rate of loss of CYP3A4 is similar to that of CYP1A2 (Figure 12.9).

It is generally believed that glucuronidation does not decline as rapidly as the decline in the activity of CYP2C19 and CYP3A4. This is based to some extent on the fact that the kinetics of diazepam and midazolam (metabolized by CYP2C19 and CYP3A4) are affected to a greater degree than oxazepam and lorazepam, which are, of course, conjugated with glucuronic acid. *In vitro* studies using liver biopsies from patients show that glucuronidation does appear to be the least affected of the conjugating enzymes (Figure 12.10). It has been reported that the kinetics of morphine are unaffected by hepatic disease but more a recent study has shown that there are changes in the kinetics (Table 12.3). The example of morphine illustrates a further problem in interpreting the effects of reduced hepatic function because there is also extrahepatic metabolism. The UGT isoforms may be affected to differing extents, for example in an *in vitro* study zidovudine glucuronidation, which is conjugated by UGT2B7, was very much more reduced in cirrhosis than was lamotrigine, the glucuronidation of which is catalysed by UGT1A3 and UGT1A4 (Furlan *et al.*, 1999).

The effect of liver disease on the expression of liver transporter proteins is interesting. Table 12.4 shows the significant changes in messenger RNA in patients infected with hepatitis C virus and who developed cirrhosis (Ogasawara *et al.*, 2010). The expression of transporters responsible for uptake into hepatocytes (see Figure 10.5) was reduced in most cases but the expression of efflux proteins was increased (Table 12.4). This pattern of change may be an adaptation to reduce the uptake of substances that may damage the cell and to increase the efflux of potentially toxic substances, such as bile acids.

12.3.3.6 Drug effects and dosage

Over and above any changes resultant from changes in pharmacokinetics, hepatic clearance and protein binding for example, there appear to be pharmacodynamic differences in patients with liver disease. Effects have been observed with β-adrenoceptor antagonists, diuretics and drugs that depress the CNS.

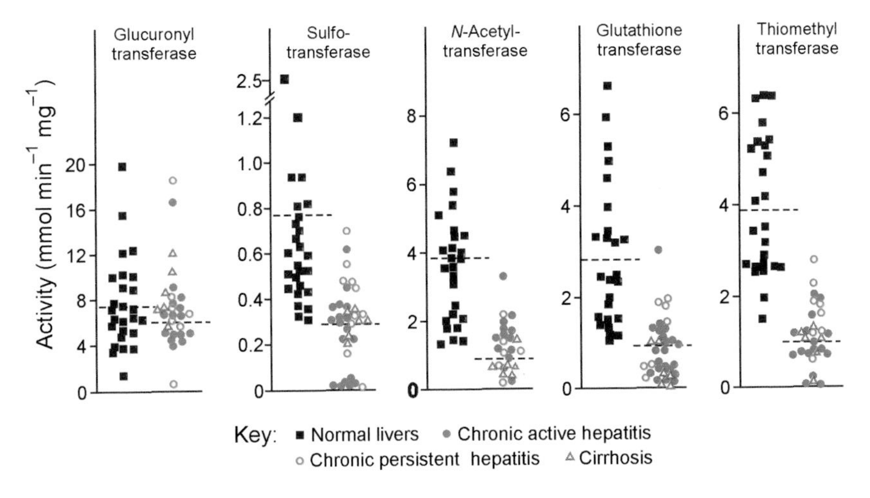

Figure 12.10 *Effects of liver disease on conjugation reactions. Broken lines represent the mean value for each group. Redrawn with permission from Pacifici* et al. *(1990).*

Table 12.4 *Changes in mRNA levels for selected transport proteins in hepatocytes from patients with hepatitis C virus and liver cirrhosis. Mean ± SD, all the differences are significant,* $p < 0.05$

Transporter	mRNA (amol/µg total RNA)		Change
	Control	Hepatitis C virus with cirrhosis	
OCT1	11.63 ± 8.91	8.87 ± 5.56	↓
OATP1B1	10.55 ± 6.71	5.92 ± 2.89	↓
OATP1B3	383 ± 2.27	1.98 ± 1.15	↓
MATE1	3.14 ± 2.04	1.34 ± 068	↓
MDR1 (P-gp)	1.15 ± 0.52	1.40 ± 0.43	↑
MRP1	0.25 ± 0.14	0.51 ± 0.53	↑
MRP4	0.09 ± 0.05	0.18 ± 0.13	↑
BCRP	0.57 ± 0.33	0.31 ± 0.68	↓

In regard to β-blocking drugs it has been proposed that a decreased therapeutic effect should be expected because there is a decrease in the sensitivity to the β-agonist effects of isoprenaline (isoproterenol) in liver disease, as well as a reduction in β-adrenoceptor density in mononuclear cells. Such a decrease in effect has been observed with propranolol.

Diuretics such as furosemide, and also triamterene, torasemide and bumetanide, show decreased effects in liver patients, characterized by a need for a relatively high concentration of the diuretic in the renal tubules in order to induce an effect. This could be caused by a reduction in the number of nephrons and by the maximum response per nephron. However, this is likely to be offset by reduced hepatic clearance, leading to higher concentrations of the drug in the kidneys, so that the effect is maintained but with a different pharmacodynamic/pharmacokinetic relationship. In renal failure it has been suggested that albumin in the tubular fluid could bind diuretics and thereby reduce their efficacy.

Patients with cirrhosis are particularly sensitive to opioids and anxiolytics. Hepatic encephalopathy (HE) may be improved with the benzodiazepine antagonist flumazenil, whereas benzodiazepines may precipitate it. It is thought therefore that GABA-A activity is increased in HE. If this is the case, then barbiturates should have similar effects.

There are no clear guidelines to dose adjustment in hepatic disease and dosage should be based on clinical assessment. However, it should be remembered that drugs that are eliminated primarily via the liver, and that have:

- *high extraction ratios* may develop very high plasma concentrations after oral doses as a result of reduced first-pass effect. Plasma concentrations may not be so elevated following parenteral administration but half-lives would be longer.
- *low hepatic extraction and high plasma protein binding* will have longer half-lives as a result of reduced metabolism. The pharmacodynamic effects may be greater. Assessment and dosage adjustment should be made on the basis of non-bound concentrations.
- *low hepatic extraction ratio and low plasma protein binding* will be affected by reduced hepatic clearance but because changes in protein binding will have only slight effects any dosage should be based on total plasma concentrations.

It seems logical to try to avoid prescribing drugs that rely on metabolism for their elimination but it should also be remembered that the excretion of drugs which are usually eliminated by the kidney may be reduced by changes in kidney function. Remifentanil, which is metabolized by plasma and tissue esterases, has been suggested as an alternative to some other opioids. Hydrophilic drugs (even those with little or no plasma protein binding) may have increased volumes of distribution in patients with oedema and even more so in patients with oedema and ascites.

12.3.4 Renal impairment

Kidney failure can be acute, with rapid decline in function but equally rapid recovery if the pathology is reversed, or chronic with slow progression and little chance of recovery. In either case there is accumulation of metabolic waste products in the blood, especially urea and creatinine. Acute kidney failure can be caused by impairment of the blood supply to the kidney (e.g. CHF, bleeding, dehydration, shock or liver failure – hepatorenal syndrome) or obstruction of urine flow (e.g. prostate enlargement or other tumours) or injury (e.g. allergic reactions within the kidney, toxic chemicals, conditions affecting glomerular filtration, blocked blood vessels or crystal deposits). Chronic kidney failure can result from high blood pressure, urinary tract obstruction, glomerulonephritis, structural abnormalities, diabetes or autoimmune disorders.

Kidney function is typically evaluated by means of creatinine clearance, which is considered elsewhere (Section 3.2.1.4). Creatinine is freely diffusible at the glomerulus, and there is approximately equal transfer in each direction across the renal tubule, so creatinine clearance effectively measures GFR. Shortcuts based on measurement of serum creatinine tend to underestimate creatinine clearance, although in patients with cirrhosis creatinine clearance may be overestimated. In some cases GFR is estimated (eGFR) without measurement of serum creatinine.

The claim that kidney disease reduces the rate of elimination of drugs cannot be challenged. However, it should be realized that drugs may be excreted unchanged to any extent, varying from 0 to 100%. Clearly, a drug that is 100% dependent on the kidney for its removal from the body might be greatly affected by kidney disease. However, a drug that is 100% metabolized may also be affected to some extent, and the excretion of the metabolites of this type of drug, as opposed to the excretion of the unchanged drug, will almost certainly be affected. The metabolites will accumulate in plasma, leading to an exaggerated response if the metabolites contribute to the pharmacological effect or, possibly, toxicity that is not seen when the metabolites are excreted normally. If the presence of large quantities of metabolites reduce the rate of metabolism of the unchanged drug, by metabolite inhibition, then there is the possibility of accumulation of the parent drug. Additionally, renal impairment is likely to lead to varying degrees of water loading and this may lead to modification of the concentrations of the drug in the fluid compartments of the body, including plasma.

12.3.4.1 Significance of changes in GFR

The effect of changes in GFR on the overall kinetics of a drug will depend on the proportion eliminated via the kidney and that eliminated via other routes. The overall rate constant of elimination (λ) can be considered as:

$$\lambda = k_R + k_{NR} \tag{12.6}$$

where k_R is the rate constant of renal elimination and k_{NR} is the rate constant of non-renal elimination (hepatic and other metabolism and miscellaneous non-renal processes). Generally, k_R is proportional to creatinine clearance, CL_{cr}, so:

$$\lambda = k_{NR} + \alpha CL_{cr} \tag{12.7}$$

where α is a factor of proportionality expressing the relationship between the k_R and CL_{cr} of the drug. For drugs such as gabapentin, which are almost entirely eliminated via the kidneys, $k_{NR} \approx 0$ and Equation 12.7 describes a straight line (slope $= \alpha$) which passes through the origin of a λ versus CL_{cr} plot (Figure 12.11(a)). Desipramine, on the other hand, is almost entirely eliminated by metabolism and so is unaffected by renal clearance ($\lambda = k_{NR}$). Digoxin is eliminated ~65% unchanged in the urine, so for such drugs Equation 12.7 will represent a straight line of slope, α, that intercepts the y-axis at k_{NR}. The relationship between the elimination half-life and GFR is more complex (Figure 12.11(b)).

The ratio of creatinine clearance in uraemic patients to normal values has been used to calculate a value of systemic clearance, CL_u, of a drug in uraemic patients:

$$CL_u = CL_{NR} + CL_R \frac{CL_{cr(u)}}{CL_{cr(N)}} \tag{12.8}$$

where $CL_{cr(u)}$ is the clearance of creatinine the in the uraemic patient and $CL_{cr(N)}$ is a 'normal' value of creatinine clearance. Thus, if the ratio of the two creatinine clearances is known, along with the drug clearances for normal subjects, the clearance in the uraemic patient can be calculated. Alternatively, if the normal total body clearance and f_e, the fraction excreted unchanged in the normal case, are known, then:

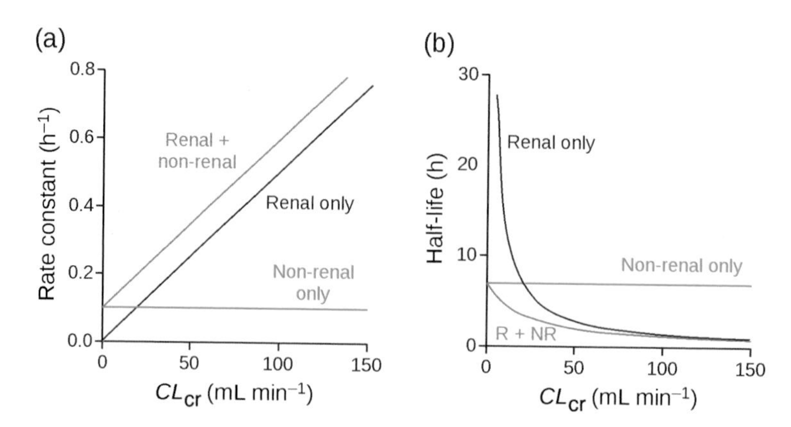

Figure 12.11 (a) Effect of renal clearance on the elimination rate constant for a drug which is only eliminated via the kidneys (red), not eliminated by the kidneys (green), and eliminated by renal and non-renal routes (blue). (b) The same data plotted as elimination half-life versus CL_{cr} calculated using $k_{NR}=0.1\ h^{-1}$, $\alpha=0.05$.

$$CL_u = CL_N(1-f_e) + f_e CL_N \frac{CL_{cr(u)}}{CL_{cr(N)}} \quad (12.9)$$

These approaches assume that there is no effect of renal impairment on non-renal elimination, which is not necessarily the case (Section 12.3.4.2).

12.3.4.2 Examples

In practice, the presence of renal impairment is a reason for care with any drug, but studies have been largely concerned with drugs excreted mostly unchanged. The majority of the work has concerned antibiotics, and a selection of values of the rate constants of elimination for these and other drugs is shown in Table 12.5.

Desipramine is included as an example with virtually no excretion of the parent drug, and for which it can be presumed that renal failure would have little effect on the half-life,

Table 12.5 Pharmacokinetic data relevant to drug elimination in anuric patients and patients with normal function

Drug	f_e	$t_{1/2(N)}$ (h)	$k_{(N)}$ (h^{-1})	$k_{r(N)}$ (h^{-1})	$k_{(anuric)}$ (h^{-1})	$t_{1/2(anuric)}$ (h)
Desipramine	0.02	22.0	0.032	0.0006	ND*	ND
Cefaloridine	0.059	1.36	0.51	0.03	0.03	23.1
Erythromycin	0.13	1.6	0.43	0.056	0.13	5.33
Sulfamethoxazole	0.14	9.8	0.071	0.0099	0.07	9.9
Chloramphenicol	0.15	3.3	0.21	0.032	0.2	3.47
Tetracycline	0.58	10.6	0.065	0.038	0.008	86.6
Digoxin	0.65	41.0	0.017	0.011	0.008	86.6
Gabapentin	1.0	6.5	0.11	0.11	ND	ND

* Not determined

while gabapentin is at the other extreme and would presumably not be eliminated at all in an anuric patient. Digoxin is an important example of a clinical need to adjust dosage in renal failure. Five of the compounds are antimicrobial drugs. One of them, chloramphenicol, is partly dependent on metabolism and excretion in bile for its removal from the body. Sulfamethoxazole is a sulfonamide which also undergoes metabolism. These two compounds were apparently the least affected of the eight compounds, perhaps less so than would be expected from the rate constant data. Generally speaking this table shows a greater impact of renal failure on drugs with the higher levels of elimination through the kidney, but cefaloridine seems to show a greater effect than would be expected from this factor alone. It is now standard clinical practice to base drug dosage with some of these compounds on creatinine clearance.

It has been suggested that various other differences arise as the result of renal failure. For example, there is apparently decreased absorption of oral iron, a carrier-mediated process, and it has also been suggested that nitrofurantoin shows impaired absorption, as there is no build-up in plasma although urinary excretion is reduced. It has been suggested that decreased absorption of drugs could arise from changes to the functioning of enterocytes as a result of oedema.

There is evidence of protein binding changes in renal failure. Generally, the trend is for serum albumin concentrations to fall, particularly when there is proteinuria, and AAG levels to rise. Thus, the binding of acidic drugs tends to fall whist the binding of basic ones may rise. However, this is not consistent. Drugs for which binding has been shown to decrease include theophylline, methotrexate and diazepam, whilst drugs showing increased binding include propranolol, imipramine, cimetidine and clonidine (Vanholder *et al.*, 1988). The exaggerated response to the antihypertensive effect of i.v. diazoxide has been attributed to reduced serum albumin (from 4.5 ± 0.07 to 3.9 ± 0.08) accompanied by reduced binding, 94.0 ± 0.4 versus 83.9 ± 0.5, in controls and renal patients, respectively. The values are mean percentage ± *SEM* (Pearson & Beckenridge, 1976).

In renal failure there is increased fluid volume, oedema and changes in acid–base balance, as well changes in protein binding, all of which may affect distribution. The volume of distribution of digoxin is reduced because of reduced tissue binding, but for phenytoin the volume is increased because of reduced plasma protein binding. The distribution of ampicillin within the kidneys of patients with chronic glomerulonephritis is different from that of control subjects, and there will clearly be difficulty in achieving therapeutic levels of drugs when the kidney is the target organ. Interestingly, it has been shown that the distribution of kanamycin is unaffected by renal failure, and this has been attributed to lack of protein binding of this drug.

It was once thought that it was unnecessary to adjust the doses of drugs that are not eliminated via the kidney, but it has been shown that acetylation (isoniazid, sulfisoxazole, hydralazine), hydrolysis (procainamide) and reduction (hydrocortisone) are reduced in renal failure. Studies in rats and the use of metabolic probes in human subjects, as described above, have shown changes in the activities of hepatic enzymes (Dreisbach & Lertora, 2003). Furthermore, infusing isolated rat livers with blood from uremic rats showed the presence of inhibitory substances. In patients, the bioavailability of propranolol was reduced the day after dialysis, supporting the idea that there are inhibitory substances in the blood of uremic patients that can be removed by dialysis. In both rat and human models there are reductions in the activities of some CYP isozymes.

The resultant fall in intrinsic clearance will affect drugs differently depending on whether their clearance is flow dependent or capacity limited, and whether or not the drug is protein bound. The reduced intrinsic clearance increases the oral bioavailability of drugs such as propranolol but does not change the kinetics of an intravenous dose, so that the half-life is unchanged in uremic subjects. The increase in the hepatic clearance of phenytoin has been ascribed to reduced protein binding. The apparent volume of distribution will also increase so that the steady-state plasma concentration falls but the concentration of non-bound drug remains the same and no dose adjustment is required. Increasing the dose to achieve the recommended steady-state values for patients without uraemia will lead to toxicity.

Thus, when attempting to rationalize the effects of renal insufficiency on the pharmacokinetics of a drug it is important to consider it in the light of the equations in Chapter 7.

12.4 Assessing pharmacokinetics in special populations

Experimentation in special populations, including the young and elderly, and those being treated for disease, raises many issues and practical difficulties. There are ethical concerns and practical issues such as the number and size of blood samples that can be taken for analysis – inadequate sampling will affect the quality of the data. The duration of the study may be limited, so that *AUC* and terminal half-lives cannot be accurately assessed. The time course of some of the studies in patients with liver disease has been barely long enough to determine an accurate half-life in control subjects and so has been woefully short for patients who had marked increases in $t_{½}$ (e.g. Table 12.3). Sampling for too short a time will nearly always underestimate the elimination half-life in i.v. studies but may overestimate it with oral administration. Also, too short a collection period will produce errors in the assessment of *AUC*, partly because of the proportion of $AUC_{(0-\infty)}$ that has to be extrapolated from the last time point and because $AUC_{(t_z-\infty)} = C_{t_z}/\lambda$, so that an error in λ as a result of the error in $t_{½}$, will produce an error in the extrapolated area. These inaccuracies will occur even when the drug has been given intravenously.

The difficulty in administering intravenous doses, either because of ethical reasons or because there is no injectable preparation available for use, has necessitated use of oral administration, which, combined with the desire to use clearance concepts, has sometimes led to conclusions that cannot be substantiated. In the populations considered in this chapter, development, aging and disease have led in many cases to changes in the GI tract and consequently to changes in oral absorption. However, observed changes in *AUC* following an oral dose can be the result of changes in bioavailability (tablet dissolution, gastric emptying, first-pass effects) and/or systemic clearance. Without further information, it is impossible to say which factors have changed, and so reported differences in apparent oral clearance (*CL*/*F*) must be interpreted with caution, particularly in those special populations in which *F* is likely to be different in control and study groups. It is not uncommon to see the use of literature values for *F* with no assessment of whether they apply to the population of subjects under study. Uncritical use of *oral clearance* must be assessed with particular care.

As with the assessment of clearance, apparent volumes of distribution should be obtained from intravenous studies whenever possible. When intravenous studies are conducted, calculating from $V = dose/C_0$ assumes that the drug confers the properties of a single compartment model. This is rarely, if ever, the case and ignoring a distributional phase means that the value will be V_{extrap}, which overestimates the apparent volume of distribution to varying degrees (Section 5.1.2.3). The best parameter to estimate and compare the distribution of drugs is V_{ss}, which can be obtained from Equation 5.31. Even the use of CL/λ is not ideal because it gives V_{area}, which changes with CL (Figure 5.5).

Interpretation of some observations requires assessment of non-bound concentrations of the drug. However, this raises practical issues. Obtaining accurate values for highly bound drugs is difficult and is best carried out with samples to which minute quantities of radioactively labelled drug have been added. This may not be possible, as the required material may not be available, and even if the material is available, facilities for radioactive counting may not be. Another practical issue is that methods of assessment (equilibrium dialysis and ultrafiltration) can perturb the binding.

Thus, the variability reported between studies may be, in part, because of differences in study design and the parameters that have been chosen for comparison. Without better experimental design and implementation, the problems of variability due to poor design will remain. Not only may these mask true biological differences, they may lead to inappropriate clinical management.

Summary

The special populations that are considered in this chapter pose particular challenges to pharmacokineticists and clinicians. The ethical and practical constraints are problems enough but the rapid physiological development in neonates and changes in some illnesses make long-term study in an individual impossible. The chapter illustrates the need to understand how the clearance of flow-limited and capacity-limited drugs, and the effects of protein binding on the latter, has such a major influence on their kinetics and hence the plasma concentrations and biological effects.

The next chapter looks at how substances can interact to change the disposition, pharmacokinetics and effects of drugs.

12.5 Further reading

Barre J, Houin G, Rosenbaum J, Zini R, Dhumeaux D, Tillement JP. Decreased alpha 1-acid glycoprotein in liver cirrhosis: consequences for drug protein binding. *Br J Clin Pharmacol* 1984; 18: 652–3.

de Wildt SN, Kearns GL, Leeder JS, van den Anker JN. Glucuronidation in humans. Pharmacogenetic and developmental aspects. *Clin Pharmacokinet* 1999; 36: 439–52.

Johnson TN, Thomson M. Intestinal metabolism and transport of drugs in children: the effects of age and disease. *J Pediatr Gastroenterol Nutr* 2008; 47: 3–10.

Kearns GL, Abdel-Rahman SM, Alander SW, Blowey DL, Leeder JS, Kauffman RE. Developmental pharmacology–drug disposition, action, and therapy in infants and children. *New Eng J Med* 2003; 349: 1157–67.

Ku LC, Smith PB. Dosing in neonates: special considerations in physiology and trial design. *Pediatric Res* 2015;77: 2–9.

McInerny TK. (Editor in Chief) *American Academy of Pediatrics Textbook of Pediatric Care*. Elk Grove Village: American Academy of Pediatrics, 2009.

Shah GN, Mooradian AD. Age-related changes in the blood-brain barrier. *Exp Gerontol* 1997; 32: 501–19.

12.6 References

Axline SG, Yaffe SJ, Simon HJ. Clinical pharmacology of antimicrobials in premature infants. II. Ampicillin, methicillin, oxacillin, neomycin, and colistin. *Pediatrics* 1967; 39: 97–107.

Birch JH, Foldes FF, Rendell-Baker L. Causes and prevention of prolonged apnea with succinylcholine. *Curr Res Anesth Analg* 1956; 35: 609–33.

Branch RA. Drugs in liver disease. *Clin Pharmacol Ther* 1998; 64: 462–5.

Branch RA, James J, Read AE. A study of factors influencing drug disposition in chronic liver disease, using the model drug (+)-propranolol. *Br J Clin Pharmacol* 1976; 3: 243–9.

Chumlea WC, Guo SS, Zeller CM, Reo NV, Siervogel RM. Total body water data for white adults 18 to 64 years of age: The Fels longitudinal study. *Kidney Int* 1999; 56: 244–52.

Davies DF, Shock NW. Age changes in glomerular filtration rate, effective renal plasma flow and tubular excretory capacity in adult males. *J Clin Invest* 1950; 29: 496–504.

Dreisbach AW, Lertora JJ. The effect of chronic renal failure on hepatic drug metabolism and drug disposition. *Semin Dial* 2003; 16: 45–50.

Foldes FF, Swerdlow M, Lipschitz E, Van Hees GR, Shanor SP. Comparison of the respiratory effects of suxamethonium and suxethonium in man. *Anesthesiology* 1956; 17: 559–68.

Fomon SJ, Haschke F, Ziegler EE, Nelson SE. Body composition of reference children from birth to age 10 years. *AmerJ Clin Nutrition* 1982; 35: 1169–75.

Frye RF, Zgheib NK, Matzke GR, Chaves-Gnecco D, Rabinovitz M, Shaikh OS, Branch RA. Liver disease selectively modulates cytochrome P450-mediated metabolism. *Clin Pharmacol Ther* 2006; 80: 235–45.

Furlan V, Demirdjian S, Bourdon O, Magdalou J, Taburet AM. Glucuronidation of drugs by hepatic microsomes derived from healthy and cirrhotic human livers. *J Pharmacol Exp Ther* 1999; 289: 1169–75.

Hasselstrom J, Eriksson S, Persson A, Rane A, Svensson JO, Sawe J. The metabolism and bioavailability of morphine in patients with severe liver cirrhosis. *Br J Clin Pharmacol* 1990; 29: 289–97.

Heaf JG. The origin of the $1 \times 73\text{-m}^2$ body surface area normalization: problems and implications. *Clin Physiol Funct Imaging* 2007; 27: 135–7.

Hepner GW, Vesell ES, Tantum KR. Reduced drug elimination in congestive heart failure. Studies using aminopyrine as a model drug. *Am J Med* 1978; 65: 371–6.

Hu OY, Tang HS, Chang CL. Novel galactose single point method as a measure of residual liver function: example of cefoperazone kinetics in patients with liver cirrhosis. *J Clin Pharmacol* 1995; 35: 250–8.

Klotz U, Avant GR, Hoyumpa A, Schenker S, Wilkinson GR. The effects of age and liver disease on the disposition and elimination of diazepam in adult man. *J Clin Invest* 1975; 55: 347–59.

Ogasawara K, Terada T, Katsura T, Hatano E, Ikai I, Yamaoka Y, Inui K. Hepatitis C virus-related cirrhosis is a major determinant of the expression levels of hepatic drug transporters. *Drug Metab Pharmacokinet* 2010; 25: 190–9.

Pacifici GM, Viani A, Franchi M, Santerini S, Temellini A, Giuliani L, Carrai M. Conjugation pathways in liver disease. *Br J Clin Pharmacol* 1990; 30: 427–35.

Pandele G, Chaux F, Salvadori C, Farinotti M, Duvaldestin P. Thiopental pharmacokinetics in patients with cirrhosis. *Anesthesiology* 1983; 59: 123–6.

Pearson RM, Breckenridge AM. Renal function, protein binding and pharmacological response to diazoxide. *Br J Clin Pharmacol* 1976; 3: 169–75.

Piafsky KM, Sitar DS, Rangno RE, Ogilvie RI. Theophylline disposition in patients with hepatic cirrhosis. *N Engl J Med* 1977; 296: 1495–7.

Thomson PD, Rowland M, Melmon KL. The influence of heart failure, liver disease, and renal failure on the disposition of lidocaine in man. *Am Heart J* 1971; 82: 417–21.

Treluyer JM, Jacqz-Aigrain E, Alvarez F, Cresteil T. Expression of CYP2D6 in developing human liver. *Eur J Biochem* 1991; 202: 583–8.

Vanholder R, Van Landschoot N, De Smet R, Schoots A, Ringoir S. Drug protein binding in chronic renal failure: evaluation of nine drugs. *Kidney Int* 1988; 33: 996–1004.

13

Drug Interactions and Toxicity

Learning objectives

By the end of the chapter the reader should be able to:

- discuss when interactions between drugs are likely to be clinically important
- describe the changes that occur with enzyme induction
- evaluate homergic interactions
- explain the effect of displacement of a drug from binding sites on plasma proteins
- discuss the interactions of phenylbutazone with warfarin
- describe how data on two or more relevant compounds can be used to predict the probability, mechanism and potential severity of drug interactions
- explain how paracetamol causes hepatotoxicity and discuss the rationale for treatment
- briefly explain how drugs cause hypersensitivity.

13.1 Introduction

In several places in this book we have alluded to the potential for a drug to interact with another drug (or substance) to produce effects that do not occur when the drugs are used individually. Generally, these are unwanted effects but occasionally two or more drugs

Introduction to Drug Disposition and Pharmacokinetics, First Edition. Stephen H. Curry and Robin Whelpton.

Companion website: www.wiley.com/go/curryandwhelpton/IDDP

may be used together because their interaction is considered beneficial. Notable interactions are those that lead to overt toxicity, usually because at least one of the drugs has a narrow therapeutic window so that relatively small increments in plasma concentrations produce unacceptable effects. Many of the most easily understood drug interactions are mediated by drug-induced changes in absorption, distribution, metabolism and excretion.

With regard to drug toxicity, it is to be hoped that most recently introduced drugs have been thoroughly tested and are free of toxic effects. Even then one cannot be sure that once in the general population untoward effects will not become apparent. The cardiovascular toxicity of the COX-2 inhibitors has led to several being withdrawn, partly to avoid expensive litigation. The thalidomide tragedy arose, in part, because at the time teratogenicity was poorly understood, and the animal models in use would not have detected the problem. It has been argued that the experience with thalidomide illustrates that animal testing is flawed, but on the contrary, more animal testing with the appropriate species would have prevented the drug reaching the market. Examples of drugs with known toxicity have been available for many years. For example, aspirin and paracetamol are well established and useful drugs, although they might not have passed current safety testing. There are several examples where the toxicity resides in the metabolites, and the effects are more apparent in special populations that produce more metabolite, for example isoniazid. Hypersensitivity reactions are difficult to predict and may not be apparent until a drug is used on a second occasion, penicillin and halothane, for example.

13.2 Drug interactions

For many drugs interactions are of little importance. It is those with a low therapeutic index, such as warfarin, where small influences caused by a wide variety of interacting drugs have led to life-threatening changes in clinical effects. Warfarin is highly bound to plasma albumin and is displaced from binding sites by other weak acids. It is eliminated by microsomal oxidation and so its metabolism is accelerated by enzyme-inducing agents or may be inhibited by competing substrates such as ketoconazole. Despite these problems it remains the most widely used anticoagulant worldwide.

13.2.1 Terminology

Drug interactions can be considered under two headings:

- *homergic interactions:* when two (or more) interacting drugs have the same effect, e.g. central nervous system (CNS) depressant drugs and ethanol
- *heterergic interactions:* when the effect of one of drugs being studied is modified by a second drug, the effects of which may also be modified.

The term *addition* (or *summation*) is appropriate to describe homergic interactions when the effect of two drugs is what would be predicted from the sum of effects of the individual drugs. If the new effect is less than that predicted from sum of the two individual contributions, then this can be classified as *antagonism*. A combined effect greater than that predicted from the sum of the individual effects may be referred to as *potentiation* or *supra-additive* (see Figure 13.6(b)).

Synergism, from the Greek 'working together', could be used to describe any interaction, but it has evolved in meaning to be synonymous with potentiation. With heterergic interactions, a reduction in the activity of the drug under consideration would be described as antagonism. The affected drug is sometimes referred to as the 'victim' drug.

13.2.2 Time action considerations

The effects of a drug *in vivo* will have an onset time, a peak intensity and a duration of action, as described in previous chapters. This can introduce added terminology issues. For example, an increase in the rate of absorption with no change in the degree of absorption will lead to an earlier onset of effect and a greater and earlier maximum effect, but a shorter duration of effect. Thus, by reference to the onset and intensity of effect, an increased rate of absorption might be termed a potentiation, but the shorter duration of effect could be classed as antagonism. Hence, observing the interaction at two times, one early and the other later, would lead to different conclusions. Given this set of circumstances, it is best to discuss the results of an interaction in terms of measured phenomena and pharmacological mechanisms, and to be careful not to be too dogmatic in the use of terminology.

For the most part, the results of interaction studies tend to be expressed in terms of changes in the descriptive pharmacokinetic parameters listed previously, such as half-life and bioavailability. However, the clearance concepts discussed in Chapter 7 make it possible to perform calculations that provide greater insight into the underlying mechanisms than is possible with descriptive pharmacokinetics alone. In some situations the *AUC* may be used to investigate the effect of the interaction on the exposure to the drug.

13.2.3 Absorption

By now the reader will be able to write his or her own list of mechanisms by which the absorption of drugs is affected. This will include substances that alter gastric emptying, transit time, gastric pH, competition for both influx and efflux transport proteins, first-pass metabolism, cardiac output and hepatic blood flow. Examples of some of these were presented in Table 2.2.

The potential effect of food on gastric emptying was mentioned in Chapter 11, and Figure 13.1(a) shows a typical pattern of food delaying absorption but having little effect on the extent, probably as result of delayed gastric emptying. C_{max} is reduced and t_{max} is increased. Ketoconazole is also an example of a drug that is better absorbed when the gastric pH is low. Figure 13.1(b) shows the effects of lowering gastric pH by administration of glutamic acid hydrochloride and simulating achlorhydria by a combination of cimetidine and $NaHCO_3$, which reduced the absorption of ketoconazole to almost zero. Thus it can be anticipated that the absorption of such drugs will be reduced by concomitant administration of antacids, H_2-receptor antagonists and proton pump inhibitors such as omeprazole. The converse will be true for acid labile drugs, for example benzylpenicillin and methicillin.

The effect of drugs altering pre-systemic metabolism will depend on whether or not the victim drug is active or is a prodrug that requires activation. A well-documented example is the interaction of grapefruit juice and terfenadine, which is normally metabolized to the antihistamine fexofenadine. A constituent of grapefruit inhibits the first-pass conversion of terfenadine to its metabolite by CYP3A4 (Figure 13.2). The raised terfenadine concentrations in plasma are sufficient to cause a severe toxic effect of terfenadine-induced

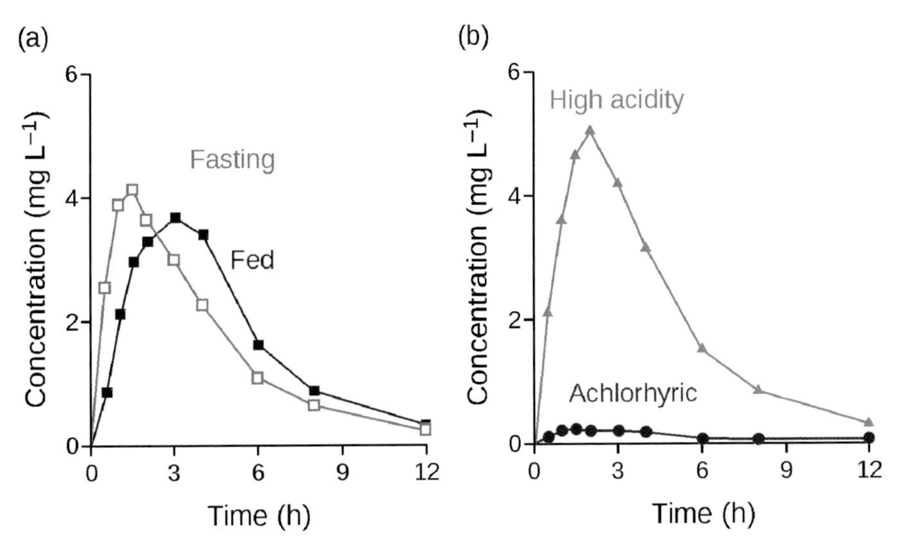

Figure 13.1 *Effect of (a) food and (b) gastric pH on the absorption of ketoconazole. The increased acidity was achieved with glutamic acid hydrochloride. Cimetidine and sodium hydrogen carbonate were given to simulate achlorhydria. Adapted, with permission, from Charman* et al. *(1997).*

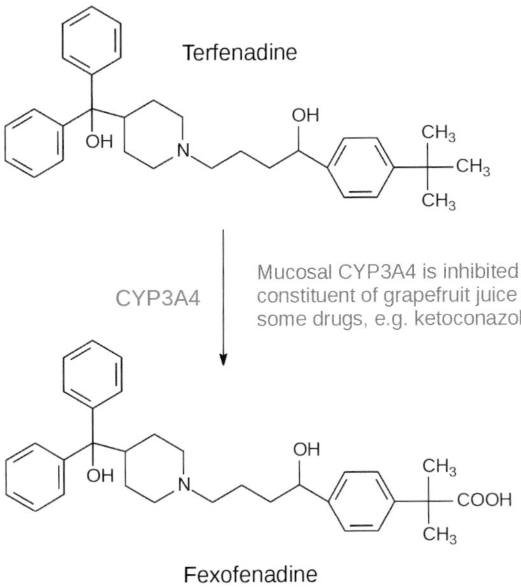

Terfenadine

CYP3A4

Mucosal CYP3A4 is inhibited by a constituent of grapefruit juice and some drugs, e.g. ketoconazole

Fexofenadine

Figure 13.2 *Inhibition of first-pass metabolism of terfenadine to fexofenadine increases plasma terfenadine concentrations, increasing the risk of toxicity.*

prolongation of the QT interval, as seen in the electrocardiogram. Both the C_{max} and the half-life of terfenadine were increased, greatly increasing the patient exposure to the drug. Other inhibitors of CYP3A4, ketoconazole and erythromycin, interact in a similar way, and as a consequence, terfenadine has been withdrawn from the market.

13.2.3.1 Transporters

The interplay between intestinal enzymes, efflux pumps and uptake transporters can produce seemingly contradictory results. For example, the renin inhibitor aliskiren is a substrate for CYP3A4, P-gp and OATP2B1. Rifampicin (rifampin), which induces P-gp, reduced aliskiren bioavailability but itraconazole, which inhibits both P-gp and CYP3A4, increased the bioavailability of aliskiren. When grapefruit juice, which inhibits all three proteins, CYP3A4, P-gp and OATP2B1, was ingested, the bioavailability of aliskiren was decreased, suggesting that inhibition of uptake is the more important interaction between grapefruit juice and aliskiren.

Plasma concentrations of digoxin are reduced by rifampicin, as assessed by *AUC* measurements, with oral doses being affected more than i.v. doses. In patients who underwent duodenal biopsies it was found that rifampicin increased intestinal P-gp three-fold. Because there were no effects on renal clearance or the half-life of digoxin, the conclusion was that the interaction was induction of digoxin efflux from enterocytes, as digoxin is not metabolized by CYP3A4. A similar mechanism may apply to St John's wort, which reduced digoxin concentrations by approximately 20% over 16 days when 900 mg d^{-1} of hypericum (St John's wort) was co-administered. Hypericum is also a CYP3A4 inducer, with potentially catastrophic effects on drugs that are substrates for this enzyme and P-gp, for example ciclosporin.

13.2.4 Distribution

Competition between different compounds for the same binding sites on albumin and other proteins can lead to the incumbent drugs being displaced. In a balanced therapeutic situation, this in turn can lead to an increased amount of displaced drug being available in plasma water and in other body fluids. The change in the volume of distribution is given by Equation 7.22. Displacement will have very little effect on the clearance of flow-limited drugs, but for drugs with capacity-limited (restrictive) clearance, Equation 7.19 applies:

$$CL = f_u CL'_{int} \tag{7.19}$$

The increase in *CL* is greater than the increase in *V*, so the half-life decreases (see Figure 7.4). For a drug that is 99% bound to plasma protein, the non-bound concentration will be 1 mg L^{-1} when the total steady-state concentration is 100 mg L^{-1}. If the binding falls to 98% when a second drug is introduced, the non-bound concentration will be momentarily 2 mg L^{-1}, resulting in a doubling of *CL* according to Equation 7.19. For drugs where the change in distribution is negligible, the average C^{ss} is inversely proportional to *CL* (Equation 4.41), so the total plasma concentration will fall to 50 mg L^{-1} and the non-bound concentration to 1 mg L^{-1}, which is what it was originally. Note that if the total drug concentration is being monitored, the dose should not be increased because this risks increasing the non-bound concentration to the point where toxicity is apparent. Thus the displacement of one drug by another leads to a transient increase in plasma concentration until the modifications in distribution and clearance have reduced the total concentration, leaving the average steady-state concentration of non-bound drug unchanged. The time for the concentration to re-equilibrate will depend on the elimination half-life. Whether, the transient increase is clinically important will depend on the circumstances and Rolan

(1994) has produced an algorithm to aid in deciding whether a transient increase in non-bound plasma concentrations is clinically important. The reduction in half-life will increase the fluctuation between peak and tough concentrations (Equations 4.38 and 4.39), which might be clinically relevant. Drugs that might be expected to be affected are warfarin (V=9 L, f_u=0.01) and tolbutamide (V=10 L, f_u=0.04), both weak acids with high plasma protein binding and low apparent volume of distribution, but, even so, drug–drug interactions with both of these drugs are more likely to be because of changes in metabolism rather than displacement. In a similar way, in neonates the displacement of bilirubin by sulfonamides is clinically important.

The situation is different for flow-limited drugs because CL will be unaffected and any change in half-life will be as a result of an increase in V as there will be more non-bound drug to diffuse into tissues. The tendency will be for $t_{½}$ to increase unless this is offset by increased renal clearance of non-bound drug. For drugs with high values of V the proportionate increase will be marginal, with little change in total plasma concentration. Potentially, this could be clinically significant as the non-bound concentration in plasma is higher than before displacement. However, when the drug is administered orally, the increase in f_u decreases F and the decrease in bioavailability reduces both the total and non-bound concentrations. So the occasions when displacement could lead to problems is when a flow-limited drug is administered parenterally. In theory this could occur with intravenous infusions of lidocaine in the presence of a displacing agent.

When quinidine interacts with digoxin, plasma concentrations of the latter are increased two- to three-fold, inducing arrhythmias as an adverse effect of the digoxin, in patients being treated for arrhythmias by the two drugs in combination. The apparent volume of distribution of digoxin is reduced – an explanation for this is that quinidine competes with digoxin for tissue-binding sites, something that has been confirmed in animal studies. However, quinidine inhibits P-gp in enterocytes and OATP1B3 in renal tubular cells (Figure 3.18), leading to increased bioavailability and reduced clearance, so displacement from tissues may be only a minor contributor to the observed increase in plasma concentrations. Aspirin overdose often leads to acidosis, which reduces the renal clearance of salicylate and increases its distribution into tissues (Section 2.5.2.1). Administration of $NaHCO_3$ increases plasma pH so that salicylate diffuses from the tissues into plasma to be cleared by the kidney.

13.2.5 Metabolism

13.2.5.1 Enzyme induction

Induction of drug-metabolizing enzymes is probably the longest established of the major pharmacokinetic mechanisms of drug interaction. This phenomenon was discovered when the influence of phenobarbital on the effects and plasma concentrations of anticoagulants, in particular, was first observed. Enhanced enzyme activity can be detected in animals after just two or three doses of phenobarbital. Affected drugs typically show shortened half-life values, reduced areas under the curve, lower steady-state concentrations on multiple dosing, increased metabolite formation and reduced pharmacological effects. As well as most of the historically significant barbiturates, other examples of drugs that induce microsomal P450 activity include rifampicin (rifampin), carbamazepine, valproic acid and phenytoin, some of which feature several times in Table 3.1.

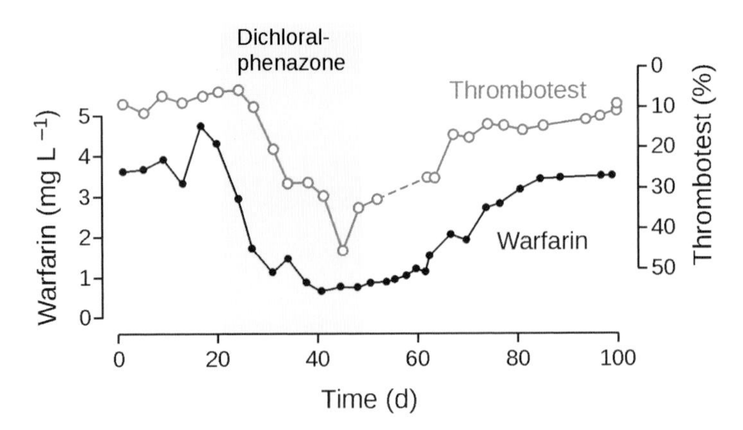

Figure 13.3 Example of enzyme induction in a patient taking warfarin (5 mg d^{-1}). Addition of dichloralphenazone reduced warfarin concentrations and anticoagulant effects, both of which returned on cessation of the enzyme-inducing drug. Adapted, with permission, from the data of Breckenridge & Orme (1971).

Because of the potential for life-threatening events, interactions with warfarin have been widely studied. Figure 13.3 shows one such interaction with dichloralphenazone, a 2:1 mixture of chloral hydrate and phenazone (antipyrine), which has been used in some migraine treatments. On introduction of the second drug, both the warfarin concentrations and anticoagulant activity (as assessed by Thrombotest, 100% equals normal coagulation) fell but rose again on cessation of exposure to dichloralphenazone. The authors showed that phenazone was the inducing drug not chloral hydrate (Breckenridge & Orme, 1971). Amobarbital (amylobarbital) and secobarbital, but not nitrazepam, also caused significant falls in warfarin concentrations and increased clotting times.

Cigarette smoking is known to, or has the potential to, affect the properties of approximately one-third of the 100 most prescribed drugs. The most common mechanism is enzyme induction by polycyclic aromatic hydrocarbon (PAH) compounds found in cigarette smoke. These compounds induce the activity of CYP1A1, CYP1A2 and possibly CYP2E1. One model PAH is 3-methylcholanthrene, which was one of the first inducers of drug metabolism to be discovered. This compound is still used as a tool in drug metabolism and carcinogenicity research. Nicotine is primarily metabolized by CYP2A6, and it is also an inducer, increasing the activity of the same enzymes as those affected by the PAH compounds, and also inducing CYP2B1 and CYP2B2, but this effect is probably not clinically important. Examples of drugs affected by smoking include theophylline, propranolol, diazepam and chlordiazepoxide. The induction of propranolol metabolism is associated with a decreased effect on blood pressure and heart rate. Examples of other drugs that are affected include caffeine, imipramine, haloperidol, pentazocine, flecainide and oestradiol. The effects on oestrogen kinetics can negate the efficacy of oral contraception. When smokers give up smoking, this induction dissipates at a rate related to turnover of microsomal protein. Clozapine plasma levels have to be carefully monitored and a change in smoking habits is one of reasons that the dose might have to be adjusted. Other variables that need to be considered include age, sex and metabolic activity (Rostami-Hodjegan *et al.*, 2004). Regular alcohol consumption induces alcohol dehydrogenase so that drinkers become tolerant to the effects. Ethanol induces several drug-metabolizing enzymes, but interest has been focused on induction of CYP2E1, which is thought to increase the oxidation of paracetamol to the toxic metabolite, NAPQI (Figure 3.16, Section 13.3.4.2). Thus smokers and imbibers of alcohol should be added to our list of special populations.

The mechanism of enzyme induction has received considerable attention. When enzyme induction occurs, the liver increases in weight and the smooth membranes of the endoplasmic reticulum (microsomes) increase in quantity and in protein content, including NADPH-cytochrome *c* reductase, cytochromes P-450 and cytochromes b_5. Treatment with phenobarbital increases liver blood flow. Activation is reduced by compounds which inhibit protein synthesis: ethionine, puromycin and actinomycin D. It is believed that the mechanism of enzyme induction involves an increase in the quantities of drug-metabolizing enzymes, rather than changes in the availability of cofactors. Because the enzymes are relatively non-specific, induction by one substrate often leads to other substrates being more rapidly metabolized, thus providing a means of one drug affecting the response of another. Auto-induction, that is, the increased rate of metabolism of the inducing agent, usually occurs to some extent. In the case of carbamazepine, plasma concentrations are monitored and the dose adjusted upwards if auto-induction has reduced the value to below the target range.

13.2.5.2 Enzyme inhibition

Drugs may bind irreversibly to drug-metabolizing enzymes, so that activity will not return until more enzyme has been synthesized. Although most barbiturates are enzyme inducers, allobarbital binds irreversibly to the enzyme, thereby destroying it. First-generation monoamine oxidase inhibitors (MAOIs), phenelzine and iproniazid for example, bind covalently, and because the enzyme has been depleted, MAOI activity continues after the drug can no longer be measured, which led to them being referred to as 'hit-and-run' drugs. Second-generation MAOIs (moclobemide) are reversible inhibitors that are considered safer to use. Ketoconazole binds to CYP450 14α-demethylase to block ergosterol synthesis in fungi, so it is hardly surprising that it inhibits other microsomal oxidases.

Although mutual inhibition of metabolism by various substrates for the P450 system had been known for many years previously, it was the observation in the early 1980s that cimetidine caused major changes in the responses to a wide variety of drugs that focused attention on this mechanism of drug interactions. The list of drugs that inhibit CYP3A4 and/or CYP2D6 in particular now includes ketoconazole, fluoxetine, fluvoxamine, quinidine, theophylline and erythromycin (Table 3.1), with several examples causing interactions that affect prescribing practices. Concentrations of CYP2D6 in the liver are relatively low and so some of the interactions that have been observed for substrates of this enzyme, phenothiazines and SSRIs for example, may be competitive. *In vitro* studies have shown that inhibition of CYP2D6 by cimetidine is not competitive, but mixed, but this may not be the case *in vivo*. Whatever the mechanism, cimetidine is capable of inhibiting a large number of drug-metabolizing enzymes. The effect of cimetidine on plasma concentrations of amitriptyline and its *N*-desmethyl metabolite are shown in Figure 13.4. The increased amitriptyline *AUC* may be a result of inhibition of CYP3A4, leading to a decrease in pre-systemic metabolism and/or inhibition of other enzymes such as CYP2D6 with a subsequent reduction in clearance.

The displacement interaction between warfarin and phenylbutazone, which causes a transient increase in the unbound concentration of warfarin, was discussed above, but the interaction with phenylbutazone is more complex than a simple displacement interaction.

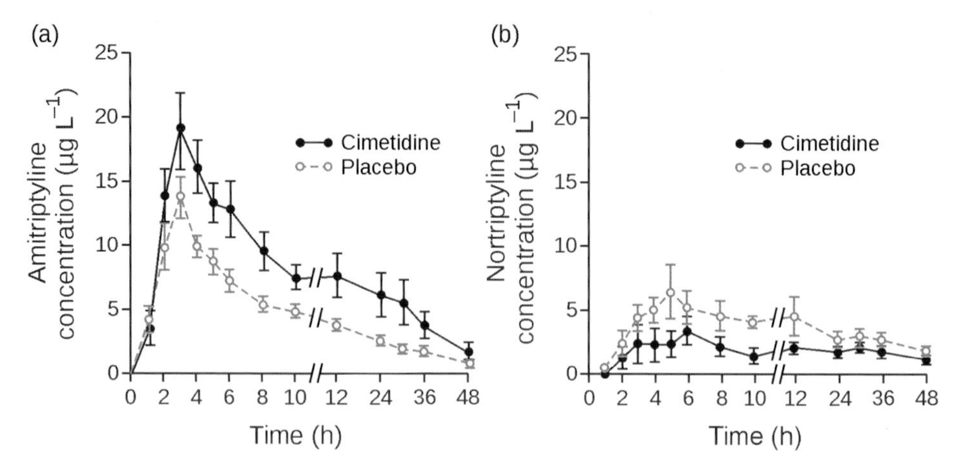

Figure 13.4 Interaction of cimetidine and amitriptyline. (a) Co-administration of cimetidine increased amitriptyline concentrations compared with placebo, whereas (b) the corresponding concentrations of the metabolite, nortriptyline, were reduced. Note that the area under the curve for amitriptyline was increased and that for nortriptyline was reduced. From the data of Curry et al. *(1985).*

Warfarin is marketed as the racemate and the *S* isomer, which is some three to four times more potent than the *R* isomer, is metabolized to 7-hydroxywarfarin by CYP2C9. The *R* isomer is metabolized by CYP1A2 to 6- and 8-hydroxywarfarin and by CYP3A4 to 10-hydroxywarfarin. Normally the *S* isomer is more rapidly metabolized than the *R*, but phenylbutazone inhibits CYP2C9 so that the production of 7-hydroxywarfarin is reduced and the concentration of the active *S* isomer is increased. By a remarkable coincidence, phenylbutazone induces CYP3A4 so that the concentration of *R*-warfarin is decreased, and the 'total' warfarin remains more or less the same (http://www.jci.org/articles/view/107711). It was not until the development of stereospecific assays that the interaction could be explained satisfactorily (Lewis *et al.*, 1974).

Historically, the greatest clinical danger with the enzyme induction interaction has been when a barbiturate was withdrawn from an established regimen of barbiturate plus warfarin, such as occurred when patients admitted to hospital for anticoagulation therapy were also given sleeping tablets containing barbiturates, and became stabilized on the combination. On returning home, the patients no longer required the hypnotic drugs and their enzymes reverted to lower activity, leading to the warfarin then being metabolized more slowly. The increased concentrations of warfarin could lead to haemorrhage. Today's hypnotics do not affect anticoagulation therapy in the same way, but this example provides a lesson of lasting significance in the history of, and potential for, drug interactions. It is good advice to those stabilized on warfarin not to change their eating and drinking habits.

13.2.6 Excretion

Drug-induced changes to hepatic and renal blood flows would be expected to affect excretion. Substances that cause acidosis or alkalosis will affect the renal clearance of some weak electrolytes and changes in plasma protein binding can affect the amount of drug filtered at the glomerulus.

Approximately 70–80% of hepatic blood flow arises from the portal vein, and so changes to uptake by hepatocytes can have a marked effect on first-pass metabolism. Statins are of particular interest because these drugs reduce lipoprotein concentrations by inhibiting 3-hydroxy-3-methylglutaryl-coenzyme A, principally in the liver, and also several statins are prodrugs and so have to be metabolized to be active. Most of them are eliminated by metabolism. Thus, reducing the amounts reaching the liver will not only reduce their efficacy but result in higher plasma concentrations, which are associated with increased risk of muscle degeneration and even life-threatening rhabdomyolysis. Ciclosporin interacts with several organic anion transporter polypeptides (Figure 10.5) and the interaction with statins results in increased *AUC* values of up to 20-fold in some cases. Some of the increase may be as a result of enzyme inhibition, but the increases in *AUC* for pravastatin and rosuvastatin, which are eliminated unchanged, were five to seven times and nine times, respectively. Animal studies using administration of rosuvastatin to the jejunum demonstrated the major mechanism was inhibition of hepatic uptake. Not only were C_{max} and *AUC* increased but inhibition of hepatic uptake reduced the apparent volume of distribution. Similarly, the hepatic efflux transporters in the canalicular membrane are inhibited. Quinidine and verapamil inhibit P-gp-mediated efflux of digoxin, whilst it is increased by rifampicin, which induces the transporter. Some of the toxicity associated with inhibitors of canalicular efflux may be because of accumulation of bile salts.

Renal transporters were discussed in Section 3.3.1.2. Probenecid inhibits OAT1 and OAT3 (Figure 3.18), thereby blocking the excretion of OAT substrates: penicillins, cephalosporins, furosemide and methotrexate, for example. When penicillin was first introduced probenecid was co-administered to prolong the half-life of the antibiotic. Because it blocks uric acid uptake by renal tubular cells probenecid has been used as a uricosuric. Probenecid is now used to inhibit OAT-mediated transport of cidofovir into PCT cells to reduce the nephrotoxicity associated with this antiviral agent. Cimetidine inhibits the renal excretion of several drugs by blocking OCT2. The affected drugs include ranitidine, amiloride and metformin. It has been suggested that cimetidine might also block the efflux of metformin by MATE1. Whatever the mechanism, cimetidine can increase the *AUC* of metformin to the point that the dosage should be adjusted.

13.2.7 Homergic interactions

Because homergic interactions occur between drugs having the same pharmacological effects, such interactions are often more pharmacodynamic in nature than pharmacokinetic. This does not mean that pharmacokinetic interactions do not occur. The increased effect when ethanol and CNS depressant drugs interact may be a mixture of PK and PD effects because ethanol can affect drug absorption and metabolism. Interactions with alcohol often lead to concentration changes in plasma and appear to occur to some extent with barbiturates, other depressant drugs, such as phenothiazines and antihistamines, phenytoin and some antidepressants. There is a significant interaction with benzodiazepine drugs, but any changes resulting from mutual influences on the plasma concentrations of benzodiazepines and ethanol appear to be minimal, although the pharmacological consequences of this interaction are considerable.

One particular technique especially applicable to homergic interactions is isobolography. An isobol is a line linking equipotent doses of two drugs on a graph in which the two axes are doses of the two drugs involved in the interaction. Thus, it is necessary to define

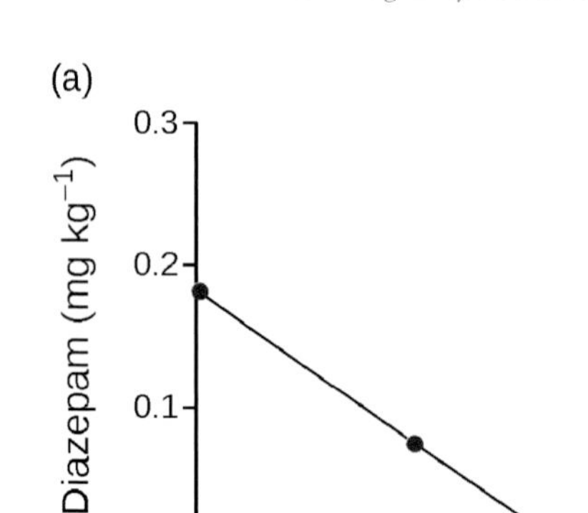

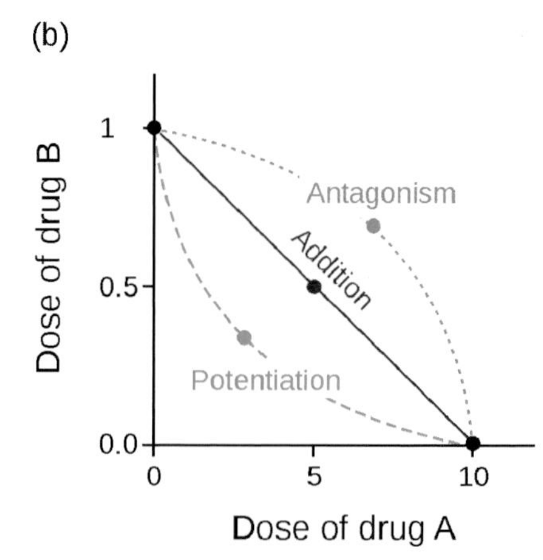

Figure 13.5 (a) Isobologram depicting effects of diazepam, ethanol and a combination of the two on the digit substitution test. From the data of Curry & Smith (1979). (b) Isobologram showing differentiation of potentiation, addition and antagonism for homergic drug interactions.

an 'end-point' or 'criterion-effect' such as ED_{50}, or minimum inhibitory concentration, in the case of antimicrobial agents. Figure 13.5(a) is an isobologram for the interaction of diazepam and alcohol, studied by means of a digit symbol substitution test. Dose–response curves were constructed to determine the dose at which each drug produced the criterion effect. Combination doses that were expected to give the same criterion effect were calculated, and one was tested. A three-point isobologram was constructed using the individual doses and the combination dose, producing a straight line, indicative of addition of the effects (Figure 13.5(a)). A concave line would have indicated potentiation, while a convex line would have indicated antagonism (Figure 13.5(b)). More complex isobolograms have been constructed in situations where the pharmacology was more complicated, and were particularly valuable in designing drugs such as co-trimoxazole, a mixture of sulfamethoxazole and trimethoprim, which show marked potentiation when used combination.

13.2.8 When drug interactions are important

There is an immense literature on drug interactions. Almost any combination of drugs can be employed to demonstrate an interaction. Thus, the literature is packed with reports that one drug affects the plasma levels of another in humans, the metabolism of another in mouse liver homogenates, the concentrations of another in plasma, the action of another in isolated enzyme preparations, etc. Many of these interactions are benign. In practice, drug interactions are important when one or more of the drugs concerned has a low therapeutic index, particularly when the effect is on delicately balanced physiological control mechanisms, such as those concerned with the heart, maintenance of blood pressure and blood coagulation. Thus, clinically significant interactions are those involving anticoagulants, cardioactive drugs, antihypertensive drugs, antidepressants, local vasoconstrictors, older hypnotics, anticonvulsants and ethanol. In other words, the usual suspects: warfarin, heparin, digoxin, quinidine, lidocaine, phenytoin and diuretics, particularly if they affect serum potassium concentrations, ethanol and CNS depressants. Overdoses of benzodiazepines are rarely fatal, but deaths have been known when they have been taken with ethanol.

Changes in enzyme concentration that occur during induction or inhibition occur in proportion to the concentration of the inducing ligand and over a period of time, thus rarely do they precipitate an *acute* medical emergency. Rather, they cause a gradual change in drug response that can be recognized and controlled in the patient by appropriate dosage adjustment. However, interactions with warfarin have caused acute haemorrhage, and possibly this drug will one day be replaced by the emerging directly-acting anticoagulants (Gomez-Outes *et al.*, 2015). However, the one advantage of warfarin is that its actions can be reversed by injecting vitamin K.

13.2.9 Desirable interactions

The term 'drug interaction' is often associated with alarm, as it engenders feelings related to potential lack of control over the effects of drugs in the body. There is some validity in this, as some drug interactions are, indeed, the cause of great difficulty in therapeutics, as described in this chapter. However, it should not be forgotten that some interactions can be turned to advantage, and far beyond the use of combination therapy. For example, the use of:

- vasoconstrictors, both epinephrine and octapressin, in prolonging the effect of local anaesthetics, thus allowing larger doses to be given safely
- carbidopa to enhance the effect of L-dopa, allowing lower doses and reduced unwanted effects
- probenecid to prolong the effects of penicillin and to reduce the nephrotoxicity of cidofovir
- cilastatin to reduce the renal metabolism of imipenem, enhancing its duration of action and preventing formation of nephrotoxic metabolites.

13.2.10 Predicting the risk of drug interactions with new chemical entities

With so many drugs in use, and with others in development, there is an obvious need for accurate and efficient evaluation of whether or not a new chemical entity (NCE) is likely to have significant effects on the actions of existing drugs when it is co-prescribed, and/or whether the NCE itself will cause unexpected effects in patients because of drug interactions. In recent years, several new drugs, as with terfenadine, have been withdrawn from the market because of drug interactions that were not anticipated before their introduction and discovered only through post-marketing surveillance, highlighting the risk of loss of the investment of time and money. It is not possible to conduct a comprehensive laboratory or clinical evaluation of this risk before approval for marketing, for reasons of both practicality and protection of the human subjects used in clinical pharmacological testing, yet the need is recognized by both research personnel and regulatory authorities.

13.2.10.1 In vitro prediction of drug interactions

The use of *in vitro* incubation of hepatocytes and microsomes was discussed in Chapter 7, and similar techniques can be employed to investigate drug interactions by introducing a NCE to the incubation medium. Inhibitor constant values, K_i, are calculated using conventional enzymology approaches:

$$v = \frac{V_{max} C}{Km\left(1 + I/K_i\right)} \qquad (13.2)$$

where I is the inhibitor concentration. The type of inhibition, such as reversible, irreversible, competitive (competition for the active sites on the enzyme), non-competitive (e.g. an allosteric change that reduces activity), can be derived from Lineweaver–Burk plots. The lower the K_i value, the more potent the inhibitor, but other factors must be taken into consideration, including:

- the concentration of inhibitors likely to occur *in vivo*
- the concentration range of potential substrates (victim drugs) occurring *in vivo*
- the therapeutic index of the victim drugs
- whether the inhibitor and victim are likely to be co-prescribed.

Livers from pre-treated animals can be used. For example, pre-treatment of rats *in vivo* for as little as 2 days with phenobarbital leads to an increase in microsomal intrinsic clearance of other compounds when measured *in vitro*. The effect of the NCE on enzyme induction can be tested in a similar manner.

To be of predictive value it is important to estimate the inhibitor concentrations occurring at the active enzyme site while recognizing that non-specific binding to liver material can affect the active concentrations in the biophase. Studies should focus the effect of the inhibitor on the important drug-metabolizing enzymes, CYP2C9, CYP2D6 and CYP3A4, but also consider the fraction of the dose of the victim drug metabolized by a particular CYP isoform as this will affect the risk of the interaction being significant *in vivo*. The absorption rate constant of the inhibitor should be incorporated into the risk assessment whenever possible because this could affect whether or not the inhibitor and victim drugs are likely to interact with endothelial and liver CYP3A4. The predicated average plasma concentration of the inhibitor is important, particularly when inhibitor and victim drug are metabolized by the same enzyme, particularly CYP2D6.

The result of this attention to detail has been a marked improvement in the prediction of interactions *in vivo* as assessed by *AUC* values in laboratory animals. In contrast, attempts to predict for protein binding and metabolism by intestinal mucosa in P450 inhibition interactions have been less successful. In practice, a potent inhibitor of CYP2D6 or CYP3A4 is unlikely to go forward into clinical trials unless the NCE is of such exceptional value that any risk of interactions would be seen as acceptable in relation to the potential clinical benefit.

Ultimately, the need is to utilize predictive information with care, as there is just as much a risk of rejection of a useful NCE with a benign level of interaction risk as there is of acceptance for further development of a NCE carrying with it serious future risk. Having said that, the sooner a NCE is identified as being unsuitable for further development the less time and money will be consumed fruitlessly.

13.2.10.2 *In vivo* predictions in human subjects

At one time the half-life of phenazone (antipyrine) was commonly used to demonstrate enzyme induction *in vivo*. Phenazone, an obsolete analgesic drug, has rapid oral absorption, an oral systemic availability close to unity, is distributed in total body water and it is a substrate for CYP1A2, CYP2C8, CYP2C9 and CYP3A4. Thus, changes in its half-life are a result of changes in enzyme activity (clearance) rather than changes in the volume of distribution. With the identification of CYP450 isozymes, phenazone has been superseded by compounds specific to particular isoforms, often using a cocktail of test drugs (Table 10.2). For example, this approach was used to show that long-term administration of St John's wort caused a significant increase in the clearance of oral midazolam, and a lesser increase in the clearance of intravenous midazolam. The activities of CYP1A2, CYP2C9 or CYP2D6 were unaffected. The clinical implications for management of patients using oral substrates of CYP3A4 are considerable.

Another approach to the study of specificity in drug interactions is the use of human subjects that have been phenotyped for their levels of the various CYP isozymes. Thus, if an evaluation of a drug for its ability to inhibit CYP2D6 in particular is sought, then subjects with known CYP2D6 hepatic activity can be used, ensuring optimal experimental design.

13.3 Toxicity

13.3.1 Terminology

Various terms are used in relation to toxicology. 'Unwanted effect' is deliberately non-specific, but does allude to the fact that drugs have 'effects', which may be 'wanted' or 'unwanted'. Furthermore, an unwanted effect in one situation may be wanted in another. So, for example, constipation arising from the use of morphine postoperatively to treat pain is an unwanted effect, particularly after bowel surgery, but the same effect is a wanted effect of morphine in anti-diarrhoeal preparations. The term 'side effect' is similarly all inclusive, and frequently used, but it has the disadvantage that it suggests that effects and side effects are different and can be separated – this is often not the case. The term 'adverse reaction' has a connotation of greater severity than does unwanted effect, but that is not always intended when this term is used. Idiosyncratic was used to describe individuals who responded differently to others, and sometimes for no apparent reason. Such people are now usually thought of as being from a definable subgroup of the population, for example ultrarapid metabolizers (Chapter 10).

There have been many attempts to classify unwanted drug effects, based variously on mechanisms, clinical manifestations and other approaches. One relatively simple and useful classification is as follows:

1. *Pharmacokinetic unwanted effects*. Effects arising from the presence of excessive concentrations of molecules that exert only desirable effects in normal quantities. This group makes allowance for overdose, be it deliberate or accidental, acute or cumulative, absolute or relative, and resulting from anything to do with drug disposition and pharmacokinetics.
2. *Pharmacodynamic unwanted effects*. Effects occurring by pharmacodynamic mechanisms at normal clinical dosage, acutely or after prolonged exposure. This group allows for anything describable as a side effect, a secondary effect, or an idiosyncrasy or intolerance where physiological or biochemical sensitivity is a factor.
3. Drug allergy. Hypersensitivity with an immunological basis.
4. Toxic drug reactions. Effects caused basically by means of covalent chemical reaction between pharmacologically active foreign molecules and biologically important materials.
5. Drug interactions.

Classes 1 and 2 are considered to be reversible, in the sense that conventional dose–response relationships apply to both wanted and unwanted effects, so that reduction in concentrations consequent on removal of the drug and/or its metabolites from the body leads to the dissipation of the effects. Classes 3 and 4 are thought of as irreversible because, although they can clearly be reversed by a variety of medical interventions, and by the normal processes of body healing, covalent chemical reactions are key to their origins, and removal from the body of the offending molecules and the molecules with which they have reacted is presumed to be needed. Class 5 was considered earlier in this chapter. It should be appreciated that the clinical outcome with any particular drug can invoke a combination of these processes.

13.3.2 Dose–response and time–action relationships with special reference to toxicology

Interpretation of toxicity studies is conducted, in part, in ways already discussed in regard to studies of the desired effects of drugs. Appropriate recognition is given to dose–response relationships, and the growth and decay curves underlying the time–action relationships at the core of pharmacological science. So, the time-course and dose relationships discussed in previous chapters, are crucial. There are, however, some variations on this theme in toxicology, for example it is not uncommon for unwanted drug effects to arise only in subgroups of the treated population, and the characteristics of those subgroups are important. Toxicity is often discovered only through epidemiological studies, for example cigarette smoking and lung cancer, thalidomide and birth defects (Kim & Scialli, 2011), and, more recently, COX-2 inhibitors and cardiac toxicity.

13.3.3 Toxicity associated with prolonged exposure to therapeutic doses

Some of the most serious toxic reactions are associated with the normal pharmacological properties of drugs, experienced over a prolonged period of time, possibly involving long-term exposure to the metabolites of the drugs concerned.

Serotoninergic syndrome is associated with prolonged use of selective serotonin reuptake inhibitors (SSRIs) and, to lesser extent, with tricyclic antidepressants (TCAs), which increase the concentrations of serotonin in the brain. The clinical picture involves confusion, autonomic instability and neuromuscular abnormalities. It is most often seen in patients taking two or more drugs at therapeutic doses that increase CNS serotoninergic activity by different mechanisms, such as MAOIs or pethidine, with SSRIs or TCAs. The combination of MAOIs and SSRIs or TCAs is discouraged but it is not unknown. The syndrome could be classed as an interaction and may be precipitated by drugs that inhibit, for example, CYP2D6. Additionally, particularly with amitriptyline, there can be a life-threatening combination of hypotension and heart block (produced by peripheral anti-adrenergic effects), parasympathetic block (rendering ineffective the normal reflex responses to a reduction in heart rate) and CNS depression associated with respiratory depression.

Opioid syndrome occurs with prolonged opioid exposure, in which the clinical manifestations are bradycardia, CNS depression, hypotension and respiratory depression, plus, with some opioids, miosis (constricted pupils) and decreased GI motility, complicated, of course, by addiction. Acute overdose can be life-threatening, more in this case, from respiratory depression than from peripheral nervous system effects.

Neuroleptic syndrome is caused by chlorpromazine and other antipsychotic drugs. When chlorpromazine was introduced it caused such a revolution in psychiatry that prescribers were overenthusiastic in their use of it. In time, it was realized that, with acute dosage, chlorpromazine, like amitriptyline, can cause nerve block, with CNS, cardiac, sympathetic and parasympathetic effects combining to produce life-threatening sedation, low blood pressure and bradycardia, all at the same time. After acute doses, recovery was complete and rapid. However, long-term dosing at dose levels just below the threshold for unwanted effects caused a combination of slowly reversible nerve block, plus tardive dyskinesia, an irreversible movement disorder similar to Parkinson's disease. This syndrome was at first thought to be a characteristic of the schizophrenia that was being treated, and in a number of cases this led to an *increase* in dosing. When it became possible to measure chlorpromazine plasma concentrations it was found that by reducing the dose and thus plasma concentrations to the minimum consistent with useful therapeutic effects, the risk of developing of neuroleptic syndrome and tardive dyskinesia was greatly reduced. This work had a considerable impact on the management of schizophrenia, which evolved from treatment with high doses of drugs over periods of years in psychiatric hospitals to an acute condition treated with smaller doses in the context of outpatient care.

13.3.4 Specific examples

13.3.4.1 Isoniazid and iproniazid

Primary amines and hydrazines undergo condensation reactions with aldehydes to form imines (Schiff's bases):

$$R\text{—}CHO + R'NH_2 \rightarrow RCH{=}N\text{—}R' + H_2O$$

and this reaction occurs between isoniazid and pyridoxal phosphate, the active form of vitamin B6 (Figure 13.6). Pyridoxal phosphate is an important co-factor for the synthesis of monoamine neurotransmitters and a reduction in its availability can produce a peripheral neuropathy and CNS effects. The neuropathy can be reversed by removal of the isoniazid, administration of pyridoxine or both. It occurs most commonly in slow acetylators, who will have higher concentrations of the parent drug (Section 10.3.1). It has been suggested that the response of tuberculosis patients to isoniazid is associated more with the magnitude of peak concentrations in plasma than with overall exposure during daily dosing. The drug is activated by bacterial catalase, which results in inhibition of the synthesis of mycolic acid required for synthesis of mycobacterial cell walls.

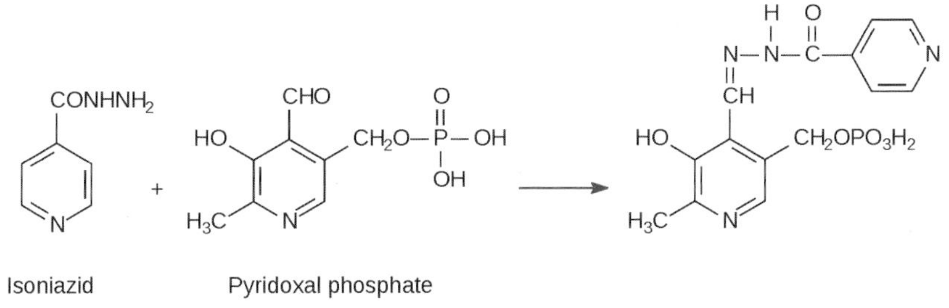

Isoniazid Pyridoxal phosphate

Figure 13.6 Condensation of pyridoxal phosphate with isoniazid.

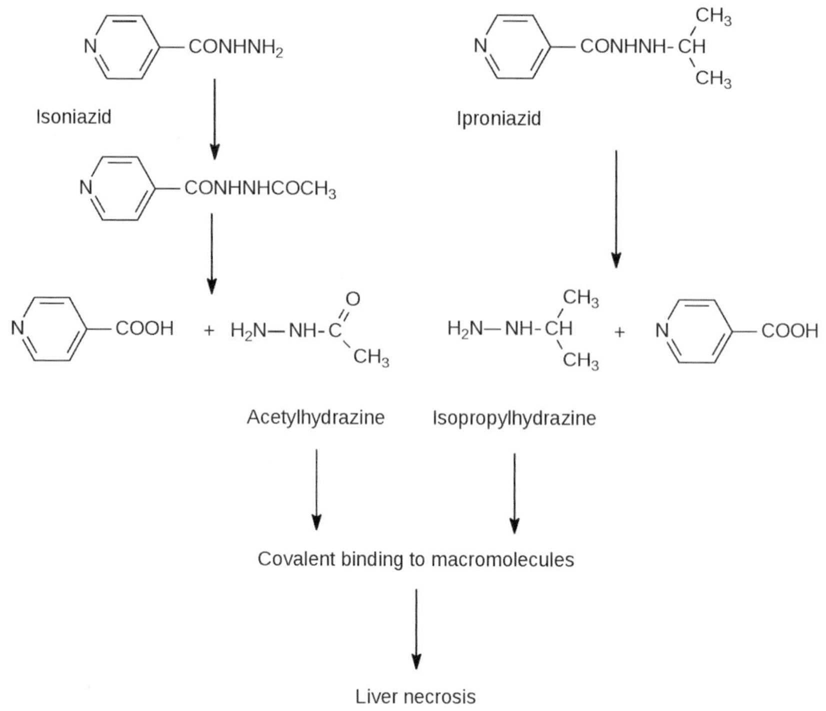

Isoniazid

CONHNH$_2$

CONHNHCOCH$_3$

COOH + H_2N—NH-C(=O)-CH$_3$

Ipronizaid

CONHNH-CH(CH$_3$)$_2$

H_2N—NH-CH(CH$_3$)$_2$ + COOH

Acetylhydrazine   Isopropylhydrazine

Covalent binding to macromolecules

Liver necrosis

Figure 13.7 Proposed mechanism of hepatotoxicity of isoniazid and iproniazid.

Isoniazid is also hepatotoxic in certain individuals, particularly fast acetylators. Iproniazid causes similar toxicity, although with this drug the isopropyl group prevents acetylation (Figure 13.7). In rats, enzyme induction with phenobarbital pre-treatment increased the incidence of liver toxicity, suggesting that a metabolite was responsible. Studies with radiolabelled isoniazid showed that the ring structure was not incorporated into liver, but the side-chain was. The conclusion was that hepatic necrosis results from cleavage of the drugs into acetylhydrazine and isopropylhydrazine, which then bind covalently to liver. These two products have been shown to be highly toxic. Fast acetylators are believed to be more susceptible than slow acetylators because of the higher concentrations of the acetyl metabolite that they produce.

Isoniazid demonstrates many of the general ideas involved in drug toxicity. Various terms, including unwanted effect and adverse reaction, may be used and because the toxicity occurs in a subgroup of the population it may be described as idiosyncratic. The toxicity is mediated via two metabolites, synthesis of the first being under genetic control, and there is a plausible physiological/biochemical explanation for the toxicity providing a basis for methods of successful treatment. Therefore isoniazid toxicity falls into classes 1 and 4 listed above.

13.3.4.2 Phenacetin and paracetamol

Phenacetin (acetophenetidin) was introduced in 1887 and was an ingredient in many over-the-counter analgesic and anti-fever remedies. When a number of people were observed to have a blue colour to their skin this was traced to methaemoglobinaemia caused by phenacetin. In their study, Brodie & Axelrod (1949) discovered a metabolite, paracetamol (acetaminophen), of even greater potential value as an analgesic. Phenacetin was notorious for its propensity to cause renal damage, and when it was also shown to be carcinogenic was withdrawn from sale in the USA in 1983.

Methaemoglobinaemia is a condition in which the normal level (1–3%) of oxidized haemoglobin in the blood is elevated. Cyanosis occurs at about 15%, and various breathing and cardiovascular problems occur as the percentage rises. Levels of 70% and above are usually fatal. When oxygen binds to haemoglobin it oxidizes the iron, removing an electron and forming superoxide (Figure 13.8). Because molecular oxygen can only bind to haemoglobin, the Fe^{3+} in methaemoglobin must be reduced back to Fe^{2+}. Several enzymes are involved, the chief one being cytochrome b_5-reductase (NADH-methaemoglobin reductase), others being NADPH-methaemoglobin reductase. Ascorbate and reduced glutathione, GSH, are also required. Direct oxidizing agents, nitrites for example, and a long list of drugs, including phenacetin, which probably interfere with the redox cycle and deplete GSH, cause methaemoglobinaemia.

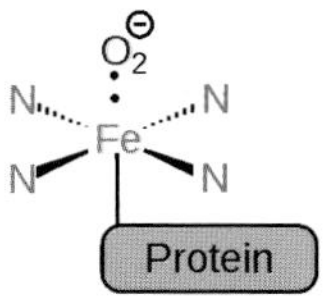

$O_2^{\ominus}$ N N Fe N N Protein

Figure 13.8 Simple representation of oxyhaemoglobin. The porphyrin nitrogens are depicted and the globular protein that binds via a histidine nitrogen. Oxyhaemoglobin is diamagnetic, that is, there are no unpaired electrons, so it is postulated that one electron from iron (giving Fe^{3+}) is paired with one from oxygen, giving superoxide.

Phenacetin is chemically related to acetanilide, which is metabolized to aniline and then to phenylhydroxylamine and nitrosobenzene (Figure 13.9). *In vivo*, acetanilide causes haemolysis and methaemoglobinaemia. Phenylhydroxylamine is highly toxic when it is administered directly. Thus, oxidizing agents, including oxidized metabolites, for example phenylhydroxylamine, which can be oxidized to a nitroso derivative and reduced back to aniline, can promote the formation of methaemoglobin from haemoglobin, either by direct oxidation or by depleting the cofactors required for the reduction pathways referred to above.

Phenacetin caused renal papillary necrosis, and probably renal pelvic carcinoma when it was widely used. Large doses also caused hepatic necrosis in animals, especially in

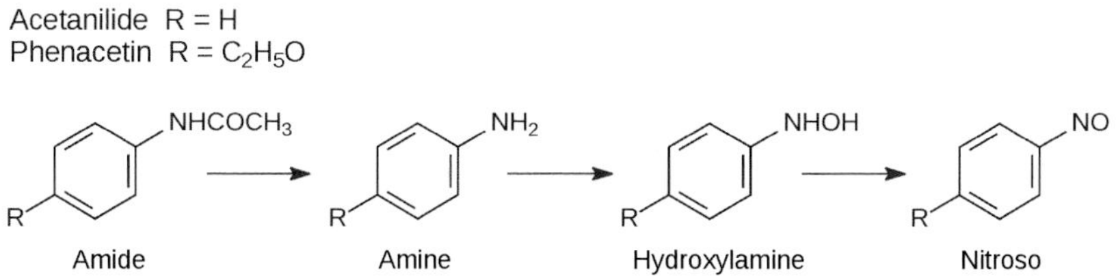

Acetanilide R = H
Phenacetin R = C_2H_5O

$NHCOCH_3$ → NH_2 → NHOH → NO

Amide Amine Hydroxylamine Nitroso

Figure 13.9 Oxidation of acetanilide and phenacetin.

species forming large quantities of *N*-hydroxylated metabolites. Paracetamol is particularly able to cause hepatic necrosis in such species and in humans taking large acute doses. With phenacetin, the intact molecule is involved in covalent binding to liver, shown by the use of ring-labelled and acetyl-labelled derivatives, which bind equally. However, elucidating a definitive mechanism is complicated by the fact these drugs may be metabolically deacetylated and reacetylated (Nicholls, *et al.*, 2006).

Paracetamol hepatotoxicity is associated with single acute doses above ~10 g – only 20 × 0.5 g tablets. Unlike phenacetin, the phenolic hydroxyl group in paracetamol can be conjugated directly without a prior phase 1 reaction. After therapeutic doses of 4–6 g per day, about 90% of the dose is excreted as glucuronide and sulfate conjugates, with the remainder excreted as the mercapurate (Figure 3.16). At higher doses, the concentration of the reactive phase 1 metabolite, *N*-acetyl-*p*-benzquinoneimine, is sufficient to deplete GSH to the point that NAPQI forms 'acetaminophen-cysteine adducts' in the liver. Immunochemical studies have shown that the cellular sites of covalent binding correlate with the toxicity. Peroxynitrite, which is a highly reactive nitrating and oxidizing species formed by reaction of nitric oxide (from endothelial nitric oxide synthase, NOS) and superoxide, produces nitrated tyrosine residues in proteins (Figure 13.10). Normally, GSH detoxifies peroxynitrite, but GSH is depleted by NAPQI. Nitration of tyrosine correlates with the degree of necrosis. It has also been hypothesised that lipid peroxidation may play a role in the toxicity.

Treatment of paracetamol overdose may be initiated with gastric washout (lavage) of unabsorbed drug and activated charcoal to reduce any further absorption. It has been recommended that charcoal not be used if antidotes are to be administered orally, but some studies have shown charcoal not to be a problem. The most successful treatments are those designed to replenish GSH, by providing glutathione precursors, or SH-donors,

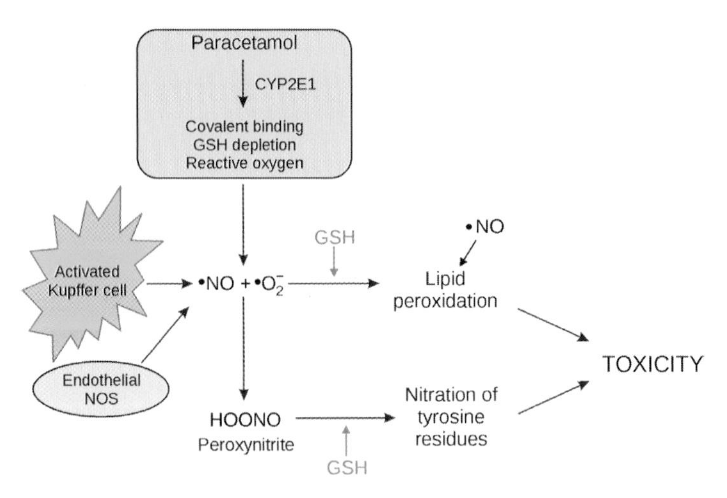

Figure 13.10 Postulated mechanisms of paracetamol-induced hepatotoxicity. Hepatocytes, Kupffer cells and endothelial cells participate in production of reactive nitrogen and oxygen species. The relative levels of nitric oxide (NO) and superoxide (O^-_2) determine whether mechanism of hepatic necrosis is dependent on protein nitrosylation or lipid peroxidation. Adapted with permission from Jaeschke et al. *(2002).*

because glutathione cannot penetrate hepatocytes. *N*-acetylcysteine (NAC), L-cysteine, cysteamine, L-methionine, D-penicillamine, and dimercaprol, have been used. NAC has emerged as the treatment of choice because it (i) is a glutathione replacement, (ii) is a free radical scavenger, (iii) binds directly to NAPQI and (iv) increases the microcirculation of oxygen. In the UK, NAC is normally administered intravenously, whereas in the USA it is usually given orally. Intravenous administration is associated with anaphylactic reactions, which although infrequent have occasionally resulted in deaths.

Treatment is guided either by the dose consumed (when ingestion is staggered over a period of time) or by plasma concentrations of the drug (when the dose was in one single event at an identifiable time). This is because use of *N*-acetylcysteine is not without complications, and clinicians are reluctant to initiate treatment unless it is deemed necessary. Nomograms have been developed to aid decision making. An example is shown in Figure 13.11. Samples are taken at times following ingestion and the paracetamol concentrations compared with the lines on the nomogram. Values above the line indicate a strong possibly of hepatotoxicity and the need for treatment. In the UK two lines may be used: 'normal' (the blue line in Figure 13.11) and 'high risk' (the red line) patients, who are expected to have low concentrations of GSH and/ or high concentrations of NAPQI. Because ethanol induces CYP2E1, chronic alcohol users fall into this category.

During treatment the clinical situation should be monitored because although the paracetamol concentrations in early blood samples may fall below the treatment lines, the concentrations may rise above the threshold for treatment at later times. This is more likely to be the case when high doses of paracetamol, or other drugs, delay absorption. Sustained-release preparations will be a particular problem, as shown in Figure 13.11,

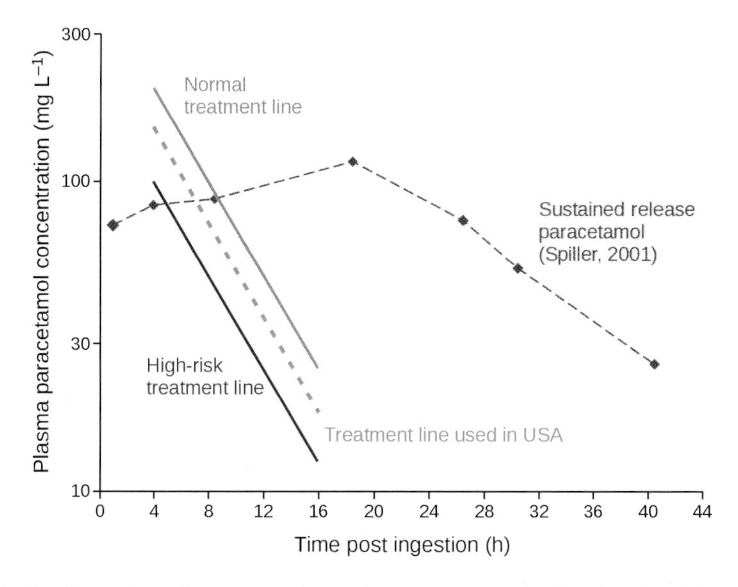

Figure 13.11 Paracetamol concentrations after paracetamol self-poisoning. Concentrations above the treatment lines indicate a high probability of hepatotoxicity occurring unless antidotal therapy is initiated. The high-risk line is for patients with low GSH concentrations. Adapted from Flanagan et al. *(2008).*

when absorption of the paracetamol was delayed such that the first three plasma paracetamol concentrations indicated no need for treatment. However, the paracetamol concentration continued to rise for 24 h after ingestion (Spiller, 2001). It has been recommended that NAC infusions should be continued until:

- plasma paracetamol concentrations are $<10\,\text{mg}\,\text{L}^{-1}$
- liver function tests (transaminase, bilirubin, blood clotting) are normal
- the patient is clinically well
- the patient receives a liver transplant
- the patient dies of fulminant liver failure.

For those who do survive paracetamol overdose the prognosis is good because usually the liver returns to normal after treatment. Children are less prone to the effects of paracetamol, possibly because of a greater capacity to sulfate the drug (Chapter 12).

13.3.4.3 Salicylate

Poisoning by salicylate and other NSAIDs is a common problem in emergency medicine. This syndrome is characterized by vomiting blood (haematemesis), difficulty breathing, which can be rapid (tachypnoea), deep breathing (hyperpnoea) or shortness of breath (dyspnoea), tinnitis, deafness, lethargy, seizures and confusion. Clinical toxicity is primarily a function of the degree and duration of acid–base disturbance, resulting from the depression of the respiratory centre and consequent carbon dioxide exchange. Aspirin toxicity can result from both acute and chronic overexposure. Occasionally, the antithrombotic effect of aspirin can cause bruising. Very rarely, there can be an allergic response exhibited as urticaria or anaphylactic shock.

Acute or chronic overexposure causes ringing in the ears and deafness as well as a broad variety of CNS effects ranging from headache to coma. Stimulation by salicylate of the respiratory centre in the medulla causes hyperventilation and hyperpyrexia, which in turn lead to sweating and dehydration. Metabolically there is an increase in oxygen consumption and carbon dioxide production, and thus metabolic acidosis plus respiratory alkalosis. The metabolic acidosis reduces the ionization so that more salicylate enters tissues, including the brain, thereby enhancing its CNS toxicity. Also, the kidney excretes more bicarbonate (along with sodium and potassium). The bicarbonate in the urine causes the urine pH to rise, which in fact accelerates the excretion of the salicylate. Thus aspirin in overdose exhibits two of the important pH effects on distribution and elimination which were discussed in Chapter 2.

Treatment involves whole-body cooling, plus cardiovascular and respiratory support, and correction of the acid–base abnormalities. Toxicokinetic intervention involves sampling to identify the toxic chemical(s) involved, administration of activated charcoal to adsorb any unabsorbed aspirin in the GI tract and administration of $NaHCO_3$ to alkalinize the urine further in order to speed the elimination of the absorbed dose. The bicarbonate also assists in the reversal of the metabolic acidosis. Forced diuresis along with the bicarbonate is no longer recommended as it has been found that the additional fluid load involved can complicate the stabilization of the electrolyte balance. Dialysis can be used as a treatment but it is not often needed.

13.3.4.4 Methanol and ethylene glycol

Methanol and ethylene glycol lead to accumulation of aldehyde and acid metabolites, and promotion of lactic acid formation due to reduced NAD/NADH ratios. This is assessed as the anion gap:

$$\text{anion gap} = \left(\left[\text{Na}^+\right] + \left[\text{K}^+\right]\right) - \left(\left[\text{HCO}_3^{2-}\right] + \left[\text{Cl}^-\right]\right)$$

If the anion gap exceeds 14 mmol L^{-1} action is required. Ethylene glycol or methanol concentrations of >500 mg L^{-1} are a possible indication for pro-active treatment with fomepizole (4-methyl pyrazole, 4MP) or intravenous ethanol and/or haemodialysis. The use of ethanol as a treatment is based on the fact that ethylene glycol and methanol poisoning are mediated by metabolites. If the formation of these metabolites can be slowed down, then the alternative pathway of renal excretion will dominate the elimination. Ethanol does just that. In being a competing substrate for the enzyme ADH, ethanol slows down the metabolism of ethylene glycol and methanol, thus prolonging the treatment (which may seem counterintuitive) but reducing exposure to the toxic metabolites. Ethanol itself is metabolized to less toxic metabolites. Fomepizole is an inhibitor of alcohol dehydrogenase that achieves a result similar to that achieved with ethanol, without the ethanol effects. Both approaches markedly increase the elimination half-lives of the ethylene glycol and methanol, thus preventing rapid production of toxic metabolites. The half-life of ethylene glycol was increased from 2.5–4.5 h to 17 h with the addition of ethanol and to 19.7 h with fomepizole. Similarly the half-life of methanol was increased from 3.0 h to 43 h and 54 h with co-administration of ethanol and fomepizole, respectively (Chu *et al.*, 2002; http://www.atsjournals.org/doi/pdf/10.1164/rccm.2108138). Ethanol is inexpensive and usually readily available in hospital.

13.3.4.5 Halothane

Halothane undergoes oxidative dehalogenation to trifluoroacetyl chloride and hydrogen bromide. The acid chloride may be hydrolysed by water to trifluoroacetic acid or may acylate neighbouring proteins to produce antigens (Figure 13.12). The trifluoroacetylated proteins are found chiefly in the endoplasmic reticulum, but some are found in the cell membrane. The toxicity arises when halothane is used on a second occasion because the antigens can trigger a potentially fatal immune response resulting in massive centrilobular liver necrosis that leads to fulminant liver failure. The prognosis for people developing

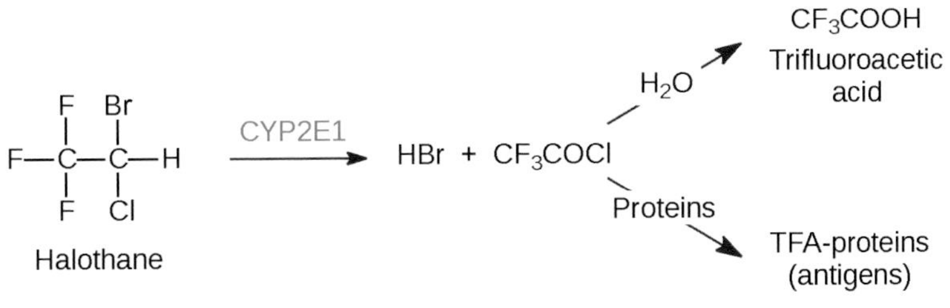

Halothane (F–C(F)(F)–C(Br)(Cl)–H) $\xrightarrow{\text{CYP2E1}}$ HBr + CF_3COCl

CF_3COCl $\xrightarrow{H_2O}$ CF_3COOH Trifluoroacetic acid

CF_3COCl $\xrightarrow{\text{Proteins}}$ TFA-proteins (antigens)

Figure 13.12 Oxidative debromination of halothane. Trifluoroacetyl chloride is hydrolysed or reacts with proteins to produce antigens capable of triggering an allergic reaction.

halothane hepatitis is poor; the fatality rate is 50%. The incidence of similar reactions to other halogenated general anaesthetics is very small as these are not as extensively metabolized as halothane. Approximately 20% of a dose of halothane is metabolized, whereas oxidative metabolism of enflurane and isoflurane accounts for 2% and 0.2% of the administered doses, respectively. It has been estimated that the incidence of fatalities because of halothane hepatotoxicity is 1:35,000.

Halothane toxicity represents an example of Class 3 of our classification of toxicity, but unlike other hypersensitivity reactions, for example with penicillin, the effect is located to the liver, presumably because that is where the antigens are concentrated.

Summary

This chapter has considered various aspects of drug interactions and toxicity. Many of the examples refer to drugs that are rarely used or have been withdrawn, such as phenylbutazone, but these observations were crucial to understanding the mechanisms of drug interactions and toxicity. Our comprehension has been refined by elucidation of CYP450 isozymes and the identification and location of transporter proteins, and the role of pharmacogenetics in drug response, but the basic principles remain unchanged. The world of drug interactions and toxicology provides a fascinating area of application of most, perhaps all, of the principles involved in this book and while treatment relies more on history, clinical assessment and interpretation of ancillary investigations, measurement and interpretation of drug concentrations remain critical in providing baseline data, and understanding of mechanisms. To put interactions between drugs into perspective, it is necessary to appreciate how appropriate data on two or more relevant compounds can be used to predict the probability, mechanism and potential severity of drug interactions from among the myriad of possibilities that exist.

13.4 Further reading

Hussar DA. *Drug Interactions*. In: Remington G, editor. *The Science and Practice of Pharmacy*. Philadelphia: Lippincott, Williams and Wilkins, 2006: 1889–1902.

Walsky RL, Boldt SE. In vitro cytochrome P450 inhibition and induction. *Curr Drug Metab* 2008; 9: 928–39.

13.5 References

Breckenridge A, Orme M. Clinical implications of enzyme induction. *Ann N Y Acad Sci* 1971; 179: 421–31.

Charman WN, Porter CJ, Mithani S, Dressman JB. Physiochemical and physiological mechanisms for the effects of food on drug absorption: the role of lipids and pH. *J Pharm Sci* 1997; 86: 269–82.

Chu J, Wang RY, Hill NS. Update in clinical toxicology. *Am J Respir Crit Care Med* 2002; 166: 9–15.

Curry SH, DeVane CL, Wolfe MM. Cimetidine interaction with amitriptyline. *Eur J Clin Pharmacol* 1985; 29: 429–33.

Curry SH, Smith CM. Diazepam-ethanol interaction in humans: addition or potentiation? *Commun Psychopharmacol* 1979; 3: 101–13.

Flanagan RJ, Taylor A, Watson ID, Whelpton R. *Fundamentals of Analytical Toxicology*. Chichester: Wiley, 2008.

Gomez-Outes A, Suarez-Gea ML, Lecumberri R, Terleira-Fernandez AI, Vargas-Castrillon E. Direct-acting oral anticoagulants: pharmacology, indications, management, and future perspectives. *Eur J Haematol* 2015; 95: 389–404.

Jaeschke H, Gores GJ, Cederbaum AI, Hinson JA, Pessayre D, Lemasters JJ. Mechanisms of hepatotoxicity. *Toxicol Sci* 2002; 65: 16676.

Kim JH, Scialli AR. Thalidomide: the tragedy of birth defects and the effective treatment of disease. *Toxicol Sci* 2011; 122: 1–6.

Lewis RJ, Trager WF, Chan KK, Breckenridge A, Orme M, Roland M, Schary W. Warfarin. Stereochemical aspects of its metabolism and the interaction with phenylbutazone. *J Clin Invest* 1974; 53: 1607–17.

Nicholls AW, Wilson ID, Godejohann M, Nicholson JK, Shockcor JP. Identification of phenacetin metabolites in human urine after administration of phenacetin-C2H3: measurement of futile metabolic deacetylation via HPLC/MS-SPE-NMR and HPLC-ToF MS. *Xenobiotica* 2006; 36: 615–29.

Rolan PE. Plasma protein binding displacement interactions – why are they still regarded as clinically important? *Br J Clin Pharmacol* 1994; 37: 125–8.

Rostami-Hodjegan A, Amin AM, Spencer EP, Lennard MS, Tucker GT, Flanagan RJ. Influence of dose, cigarette smoking, age, sex, and metabolic activity on plasma clozapine concentrations: a predictive model and nomograms to aid clozapine dose adjustment and to assess compliance in individual patients. *J Clin Psychopharmacol* 2004; 24: 70–8.

Spiller HA. Persistently elevated acetaminophen concentrations for two days after an initial four-hour non-toxic concentration. *Vet Hum Toxicol* 2001; 43: 218–9.

14

Perspectives and Prospects: Reflections on the Past, Present and Future of Drug Disposition and Pharmacokinetics

14.1 Drug disposition and fate

Having read this book you will be better placed to understand the development of drug disposition and pharmacokinetics and to reflect on the topic's origins, its role in medicine today and what might be its future, including your contribution to that future.

The origins of the use and misuse of drugs are lost in the mists of time, but we are certainly talking about thousands of years. With drug use came witchcraft, and eventually medicine as we now know it. The ancients knew of many drugs, several of which are still in use today. Most were of plant origin, containing mainly alkaloids and glycosides. These people clearly understood routes of administration: curare had to be delivered parenterally, especially in hunting and warfare on arrowheads, while the effects of cocaine could be best obtained by absorption through mucous membranes, such as by chewing coca leaves. They also understood that animals hunted with curare in this way could be made edible by a combination of the use of arrowhead poisons that were not absorbed when the meat was eaten, and/or chemical degradation in the process of cooking. They were pioneers in the study of the chemical stability of drugs, a critical type of experiment needed in high-quality well-controlled drug metabolism investigations today. Not only that, past generations knew that cocaine absorption could be enhanced by adding alkali, in the form of wood ash (potash, potassium carbonate) or lime ($CaCO_3$). Thus, the effects of pH partitioning had been discovered long before the term was ever used.

For centuries, dried plant materials were smoked, for absorption through the lungs, or applied to the nasal passages as snuff. Opium, cocaine and hyoscine (scopolamine) have

Introduction to Drug Disposition and Pharmacokinetics, First Edition. Stephen H. Curry and Robin Whelpton.

Companion website: www.wiley.com/go/curryandwhelpton/IDDP

been, and still are, administered this way. Datura stramonium, which is native to North America, but was taken to Europe and Asia, and contains hyoscyamine and hyoscine, has been smoked for its hallucinogenic properties. In the USA it grows wild and may be called Jimson weed, and its misuse leads to hospitalization and sometimes deaths, even today. Hyoscine is administered transdermally for rapid effect in motion sickness, but in the past stramonium concoctions were administered to the skin or mucus membranes. Witches were believed to administer hyoscine intravaginally or under the armpits, where possibly the moist skin aided transdermal absorption. The hydration principle is used today in some transdermal preparations, which include agents to hydrate the skin to improve drug absorption. Hyoscine is lipophilic enough to be readily absorbed across intact skin and is available today as Scopoderm TTS, which contains 1 mg in a transdermal patch for release over 72 h, which illustrates the highly potent nature of this alkaloid.

Coniine, which is chemically related to nicotine, is an alkaloid found in hemlock. It has properties similar to tubocurarine, causing muscle paralysis, but because it is a tertiary amine rather than a quaternary ammonium salt, it is active orally. It was a drink containing hemlock that was used for the execution of Socrates in about 400 BC: a judicial order to commit suicide, one might say.

Drug residues that have been found in 3000-year-old mummies give some indication of how long drugs have been used, but for more detailed information it is necessary to rely on writings and observations about isolated tribes and peoples whose lifestyles have probably not changed for hundreds of years. In 1774 Georg Wilhelm Steller writing about the Koryak (people of north-western Russia) and their use of 'magic mushrooms' (fly agaric, *Amanita muscaria*) noted that 'those who cannot afford the high price [of the dried mushrooms] drink the urine of those who have eaten it, whereupon they become as intoxicated, if not more so. The urine seems to be more powerful than the mushroom and its effects may last through the fourth or fifth man.' Clearly these people, and probably many before them, knew that some drugs are excreted via the urine and that this fact could be exploited.

Interest in drug metabolism appears to more recent, having begun in the 1800s. Friedrich Wöhler (1800–1882) studied the passage of substances into the urine when he was a medical student at Marburg University. He proposed that benzoic acid, or a metabolite of benzoic acid, was excreted into the urine, but it is Alexander Ure (1810–1860) who is credited with the first metabolism study in humans in 1841 when he showed that benzoic acid is converted to hippuric acid, the glycine conjugate of benzoic acid. Thus, the first documented metabolic reaction in human beings was a phase 2 reaction. That cinnamic acid is also excreted as hippuric acid lead to the conclusion that there must be an intermediate oxidation step, that is, a phase 1 reaction. Bernhard Naunyn (1839–1925) showed that benzene is oxidized to phenol, and toluene was shown to be oxidized to benzoic acid.

The investigations and writings of Richard Tecwyn Williams (1909–1979), his students and colleagues brought together many of the developments in drug metabolism. Julius Axelrod firstly working with liver slices and then microsomal fractions showed that many drug-metabolizing enzymes were associated with the smooth endoplasmic reticulum, and later CYP-450 was shown to be a haem-containing protein.

It had long been known that drugs could be absorbed and secreted into bile and renal tubular fluid via carriers that could be saturated or inhibited by other drugs (Chapters 2 and 3). Victor Ling discovered P-gp in 1971 whilst investigating drug-resistant cancers, and the realization of the widespread location of this and other efflux proteins having large numbers of substrates changed our thinking about how drugs are transferred across biological membranes and how their induction and inhibition may explain many drug–drug interactions.

14.2 Pharmacodynamics

The ancients probably understood something about pharmacodynamics, certainly dose–response relationships. The Incas had a good idea of how many coca leaves were required to achieve a day's work and South American Indians adjusted the amount of curare on their arrow tips so that they could either kill or incapacitate their prey, which would be captured to be eaten later. Literature is another source of contemporary thinking. Shakespeare wrote about the risk–benefit equation of alcohol in Macbeth and the duration of drug action in his version of Romeo and Juliet.

Plant alkaloids played a crucial role in formulating receptor theory. John Langley (1852–1925) investigated the effect of pilocarpine, first on the heart and then on salivary secretion. Pilocarpine (a muscarinic agonist) increased salivary secretion and this could be blocked by atropine. Furthermore the effect of atropine could be reversed by applying more pilocarpine. Langley formulated the theory that the surface of the gland contained a 'receptive substance' with which the alkaloids were reacting. It was later that Paul Ehrlich coined the expression 'receptor'. Observations of the action of pilocarpine and atropine on the heart led Otto Loewi to prove that acetylcholine is a neurotransmitter, for which he gained the Nobel Prize in Physiology or Medicine. Although the concept of receptors was accepted in pharmacology, receptor proteins were not isolated and unequivocally identified until the 1970s.

14.3 Quantification of drugs and pharmacokinetics

We noted earlier the almost complete dearth of pharmacokinetic information on new drugs until about 1960 onwards. One of the reasons for this was the lack of suitable chemical analyses. The study of pharmacokinetics requires the development of assays to quantify the drugs under investigation. The quality of the results of a pharmacokinetic investigation can be no better than the quality of the concentration–time data. The analytical method must have the required:

- *specificity*: the compound under investigation must be assayed and not some interfering substance
- *accuracy:* the measured value of concentration must be close to the actual value
- *precision:* repeated measurements of concentrations should fall within a small range of values
- *sensitivity*: is important because for each halving of the lowest limit of quantification, the decline in concentration can be followed for a further half-life.

It has been the introduction of new analytical techniques with improved selectivity and sensitivity that has led to developments in the subject.

The 1880s saw the development and documentation of many scientific discoveries, including chemical analysis. In 1836 James Marsh published an account of his test for arsenic, which was not only more sensitive than previous tests but was also quantitative, allowing the amount of the poison to be assessed. Henry Bence Jones (1813–1873), a polymath who has a protein named after him (a test for which is still used today), developed methods to measure lithium and quinine. In a letter to the Royal Society in 1865 he noted that lithium 'appeared in every tissue of the body'. This observation was made using guinea pigs but Jones also studied lithium in lenses from patients undergoing cataract operations. With regard to quinine he realized that to obtain maximum fluorescence, samples had to be irradiated with ultraviolet light, provided by sparks from an induction coil. Tissues were extracted with sulfuric acid, neutralized and extracted into diethyl ether. After evaporation the residue was dissolved in acid for measurement. And so it was that Bence Jones was able to define the time course of quinine. His approach to extraction with pH-controlled back-extraction has been widely used by countless analysts since.

Much of the development in pharmacokinetics parallels the introduction and application of new or improved analytical techniques. Some of the progress was driven by the need of the military during World War II. A team of scientists led by James Shannon at the Goldwater Memorial Hospital in New York City were tasked with finding an antimalarial drug to replace quinine, sources of which were no longer available because of the war. Bernard B. Brodie (1907–1989) and Sidney Undenfriend (1918–2001) developed a fluorometric assay for mepacrine (quinacrine). From their blood concentration data they were able to recommend an efficacious and safe dosage regimen based on an initial loading dose followed by smaller maintenance doses. This nicely illustrates the use of pharmacokinetics to achieve a better clinical outcome. Brodie published a series of six papers in January 1947 which outlined the group's rationale to drug analysis, much of which relied on selective solvent extraction to separate parent drug from metabolites and interfering substances. This must have been 'bread and butter' to an organic chemist such as Brodie.

In 1949 Shannon persuaded several people from the Goldwater group to join him at the newly created National Heart Institute (NHI) at the National Institutes of Health (NIH) in Bethesda, Maryland. Brodie became Chief of the Laboratory of Chemical Pharmacology. Brodie's contribution to explaining the time course of thiopental was discussed in Chapter 5, and this concept of redistribution as a determinant of the duration of action and safe use of general anaesthetics is still relevant today. At the time it helped rationalize the use of intravenous anaesthetics, especially in military field hospitals. Numerous scientists from some 29 countries worked alongside Brodie at some stage in their careers, many of whom became eminent in their own right. In 1972 Brodie was superseded at NIH by James R. Gillette (1921–2001), who had originally joined Bert LaDu's team. They identified the need of the mixed function oxidases for NADPH as cofactor and investigated environmental factors and enzyme induction on the activity of CYP-450. Some idea of the many contributions that were made by this group can be gained from Costa *et al.* (1989). It was a fundamental philosophy of this group that meaningful observations could only be based on data that came from application of high-quality assay methods. The role of the analytical chemist in drug metabolism and pharmacokinetics was assured.

In science it is important to recognize how innovations can be exploited to address particular problems. The introduction of gas chromatography (GC) with flame ionization detection (FID) provided selectivity with improved limits of detection. The introduction of the ^{63}Ni electron capture detector (ECD), which was developed to measure residues of halogenated pesticides, particularly DDT, in biological samples, provided greatly increased sensitivities towards electron-deficient compounds. The key that enabled chlorpromazine to be measured for the first time in plasma samples from psychiatric patients was recognizing that this phenothiazine is electron-deficient (Curry, 1968). Similarly, the application of high-performance liquid-chromatography (HPLC) with electrochemical detection resulted in the first assay for plasma concentrations of physostigmine in humans (Whelpton, 1983). Solid-phase extraction and use of a more stable coulometric detector allowed development of a physostigmine assay that was approximately two orders of magnitude more sensitive (Hurst & Whelpton, 1989).

Recent years have seen a huge increase in mass spectroscopic techniques, particularly different modes of ionization that enable coupling to GC, HPLC and thin-layer chromatography (TLC). These innovations have led to more rapid assay development and, in general, shorter analysis times, thereby increasing the numbers of samples that can be assayed. Increases in sensitivity have allowed the use of smaller sample volumes, thereby facilitating investigations in paediatric populations and multiple sampling in small animals and microdosing in human subjects. Microdosing refers to the administration of sub-therapeutic doses of drug during early development for the first determination of kinetic parameters. This permits the use of doses below the threshold of toxicological significance, reduces the amount of substance needed for human investigations, and facilitates the study of drugs with the potential for saturable kinetics. To achieve the required sensitivity a technique known as accelerator mass spectroscopy (AMS) can be used to quantify low doses of ^{14}C-labelled drug. PET scanning has also been used, but both AMS and PET require synthesis of radiolabelled analyte. It is to be predicted that increased sensitivity of non-radioactive techniques will allow these to be applied to microdosing without the need to prepare isotopically labelled materials, with a consequent cost saving.

With regard to mathematical approaches, three men are usually credited with the early innovations in this area. Erik M.P. Widmark (1889–1945) is famous for his work on ethanol kinetics, using an oxidative colorimetric assay. Using a single-compartment model with zero-order kinetics, he produced equations relating blood ethanol concentrations to the amount of alcohol consumed. He was also aware that drugs may be eliminated according to first-order kinetics and derived equations for accumulation of drugs; those with exponential elimination reaching steady-state concentrations while the concentration of those eliminated according to zero-order kinetics would never reach a maximum (Widmark & Tandberg, 1924). Torsten Teorell (1905–1992), a Swedish physiologist, published two papers with the same title in 1937, 'Kinetics of distribution of substances administered to the body', the first considering extravascular doses and the second considering intravascular dosing. Teorell considered the drug being absorbed, circulated in the blood, distributed into tissues and delivered to the eliminating organs, and ascribed rate constants to these processes. He derived the appropriate differential equations, and by giving the rate constants and blood and tissue volumes certain possible values, he calculated the time course of the amount of drug in the body, expressed as a percentage of the dose. This may seem obvious to us now but it was revolutionary, particularly the fact that changing the rate of absorption could change the pattern of the kinetics. Teorell's contribution has been described as the first multi-compartment model but his approach was more akin to PBPK modelling (Chapter 7) so he can also be credited with this. This seminal work was largely ignored, possibly because other clinicians were unsure as to how to utilize the work. The term 'pharmacokinetics' was first coined by Friedrich Hartmut Dost (1910–1985) in his 1953 monograph *Der Blutspiegel* (*Blood levels*); a second edition, *Grundlagen der Pharmakokinetik* (*Fundamentals of Pharmacokinetics*), was published in 1968. As a professor of paediatrics his goal was to derive methods for scaling from adult to paediatric doses, and he appears to have been the first to consider the importance of *AUC* as a measure of exposure to drugs. In many ways, Hans Theodor Schreus (1892–1970) had addressed the issue of multiple dosing in 1926 when he proposed a dosage regimen for arsphenamine (a treatment for syphilis) based on his observation that by 15 minutes the blood concentration had fallen by 35%, and therefore to maintain effective concentrations with minimum toxicity only 65% of the original dose should be injected at this time. In his paper he drew concentration time curves that we would now describe as representing a loading dose followed by maintenance doses, as discussed in Chapter 4. Thus these investigators, and Brodie with his work on thiopental, had alluded to the compartment concept, but it fell to Riegelman and Rowland in the 1960s and 1970s to demonstrate to us the significance of mathematically describing this critical concept. Riegelman and Rowland were originating leaders of the European and North American wave of interest in pharmacokinetics that arose in colleges of pharmacy in response to a then newly perceived need to quantify pharmaceutically modified dosage forms of drugs. The first English language review of pharmacokinetics appears to be that of Eino Nelson (1961), who, with many of those cited in this article, was also a part of this explosive growth in research.

The 1970s saw a move away from using elimination half-life as a descriptor for changes in pharmacokinetics to increasing use of clearance concepts and later to the use of *in vitro* clearance, as described in Chapter 7. It has been suggested that a more useful acronym than ADME is ABCD: absorption, bioavailability, clearance and distribution. James Gillette, along with others, challenged the concept of protein binding always delaying metabolism, and David Shand and Grant R. Wilkinson (1942–2006) derived equations for the effects of protein binding on flow-limited and capacity-limited drugs (Chapter 7).

In the 1980s extrapolation of results between different species, including from animals to humans, became more formalized with the introduction of what was termed 'allometric scaling'. This was a great improvement on what could be described as a 'scientific leap of faith' from data collected in animals to prediction of results of phase 1 studies in humans. Allometric scaling recognizes that clearance, and consequently elimination half-life, is related to the weight of the animal and that large animals live longer than smaller ones. Thus, both concentration and time need to be scaled for comparison between different species. Robert Dedrick gave his name to such plots. The term 'translational science' has been introduced for the study of the transition of data from animals to man.

14.4 The future

It is a brave or foolish person who attempts to predict the future, but it is reasonable to suppose that many recent innovations will be further developed and expanded. More examples for study are to be expected, rather than novel methods of study. No doubt the quest to find the Holy Grail of an oral insulin preparation will continue, although presumably it will still be necessary to take additional doses as dictated by blood glucose levels. Macromolecules, including enzymes, monoclonal antibodies and antisense nucleotides will continue to be explored. Gene manipulation for potential new therapies and as a tool for investigating basic science will be important.

In vitro studies, for example of intrinsic clearance and extrapolation to *in vivo* pharmacokinetics, is likely to continue to grow in importance in the quest for lower costs. The results will be incorporated into increasingly complex models, so-called investigations, *in silico*, to predict behaviour *in vivo*. Drug disposition and pharmacokinetics will be supported by increasingly sensitive assays, which in turn will allow more widespread use of microdosing. Application of microdosing to proteins (Rowland, 2016) should expedite development of these large molecules as drugs (Chapter 9). It may sound counterintuitive, but an aim in the development of new chemical entities (NCEs) must be to identify unsuitable candidates as soon as possible as it is in no one's interest to nurture a NCE only for it to fail at the last moment; that is the road to bankruptcy. Microdosing, without the need to use isotopically labelled drug, and *in vitro* studies should increase the throughput of compounds that can be tested and rejected, if necessary, at minimal cost.

One part of the future is certain: the development of what is currently referred to as 'personalized medicine' will continue. It has always been the case that pharmacologists and doctors have sought to treat people with the right drug, at the right dose, in the right place, at the right time, largely rejecting the 'one size fits all' approach, which was probably never really practised to any significant extent (Woodcock, 2016). Early biomedical scientists such as Claude Bernard, Paul Ehrlich and A.J. Clarke undoubtedly recognized the significance of individual variation. In recent years the quest for individualization of treatment has been supported by the use of pharmacogenetic information. Major progress has been made during the last 20 years in regard to acquisition of knowledge concerning the genetic basis of some diseases; the design of pharmacological interventions can now, sometimes, reflect this basis. Recently, the term 'precision medicine' has been introduced. Personalization refers to the idea that genetic make-up is indeed personal, and so treatment should be personal, whereas *precision* refers to the idea that close attention to detail will lead to targeting the right gene and only the right gene, or its gene products. These concepts, built on the foundations provided by our predecessors, promise significant improvements in therapeutics. This approach to targeting treatment has led to the recognition that both highly specific diagnostic tests and highly specific drugs can be developed from the same research, and regulatory authorities are beginning to approve combinations, of test and treatment, as a package (Pacanowski & Huang, 2016). The best known of these is the test for HER2 positive breast cancer (HER2 FISH PharmDx Kit, HERCEPTEST) and trastuzumab (Chapter 9). Most of the examples of this kind are in the field of cancer chemotherapy, with some from other fields, for example thallasaemia.

The examples quoted by Panakowski and Huang do not invoke concepts of genetic control of pharmacokinetics. However, there are two notable examples of this in the literature. Sjoqvist & Eliasson (2007) have promoted the idea of phenotyping for CYP2D6 as a mechanism of fine tuning therapy with tricyclic antidepressants and, to a lesser extent, selective serotonin reuptake inhibitors. Phenotyping has historically been done indirectly by studying the metabolism of 'probe' substances, as described in Chapter 10, while genotyping is increasingly being used to determine personalized dosing (Section 10.12). In a field similar to that with the antidepressants, Mahajan *et al.* (2013) have reviewed published studies of gene-related monitoring of warfarin metabolism in patients at risk from too little or too much effect on the coagulation of their blood, when treated for increased risk of clotting disorders. Multiple studies have demonstrated modest improvements in initial dosing and clotting control, frequency of clinic visits and cost-effectiveness when pharmacogenomic data is used in defining and controlling therapy. Invasive methods are not required, as direct control of anticoagulation therapy can be achieved by *in vitro* coagulation monitoring of blood samples.

There can be no doubt that the current trend towards development of drugs, supported by the science of pharmacogenomics, will continue for a considerable period of time into the future. Mostly, these drugs will continue to be administered by injection. This will almost certainly provide challenges for generations of drug metabolism and pharmacokinetics (DMPK) scientists to come, especially the analytical chemists and formulation scientists who support them.

14.5 Postscript

Our intention with this book has been to introduce you, the reader, to the terminology and facts of drug disposition and pharmacokinetics, and to assist you in the development of your ability to visualize how the time course of a drug in the body is controlled, and how it affects responses to the drug. For that reason, we have emphasized the stereotypical concentration/time patterns in Figures 4.1 and 4.2, and illustrated useful methods for describing them mathematically for any particular drug, and for any particular set of circumstances, governed by such variables as route and mode of administration, species treated, environmental, physiological and pathological variables, and interacting drugs. Thereby, the text should prove valuable to dosage formulation scientists developing formulations designed to optimize existing drugs, to those selecting from multiple candidates in the search for new therapeutic agents in universities, research institutes and industry, and to those primarily seeking to understand the individual patient response to drugs, a key component of personalized medicine. As an example to put this in perspective, no one set of published data will illustrate the expected time course of a chosen drug in a smoking patient treated with a controlled-release preparation of that drug, for which there is a significant influence on response related to age, and for which there is an anticipated effect of partial renal failure but no definitive relevant published study for these combined influences. Such a case requires the use of building blocks by an expert professional recognizing the multiple contributing influences that apply in a particular case. We hope that this book will provide an introduction to the method of using building blocks in this way. We also hope that this book will help those involved in corporate and clinical decision making to utilize relevant specific information for any particular case with full recognition of the principles involved.

You may feel that you have an adequate level of knowledge of this subject matter after reading this book, and that it can now go into your personal library as a future reference work. Alternatively, you may wish to develop your expertise in this area beyond the scope of this book. Hopefully, you are now in a position to evaluate the many higher level texts that exist, as well as information of this type in learned scientific journals that will help you develop your knowledge further, for example by delving into mathematical pharmacokinetics at a higher level, or by developing expertise in relation to the multiple species encountered in veterinary practice, or by becoming an expert in bioanalysis, or by working with patentable ideas, or by focusing on unanswered medical problems that can be solved by fusing knowledge of pharmacokinetics with knowledge of disease processes at the highest levels. Or you may want to focus on the minutiae of understanding why individual groups of compounds within groups vary so dramatically in their pharmacokinetic properties – properties governed by chemical structure. Whatever you decide to do, we hope that this book proves to be a valuable aid to you realizing your ambitions and you feel it is worthy of a place on your bookshelf.

14.6 Further reading

Andreasson R, Jones AW. Erik M.P. Widmark (1889–1945): Swedish pioneer in forensic alcohol toxicology. *Forensic Sci Int* 1995; 72: 1–14.

Costa E, Karczmar AG, Vesell ES. Bernard B. Brodie and the rise of chemical pharmacology. *Annu Rev Pharmacol Toxicol* 1989; 29: 1–21.

International Society for the Study of Xenobiotics. About James R. Gillette; http://www.issx.org/?page=Gillette, accessed 17 March 2016.

International Society for the Study of Xenobiotics.Woehler: The birth of metabolism research; http://www.issx.org/?Woehler, accessed 17 March 2016.

Paalzow LK. Torsten Teorell, the father of pharmacokinetics. *Ups J Med Sci* 1995; 100: 41–6.

Rosenfeld L. Henry Bence Jones (1813–1873): the best 'chemical doctor' in London. *Clin Chem* 1987; 33: 1687–92.

14.7 References

Costa E, Karczmar AG, Vesell ES. Bernard B. Brodie and the rise of chemical pharmacology. *Annu Rev Pharmacol Toxicol* 1989; 29: 1–21.

Curry SH. Determination of nanogram quantities of chlorpromazine and some of its metabolites in plasma using gas-liquid chromatography with an electron capture detector. *Anal Chem* 1968; 40: 1251–5.

Hurst PR, Whelpton R. Solid phase extraction for an improved assay of physostigmine in biological fluids. *Biomed Chromatogr* 1989; 3: 226–32.

Mahajan P, Meyer KS, Wall GC, Price HJ. Clinical applications of pharmacogenomics guided warfarin dosing. *Int J Clin Pharm* 2013; 35: 359–68.

Nelson E. Kinetics of drug absorption, distribution, metabolism, and excretion. *J PharmSci* 1961; 50: 181–92.

Pacanowski M, Huang SM. Precision medicine. *Clin Pharmacol Ther* 2016; 99: 124–9.

Rowland M. Microdosing of protein drugs. *Clin Pharmacol Ther* 2016; 99: 150–2.

Sjoqvist F, Eliasson E. The convergence of conventional therapeutic drug monitoring and pharmacogenetic testing in personalized medicine: focus on antidepressants. *Clin Pharmacol Ther* 2007; 81: 899–902.

Whelpton R. Analysis of plasma physostigmine concentrations by liquid chromatography. *J Chromatogr* 1983; 272: 216–20.

Widmark E, Tandberg J. Theroetische Bereckerunger Über die Bedingungen für die Akkumulation indifferenter Narkotika *Biochem Z* 1924; 147: 358–69.

Woodcock J. 'Precision' drug development? *Clin Pharmacol Ther* 2016; 99: 152–4.

Appendix 1

Mathematical Concepts and the Trapezoidal Method

A1.1 Algebra, variables and equations

Algebra is a way of describing relationships in general terms, usually as equations. For example, Wilhelm Beer observed that the optical absorbance, A, of a dilute solution was directly proportional to the concentration, C, of the solute:

$$A \propto C \tag{A1.1}$$

To write the relationship as an equation we need a constant of proportionality, k:

$$A = kC \tag{A1.2}$$

If C is in mol L^{-1}, and the path length is 1 cm, then $k = \varepsilon$, the molar absorptivity of the solute. If the solvent also absorbs light, the background absorption, b can be added to the equation:

$$A = kC + b \tag{A1.3}$$

A is known as the *dependent* variable because it changes as a result of changes in C, the *independent* variable. Equation A1.3 can be represented graphically by plotting A against C. The independent variable is plotted along the bottom (x-axis) and the dependent variable along the y-axis (Figure A1.1)

Introduction to Drug Disposition and Pharmacokinetics, First Edition. Stephen H. Curry and Robin Whelpton.

Companion website: www.wiley.com/go/curryandwhelpton/IDDP

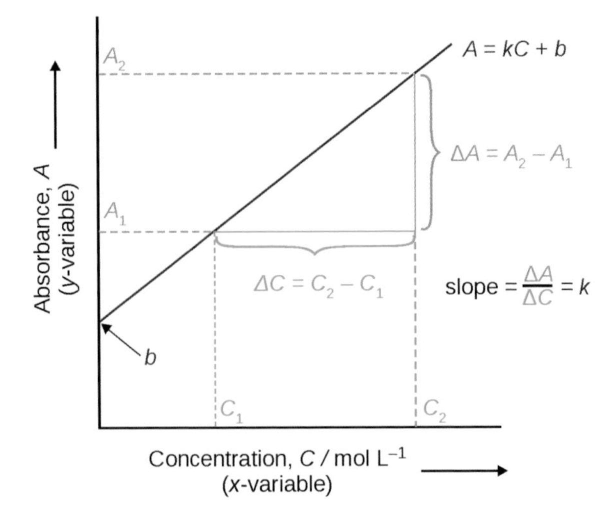

Figure A1.1 *Straight line representation of $A = kC + b$.*

When $C = 0$, $y = b$, so the value of b can be obtained from the intercept of the line with the y-axis. The value of k is obtained from the slope of the line. Estimates of the intercept and slope are best derived from least squares regression analysis, which can be done on many hand-held calculators, an Excel spreadsheet, as well as specifically designed regression software packages. If the line is a calibration line then it usual to rearrange Equation A1.3 so that concentrations can be calculated from measured absorbance values. Subtracting b from both sides and dividing both sides by k gives:

$$C = \frac{A - b}{k} \tag{A1.4}$$

A1.2 Indices and powers

When a number, a, is multiplied by itself a number of times, n, the product can be written a^n. For example:

$$a \times a = a^2 \tag{A1.5}$$

and, in this case, a is said to be 'squared'. Similarly,

$$a \times a \times a = a^3 \tag{A1.6}$$

and the result is called 'a cubed' or 'a to the 3'. Multiplying a^2 and a^3:

$$a \times a \times a \times a \times a = a^5 \tag{A1.7}$$

Note that the result is $a^{(2+3)}$ and the general rule can be written:

$$a^n \times a^m = a^{(n+m)} \tag{A1.8}$$

Using similar logic it can be shown that:

$$\frac{a^n}{a^m} = a^{(n-m)} \tag{A1.9}$$

Thus:

$$\frac{1}{a^n} = a^{-n} \tag{A1.10}$$

and:

$$\left(a^n\right)^m = a^{nm} \tag{A1.11}$$

A fractional index indicates that a root should be taken:

$$a^{\frac{1}{n}} = \sqrt[n]{a} \tag{A1.12}$$

A1.3 Logarithms

Tables of logarithms and antilogarithms are used to simplify multiplication and division of complex numbers using only addition and subtraction. With common logarithms (base 10), each number is expressed as 10 raised to the appropriate power. Logarithms have additional importance in pharmacology for contracting the range of numbers and in the linear transformation of data. The base of the logarithm can be indicated, for example $\log_{10}$, but generally common logarithms are written log. Logarithms to the base e ($e \approx 2.718...$) are referred to as natural logarithms, and can be written $\log_e$ or, more commonly, simply ln. A word of warning: in many computer languages, and in engineering parlance, log means $\log_e$. To convert a logarithm to a number the base of the logarithm is raised to the logarithm. For example,

$$\log 2 = 10^2 = 100$$

and if

$$x = \ln y, \text{ then } y = e^x \tag{A1.13}$$

Negative numbers cannot have logarithms, but note that

$$\log\frac{1}{x} = -\log x \tag{A1.14}$$

Furthermore, it should be noted that although logarithms are dimensionless, it is convention to indicate the units of the original number. So for a concentration of C mg L^{-1}, the logarithm should be written: log (C/mg L^{-1}). Note the position of the brackets, the units refer to the concentration not the logarithm; it *must not* be written log C (mg L^{-1}).

A1.4 Calculus

A1.4.1 Differentiation

Calculus was invented to deal with slopes of curves and the areas under them. When the rate of change of a measurement (e.g. y as a function of x) is changing, we may wish to determine the average rate of change of y with respect to x. This is done by differentiation. Suppose δy represents a small increase in y which occurs while x increases by δx, then $\delta y/\delta x$ is the mean gradient of the graph over the small range examined. If we reduce the values of δy and δx towards zero, then the line showing the gradient tends towards a tangent. The limiting value of $\delta y/\delta x$ is called $\mathrm{d}y/\mathrm{d}x$ or the differential coefficient of y with respect to x. The process of finding the limit of $\mathrm{d}y/\mathrm{d}x$ is *differentiation*, and $\mathrm{d}y/\mathrm{d}x$ is a measure of the slope of the tangent at a given value of x (Figure A1.2(a)). If the y variable is time, t, and the x variable is concentration, C, then $\mathrm{d}C/\mathrm{d}t$ is the instantaneous rate of change in concentration at that time.

For equations of the type: $y = ax^n + b$

$$\frac{\mathrm{d}y}{\mathrm{d}x} = anx^{(n-1)} \qquad \text{(A1.15)}$$

Thus for a straight line, $y = ax + b$, that is, $n = 1$, so

$$y = ax^1 + b$$
$$\frac{\mathrm{d}y}{\mathrm{d}x} = ax^{(1-1)} = ax^0 = a \qquad \text{(A1.16)}$$

confirming that a is the slope of the straight line.

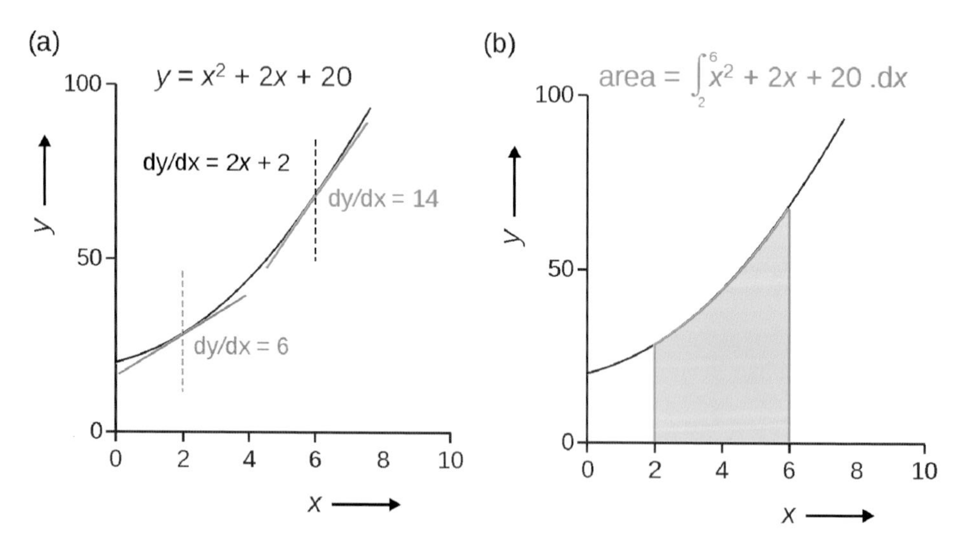

Figure A1.2 (a) Differentiation allows the slope at any value of x to be calculated. (b) Integration between $x = 2$ and $x = 6$ allows the area depicted by the shading to be calculated (see text for details).

Reciprocals of x are treated the same way:

$$y = \frac{1}{x^n} = x^{-n}$$
$$\frac{dy}{dx} = -nx^{-n-1} = -nx^{-(n+1)} \tag{A1.17}$$

However the differential of ln x is a special case:

$$y = \ln x$$
$$\frac{dy}{dx} = \frac{1}{x} \tag{A1.18}$$

A1.4.2 Integration

The reverse of differentiation is integration. If dy/dx is known we may wish to find y in terms of x. Thus, if dy/dx is some function of x, written as $f(x)$, then y is the integral of $f(x)$ with respect to x:

$$y = \int f(x).dx \tag{A1.19}$$

If $y = ax^n$, then:

$$y = \int ax^n.dx = a\int x^n.dx = \frac{ax^{(n+1)}}{n+1} + c \tag{A1.20}$$

Note the appearance of c, sometimes referred to as a constant of integration, which is necessary because constants are 'lost' on differentiation, see Equation A1.16. The value is found by substituting $x = 0$, when $y = c$.

The integral of $1/x$ is of particular importance because of the form of the rate equation of first-order reactions:

$$y = \int \frac{1}{x}.dx = \ln x + c = ce^x \tag{A1.21}$$

A quantity written as a power of e is an exponential function, and the quantity increases more and more rapidly as the power increases: *exponential growth*. If the power is negative, for example $y = e^{-x}$, then this represents *exponential decay* and y becomes ever nearer to 0 as x increases, but only reaches 0 when x is infinite; y is said to *asymptote* to 0.

A1.4.2.1 Areas under curves

Integration is important for calculating areas under curves. Using the equation of Figure A1.2 as an example, the area under the curve (AUC) from 2 to 6 is obtained by integrating between the limits:

$$AUC_{(2-6)} = \int_2^6 x^2 + 2x + 20.dx = \frac{x^3}{3} + x^2 + 20x\Big|_2^6 \tag{A1.22}$$

$AUC_{(2-6)}$ is the difference between the value obtained by substituting $x = 6$ and $x = 2$.

$$AUC_{(2-6)} = \left[\frac{6^3}{3} + 6^2 + 20(6)\right] - \left[\frac{2^3}{3} + 2^2 + 20(2)\right]$$

$$AUC_{(2-6)} = 228 - 46.7$$

$$AUC_{(2-6)} = 181.3$$

Note that because we are using the difference, the constant of integration cancels and need not be included. In this example, the numbers were dimensionless and so *AUC* has no units. However, in most practical instances the *x* and *y* values will represent variables with units and so anything derived from them should have the appropriate units. The slope of a ln (concentration) against time plot has units of time to the minus 1 (T^{-1}). The area under the curve of a plasma concentration (mg L^{-1}) against time (h) plot has units of mg h L^{-1}, for example.

A1.5 Calculating *AUC* values: the trapezoidal method

Probably the simplest way to obtain the area under a curve is the *trapezoidal method* The plasma concentration–time curve is plotted and each segment between adjacent collection time points treated as a trapezium (Figure A1.3(a)). The area of the trapezium is the average length of two sides multiplied by the width between them, so for the nth trapezium the area, A_n, is:

$$A_n = \frac{C_n + C_{n+1}}{2}(t_{n+1} - t_n) \tag{A.23}$$

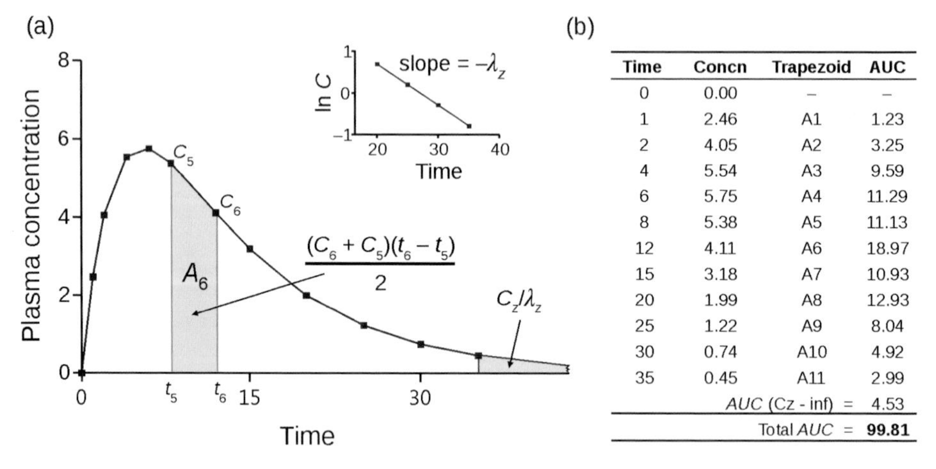

Time	Concn	Trapezoid	AUC
0	0.00	–	–
1	2.46	A1	1.23
2	4.05	A2	3.25
4	5.54	A3	9.59
6	5.75	A4	11.29
8	5.38	A5	11.13
12	4.11	A6	18.97
15	3.18	A7	10.93
20	1.99	A8	12.93
25	1.22	A9	8.04
30	0.74	A10	4.92
35	0.45	A11	2.99
		AUC (Cz - inf) =	4.53
		Total *AUC* =	**99.81**

Figure A1.3 (a) Example of the trapezoid method of calculating the area under a concentration versus time curve. The rate constant is calculated from the slope of the terminal points of ln C versus time plot (inset). (b) Printout from an Excel spreadsheet showing the areas of individual trapeziums and the total, which is good agreement with the theoretical value of 100 for this example.

The area from $t=0$ to the time of last plasma sample, $AUC_{(0\text{-}t_z)}$, is obtained by adding the areas of all the trapeziums. The remaining area, from the time of the last collected sample to infinity is extrapolated using C_z/λ_z, where λ_z is the rate constant of the terminal decay phase. This can be estimated from a ln(concentration) versus time plot of the terminal data (Figure A1.3, inset). The areas are conveniently calculated from the concentration–time data using a spreadsheet (Figure A1.3(b)).

Some points should be noted:

- It is the area under the concentration–time plot, *not* the ln(concentration)–time plot that must be used.
- Although the first segment on the oral plot is a triangle the formula for a trapezium gives the correct area because length of one side is 0.
- The greater the number of trapeziums the greater the accuracy of the calculation.
- Ideally the extrapolated area should be <5% of the total.
- The units of *AUC* are concentration × time, for example mg h L^{-1}.

Appendix 2
Dye Models to Teach Pharmacokinetics

A2.1 Introduction to the dye model

Single-compartment pharmacokinetic models can be illustrated by pumping water into and out of a suitable vessel, such as a beaker, or an even simpler system, which is to run tap water into a Büchner flask and to use the side arm for the outflow, as depicted in Chapter 4 (Figure 4.3). The advantages of this approach to teaching pharmacokinetics include the following:

- Concentration–time relationships can be demonstrated empirically without recourse to mathematics.
- It avoids using the mathematical models to generate data for analysis, only to use those same equations to derive the parameters.
- It provides good, but not perfect, quality data for analysis.
- The derived parameters can be compared with the known volumes of the apparatus and the measured flow rate of the water.
- The concentrations in the peripheral compartment of a two-compartment model can be viewed and *measured* directly.

Demonstrating multi-compartment models is technically more difficult. To facilitate visualization of a two-compartment model, a bespoke Perspex box was constructed so that the path length through the box was the same for both compartments, thereby allowing direct visual comparison of the concentrations in each (Figure 5.4).

Introduction to Drug Disposition and Pharmacokinetics, First Edition. Stephen H. Curry and Robin Whelpton.

Companion website: www.wiley.com/go/curryandwhelpton/IDDP

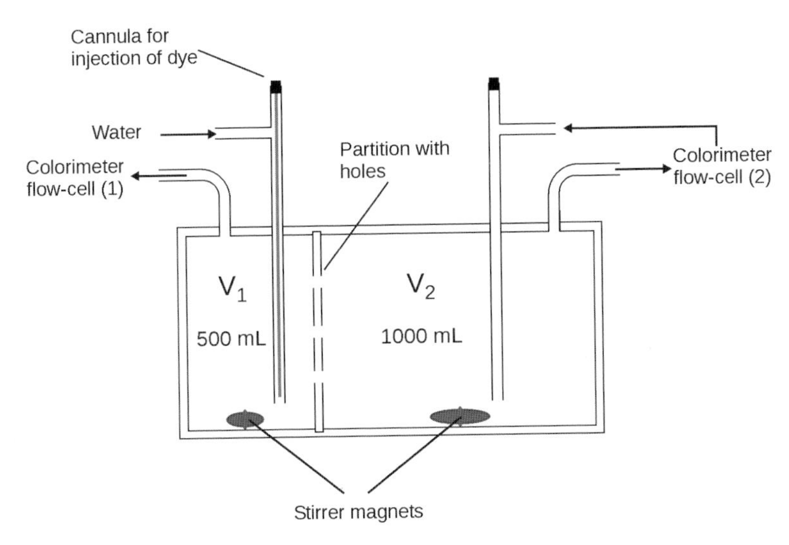

Figure A2.1 *Apparatus configured for demonstrating intravenous injection into a two-compartment, well-stirred, open model.*

A2.2 Arrangement for i.v. bolus or zero-order infusions

Inlet and outlet tubes carry water into and out of the compartments, which contain magnetic stirrer-followers to ensure thorough mixing of the contents of each volume (Figure A2.1). Constant flow rates are maintained using peristaltic pumps.

The concentrations of dye are measured by passing the effluent from the compartments through a colorimeter flow cell. When two-compartment models are being demonstrated, a partition with holes allows dye to diffuse from the smaller volume (V_1=central compartment) to the larger one (V_2=peripheral compartment). If the model is depicting elimination from the central compartment then the inlet and outlet tubes are used for the flow of water to the central compartment (V_1) and to flow out of the same compartment to waste, via a colorimeter to measure the concentration of dye. The set of tubes for the peripheral compartment are used to circulate dye though a colorimeter cell to measure the concentrations in the peripheral compartment (V_2). This arrangement was used to generate data for Figures 5.2, 5.4, 5.5 and 5.14 (i.v. bolus injection) and Figure 5.8 (i.v. infusion).

A2.3 Arrangement for first-order absorption

The first-order output from a dye model can be used as the first-order input into the Perspex box. This was achieved by connecting the outlet of a flask to the inlet of V_1 (Figure A2.2). The value of the absorption rate constant was varied by varying the volume of the flask. This arrangement was used to produce Figure 5.6. The rate constants are indirectly proportional to the volumes of the flask and compartments, so by making the input flask larger than the volume of V_2, it is possible to simulate the situation where $k_{21} > k_a$, and the concentration–time data appear to be from a single compartment, even though there are clearly two compartments.

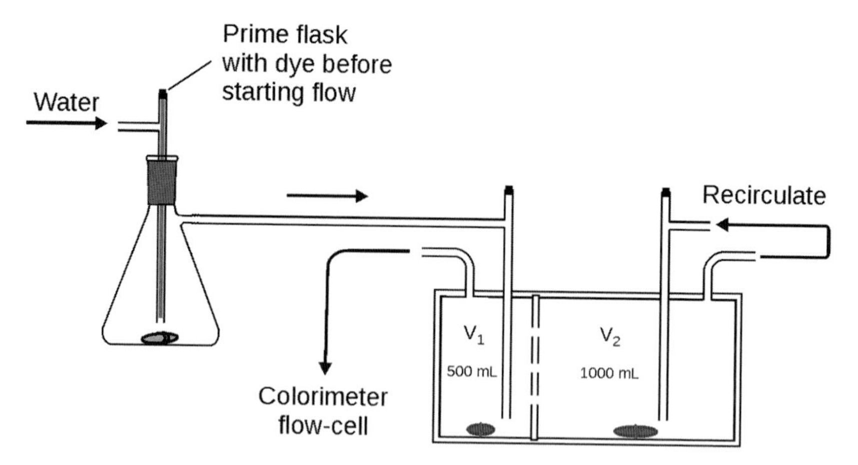

Figure A2.2 Apparatus arranged for first-order input into a two-compartment model.

Appendix 3
Curve Fitting

Reference has been made to resolving or fitting concentration–time data to obtain estimates of the pharmacokinetic parameters. Not all our readers will want to perform calculations and so the details have been reserved for this appendix.

A3.1 Graphical solution: method of residuals

Before the ready availability of personal computers and relatively inexpensive curve-fitting software, pharmacokinetic parameters were often obtained graphically. Although rarely used these days, an understanding of the approach is important when assessing the quality of the data to be analysed. The method of residuals, as it is known, is most easily understood from consideration of an intravenous bolus injection into a two-compartment model. Because $-\lambda_1$ is the slope of the steeper initial phase, λ_1 is greater than λ_2 and so the term $C_1\exp(-\lambda_1 t)$ approaches zero faster than $C_2\exp(-\lambda_2 t)$. Therefore at later times the contribution from the first exponential term is negligible and Equation 5.2 approximates to:

$$C \cong C_2 \exp(-\lambda_2 t) \tag{A3.1}$$

Provided that the plasma concentration–time curve is monitored for long enough, the terminal portion of the ln C versus t curve approximates to a straight line so C_2 and λ_2 can be estimated from the y-intercept on the y-axis and slope of the terminal phase. Values of $C_2\exp(-\lambda_2 t)$ are calculated for earlier time points (i.e. when $C_1\exp(-\lambda_1 t)$ is making a significant contribution to the plasma concentration) and subtracted from the experimental

---

Introduction to Drug Disposition and Pharmacokinetics, First Edition. Stephen H. Curry and Robin Whelpton.

Companion website: www.wiley.com/go/curryandwhelpton/IDDP

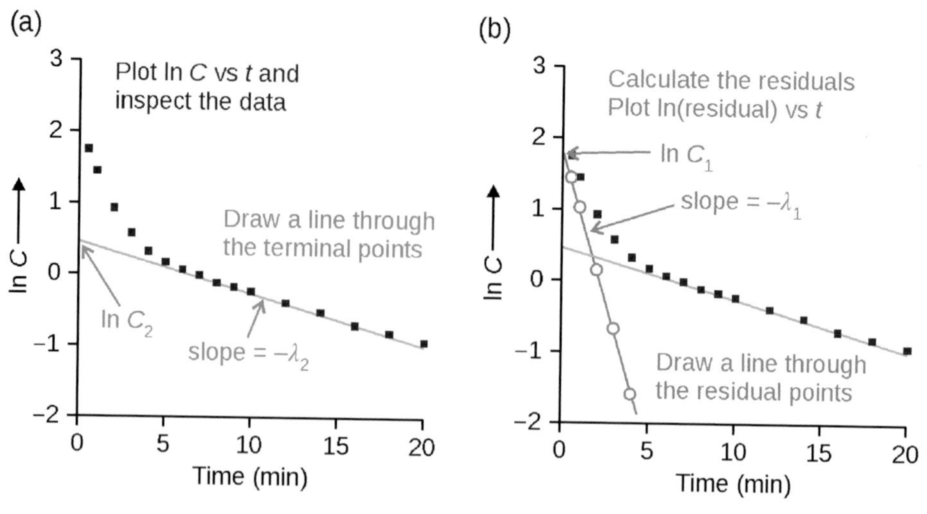

Figure A3.1 Method of residuals.

values at those times to give estimates of $C_1\exp(-\lambda_1 t)$, which are referred to as *residuals*. A plot of ln(residual) versus t should give a straight line of slope $-\lambda_1$ and intercept $\ln C_1$. Concave curvature of the residual line indicates that the first estimate of λ_2 is too low, whilst convex curvature indicates that it is too high, assuming, that is, that the data fit a two-compartment model.

In summary:

1. The data are plotted as $\ln C$ versus t (Figure A3.1(a)).
2. A construction line is drawn through (or close to) the terminal points, and the slope ($-\lambda_2$) and the intercept ($\ln C_2$) of the line determined.
3. For each time point, t, values of $C_2\exp(-\lambda_2 t)$ are calculated. These represent points on the construction line at each experimental time point.
4. For each time point, the residuals, $C_t - C_2\exp(-\lambda_2 t)$, are calculated and plotted on the graph (Figure A3.1(b)).
5. A line through the residuals allows computation of C_1 and λ_1.

The method of residuals can be applied to the majority of compartment models, and also first-order input into single or multiple compartments, and the post-infusion phase of zero-order infusions.

A3.2 Iterative curve-fitting

There are several commercially available curve-fitting programs that are sold specifically for pharmacokinetic analyses. However, other packages may be adapted to derive pharmacokinetic parameters. Concentration–time data for an intravenous bolus injection into a single compartment can be transformed to $\ln C$ versus t and solved using linear regression programs available on many handheld calculators (but see below). Many relationships do not have a mathematical solution and have to be solved iteratively, usually by computing the equation that gives the lowest residual sum of squares. On this occasion 'residual' refers to the difference between the observed value and the value calculated from the derived parameters. The square of the residuals is used because the residuals can be positive or negative. A useful source code for iterative curve fitting is that written in BASIC by Neilsen-Kudsk (1983). This code can be modified to run under different forms of BASIC and it is relatively easy to modify so that it can used to fit a wide variety of equations, including calibration curves and most compartmental pharmacokinetic relationships. With this robust algorithm several sets of data can be fitted to common parameters, for example concentration–time data following oral and i.v. administration of a drug, *simultaneously*.

A3.2.1 Weighted regression

Experimental data will have associated errors. Linear regression assumes that the errors are the same irrespective of the concentration, that is, the data are homoscedastic. However, the size of the error generally increases with concentration – indeed most assays are developed to ensure that the relative standard deviations (RSD) cover a limited range and never exceed a predefined limit, say 10%. Therefore the errors are approximately proportional to the measured concentration and so the errors associated with high concentrations are higher and may unduly affect the way in which the data are fitted. Because unweighted regression treats all points equally, the fit will be biased towards the higher concentrations at the expense of the lower concentrations. This is particularly a problem with pharmacokinetic data because the concentrations often extend over several orders of magnitude. The answer is to weight the data and to minimize the sum of weighted squares:

$$\text{weighted sum of residual squares} = \Sigma\left[(\text{weight})(\text{residual})^2\right] \tag{A3.2}$$

The size of the random errors is assessed by the variance (s^2) and, ideally, the data should be weighted by $1/s^2$, so Equation 5.29 can be written:

$$\text{weighted sum of residual squares} = \Sigma\left[1/s^2\,(\text{residual})^2\right] \tag{A3.3}$$

Thus, the points with the smallest errors assume more importance than those with the largest errors. Ideally, the points should be weighted by 1/variance, but it is unlikely that this will have been measured at every concentration because that would require replicate assays of every sample. However, if the RSD is (approximately) constant over the concentration range, then the standard deviation s is proportional to concentration and the data can be weighted 1/(concentration)^2. Many commercially available statistical programs allow the option of weighting data by $1/y^2$.

Weighting data is not some method of manipulating the result to make it appear more acceptable, *it is the correct statistical treatment for heteroscedastic data*. However, this being said, when the errors are small and the concentration range is limited, then there may be little difference in the values obtained by using non-weighted or weighted data. It should be noted that if ln C values are fitted, the transformation will weight the data and the results are likely to be different from those when C versus t is fitted.

A3.2.2 Choice of model

The number of compartments required to fit the data is given by the number of exponential terms that describe the *declining* portion of the curve. The choice of how many compartments to fit should be dictated by the data. Statistical fitting allows the various equations to be compared. Simply choosing the equation which gives the lowest residual sum of squares (*SS*) is unhelpful because this will be the equation with the largest number of parameters. Consequently, most statistical packages compute 'goodness of fit' parameters, which take into account the number of parameters in the equation. Some programs will fit two equations simultaneously and compare them using the F-test. Neilsen-Kudsk (1983) used the Akaike information criterion (*AIC*):

$$AIC = N \ln(SS) + 2M \tag{A3.4}$$

where N is the number of data points and M is the number of parameters. The equation with the lowest *AIC* is statistically the most appropriate.

The appropriateness of the model, including the weighting, should be tested by plotting the residuals as function of concentration. These should be randomly distributed about zero. The correlation coefficient, r, is often a poor indicator of the goodness of fit, unless $r^2 = 1$!

A3.3 Reference

Neilsen-Kudsk F. A microcomputer program in BASIC for interative, non-linear data-fitting to pharmacokinetic functions. *Int J Biomed Computing* 1983; 14: 95–107.

Appendix 4

Pharmacokinetic Simulations

Readers are free to use the simulations that have been provided on the web site as they wish. However, without being too prescriptive, some suggestions are given below. These have been chosen to reinforce some of the material in the book and have been arranged by relevant chapter. For simplicity and maximum flexibility, the output does not display units so it is incumbent on the user to note the appropriate units.

A4.1 Simulations for Chapter 4

SIM1 IDDP One-compartment i.v. bolus illustrates the role of apparent volume of distribution (V) and systemic clearance (CL) as determinants of the elimination rate constant, λ, and, as a consequence, the elimination half-life, $t_{½}$.

To illustrate the mono-exponential nature of the decay the output is presented on linear and semi-logarithmic coordinates. For simplicity, no units are given so that if the dose is in μg and the volume in mL, then the concentration will be μg mL^{-1}.

The input variables are dose, D, V and CL, and the duration of the plot can be varied as appropriate. Suggested starting values are below:

	D	*V*	*CL*
Set 1	1000	100	10
Set 2	1000	100	20
Set 3	1000	200	20

Introduction to Drug Disposition and Pharmacokinetics, First Edition. Stephen H. Curry and Robin Whelpton.

Companion website: www.wiley.com/go/curryandwhelpton/IDDP

Note how for a given dose the curves must intercept at the same value of C_0 when the apparent volume is the same in both cases, but the slopes ($-\lambda$) vary with varying clearance. Note what happens when the clearance value for data Set 3 is changed to 40.

SIM2 IDDP One-compartment i.v. infusion illustrates how when a drug is infused at a constant rate, R_0, the plasma concentration asymptotes to a constant value, C^{ss}. This must be the case for drugs that are eliminated according to first-order kinetics because as the plasma concentration increases the rate of elimination increases until it equals the rate of input. Although steady state only occurs at infinite time, >99% of steady state occurs after seven elimination half-lives.

With duration of infusion set to 300 and total time to 500, use initial values of $R_0=1000$, $CL=100$ and $V=7000$. Vary CL to note the effect of the elimination half-life on the shape of the curves, the rate of attainment of steady state and the value of C^{ss}. With $CL=100$, observe the effects of varying V. Why does C^{ss} not change when V changes?

Use Equation 4.38 to calculate a suitable loading dose to achieve instantaneous steady state. Is there a slightly easier way of calculating the loading dose? Vary CL and V as above, calculating a loading dose in each case. Note how CL influences the steady-state concentration and V defines the size of the loading dose.

SIM3 IDDP One-compartment multiple i.v. doses illustrates the accumulation of drug on repeated dosing and the effect of the kinetic parameters on the fluctuation between doses. Instantaneous steady-state conditions can be obtained with appropriately calculated loading doses.

Set all the values to that shown for Set 1 below, so that the three curves are superimposed. Now adjust the values of Set 2 and Set 3 as shown below.

	D	*V*	*CL*	*LD*	*Tau*
Set 1	1000	500	69.3		10
Set 2	1000	500	34.65		10
Set 3	1000	500	17.325		10

Note how the drug with the lowest clearance, and hence longest half-life, accumulates more slowly but to higher values. It reaches pseudo steady state at about 100 h ($5\times t_{1/2}$) whereas the one with the highest clearance reaches steady-state by about 25 h.

To observe the fluctuations more clearly adjust the doses for Sets 1 and 2 to give the same average steady-state concentration as Set 1 (see Equation 4.42). *Tip:* Use $AUC=C_0/\lambda$ or note that the average concentration is directly proportional to D and inversely proportional to λ.

With $D=4000$, 2000 and 1000 for the data sets respectively, reduce the fluctuation of Set 1 by reducing the dose to 2000 and the dose interval to 5. What effect does this have on the average steady-state concentration? Note that the average steady-state values for C_{max} and C_{min} are the same for Set 1 and Set 2.

If frequent dosing is a problem, the lag to steady state can be overcome by the use of a loading dose. Calculate the appropriate loading dose for Set 3 (Equation 4.41). The input data should now be:

	D	*V*	*CL*	*LD*	*Tau*
Set 1	2000	500	69.3		5
Set 2	2000	500	34.65		10
Set 3	1000	500	17.325	3414.8	10

Note how the use the drug with the longest half-life, plus a loading dose, gives instantaneous pseudo steady state, with the least fluctuation and most convenient dosing.

SIM4 IDDP One-compartment single oral dose allows the reader to investigate several fundamental features of pharmacokinetics, including:

- bioavailability (F)
- relative sizes of absorption and elimination rate constants
- effects of clearance.

To investigate the effect of changing the value of F, set up the three curves as below:

	F	D	V	CL	k_a
Set 1	1	1000	100	50	3
Set 2	0.5	1000	100	50	3
Set 3	0.1	1000	100	50	3

with duration = 10. Note how the y-intercept (C'_0), AUC and C_{max} are directly proportional to F. The time of the peak (t_{max}) is unchanged.

To compare the AUCs relative to that of an intravenous dose, increase the value of k_a in Set1 to 1,000,000 (this is a good approximation to the i.v. situation) and vary the values of F in the other sets.

To investigate the effect of k_a on t_{max} and C_{max} set up the plots with:

	F	D	V	CL	k_a
Set 1	1	1000	100	50	10
Set 2	1	1000	100	50	2
Set 3	1	1000	100	50	1

Note how the size of peak is higher and the time of the peak is earlier (increased C_{max}, reduced t_{max}) with higher values of k_a.

Equation 4.19 cannot be used when $k_a = \lambda$ because it reduces to zero. When the two rate constants are similar it is much more difficult to discern a linear phase on the semi-logarithmic plot, and this is why it is necessary to monitor plasma concentrations for sufficient time to obtain an accurate estimate of λ. With the data in the table above $\lambda = CL/V = 0.5$. Change k_a to a value close to λ and note the marked curvature in the semi-logarithmic plots, and the increasing value of the y-intercept:

	F	D	V	CL	k_a
Set 1	1	1000	100	50	2
Set 2	1	1000	100	50	0.51
Set 3	1	1000	100	50	0.501

Equation 4.19 is valid for situations where k_a is less than λ, when the absorption rate constant is rate determining – the principle of sustained release preparations (Section 4.4.4). In the example of Figure 4.11, the half-life of fluphenazine is ~0.5 d, but appears to be seven times longer after administration of the enanthate ester (3.5 d). Because these were

i.m. injections assume that $F \sim 1$. The dose was 25 mg (25,000 μg). Enter 3000 L for V and 4200 L d^{-1} for CL. Absorption after injection of the non-esterified drug was rapid, so enter $k_a = 20$. Enter a value for k_a for the second set to give an apparent half-life of ~3.5 d. This can be done empirically or a value can be calculated.

In the UK, the longer lasting decanoate ester is used, rather than the enanthate. The apparent half-life of fluphenazine after this depot injection is ~12 d. Enter an appropriate value for k_a for data Set 3. The data sets should now be:

	F	D	V	CL	k_a
Set 1	1	25000	3000	4200	20
Set 2	1	25000	3000	4200	0.198
Set 3	1	25000	3000	4200	0.058

To see the difference in time course for the two esters, enter $F = 0$ for Set 1 and duration 28. Note that after 28 days fluphenazine is still present following the decanoate injection and so there will be some degree of accumulation if it is injected every 28 days (4 weeks).

SIM5 One-compartment multiple oral dose can be used to show how sustained-release preparations can be used to reduce fluctuations between doses. Set the values for Set 1 as shown below, this gives a value for $\lambda = CL/V = 0.5$. Note the peak to trough fluctuations.

	F	D	V	CL	k_a	Tau
Set 1	1	1000	100	50	2	8
Set 2	1	1000	100	50	0.1	8

A4.2 Simulations for Chapter 5

SIM6 IDDP Two-compartment i.v. bolus. The input for this simulation is the data that would be collected following an i.v. bolus experiment: the dose, D, the y-intercepts, C_1 and C_2, and the rate constants, λ_1 and λ_2, derived from the slopes of the semi-logarithmic plot. Use the values from Table 5.1. The program calculates the additional parameters that are required for the simulation. The values of V_1 and V_2 will be in litres.

Note how the shapes of the curves (linear and semi-logarithmic) are the same as those in Figure 5.2. Enter 1 in the box to reveal the concentrations in the peripheral compartment and compare the results with Figure 5.4(b), in which the peripheral concentrations were *measured* rather than calculated. Increase the duration of the plot and note how the decay curves of the semi-logarithmic plot remain parallel.

To see the effect of changing the clearance, enter the values from Figure 5.5, with dose = 4. With the higher clearance the concentrations fall more rapidly and the terminal rate constant is greater, as would be expected. Note how at higher clearance the post-equilibrium concentrations in the peripheral compartment are relatively higher, resulting a higher value of V_{area} (see Figure 5.5). Reduce the value of λ_2 to exaggerate the effect as clearance approaches zero, or increase it to see the effects of very high values of clearance.

SIM7 IDDP Two-compartment i.v. infusion demonstrates the difficulty in calculating a suitable loading dose for this situation. The model replicates the empirical results, obtained using the dye model shown in Figure 5.8. Set the total time to 90 (min) and the infusion rate to 5 ($mg\,min^{-1}$). Use a duration of 60 (min) for the infusion time. Calculate the input loading doses using R_0/k_{10} and R_0/λ_2.

The simulation can be used to illustrate the situation depicted by Figure 5.7 when different infusion times are used. The semi-logarithmic output illustrates how it can be difficult to define a redistributional phase after cessation of the infusion as one approaches steady-state conditions. The redistributional phase is more apparent if shorter infusion times are chosen.

SIM8 IDDP Two-compartment multiple i.v. bolus. This simulation allows investigation of the shapes of the plasma concentration–time curves for different relative values of λ_1 and λ_2, and whether the repeat injections are given during the redistribution phase or post equilibration.

The data for thiopental (Figure 5.11) were fitted to $C=15\exp(-6t)+5\exp(-0.025t)$. With the duration = 0.33 h (20 min) enter the appropriate values for Set 1. To compare this to a single-compartment model, enter $C_1=20$, $C_2=0$, $\lambda_1=3.2$ and $\lambda_2=3.2$ for Set 2. Compare your plots with Figure 5.12.

SIM9 IDDP Two-compartment single oral dose allows comparison of plasma concentration curves for differing relative values of k_a. Use the data of Table 5.1 and $F=1$. In the experiment the input rates were varied by using different sized flasks: 250 mL and 1000 mL. The flow rate (= *CL*) was 130 $mL\,min^{-1}$. Calculate the absorption rate constants and use them in the simulation. Compare your plots with Figure 5.6.

A4.3 Simulations for Chapter 7

This simulation uses *SIM4 IDDP One-compartment single oral dose.*

Increasing *V* or *CL* will reduce the plasma concentration and *AUC* values, but they affect the elimination half-life in opposite ways. With duration 10, enter the following values:

	F	*D*	*V*	*CL*	k_a
Set 1	1	1000	100	20	3
Set 2	1	1000	100	50	3
Set 3	1	1000	100	100	3

Note how *AUC* is inversely proportional to *CL*, and λ is directly proportional to *CL*. Also note how the increases in λ results in reduced t_{max}, even though k_a is unchanged (see Equation 4.21).

Index

Introduction to Drug Disposition and Pharmacokinetics, First Edition. Stephen H. Curry and Robin Whelpton.

Companion website: www.wiley.com/go/curryandwhelpton/IDDP